NEUROMETHODS

Series Editor
Wolfgang Walz
University of Saskatchewan
Saskatoon, SK, Canada

For further volumes:
http://www.springer.com/series/7657

Cell Culture Techniques

Second Edition

Edited by

Michael Aschner

Department of Molecular Pharmacology, Albert Einstein College of Medicine, Bronx, NY, USA

Lucio Costa

Department of Environment/STE 100, University of Washington, Seattle, WA, USA

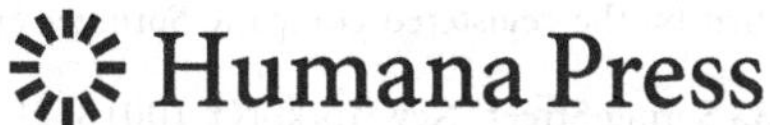

Editors
Michael Aschner
Department of Molecular Pharmacology
Albert Einstein College of Medicine
Bronx, NY, USA

Lucio Costa
Department of Environment/STE 100
University of Washington
Seattle, WA, USA

ISSN 0893-2336 ISSN 1940-6045 (electronic)
Neuromethods
ISBN 978-1-4939-9230-0 ISBN 978-1-4939-9228-7 (eBook)
https://doi.org/10.1007/978-1-4939-9228-7

This Humana Press imprint is published by the registered company Springer Science+Business Media, LLC, part of Springer Nature.
The registered company address is: 233 Spring Street, New York, NY 10013, U.S.A.

Preface to the Series

Experimental life sciences have two basic foundations: concepts and tools. The Neuromethods series focuses on the tools and techniques unique to the investigation of the nervous system and excitable cells. It will not, however, shortchange the concept side of things as care has been taken to integrate these tools within the context of the concepts and questions under investigation. In this way, the series is unique in that it not only collects protocols but also includes theoretical background information and critiques which led to the methods and their development. Thus, it gives the reader a better understanding of the origin of the techniques and their potential future development. The Neuromethods publishing program strikes a balance between recent and exciting developments like those concerning new animal models of disease, imaging, in vivo methods, and more established techniques, including immunocytochemistry and electrophysiological technologies. New trainees in neurosciences still need a sound footing in these older methods in order to apply a critical approach to their results.

Under the guidance of its founders, Alan Boulton and Glen Baker, the Neuromethods series has been a success since its first volume published through Humana Press in 1985. The series continues to flourish through many changes over the years. It is now published under the umbrella of Springer Protocols. While methods involving brain research have changed a lot since the series started, the publishing environment and technology have changed even more radically. Neuromethods has the distinct layout and style of the Springer Protocols program, designed specifically for readability and ease of reference in a laboratory setting.

The careful application of methods is potentially the most important step in the process of scientific inquiry. In the past, new methodologies led the way in developing new disciplines in the biological and medical sciences. For example, physiology emerged out of anatomy in the nineteenth century by harnessing new methods based on the newly discovered phenomenon of electricity. Nowadays, the relationships between disciplines and methods are more complex. Methods are now widely shared between disciplines and research areas. New developments in electronic publishing make it possible for scientists who encounter new methods to quickly find sources of information electronically. The design of individual volumes and chapters in this series takes this new access technology into account. Springer Protocols makes it possible to download single protocols separately. In addition, Springer makes its print-on-demand technology available globally. A print copy can therefore be acquired quickly and for a competitive price anywhere in the world.

Saskatoon, SK, Canada *Wolfgang Walz*

Preface

Research on the fundamental mechanisms underlying neurodevelopment and optimal brain function requires comprehensive mechanisms. Some research is carried out in animal models, but recent efforts have been directed at optimizing tissue culture methods to allow for the reduction of animal usage. Thus, neuromethodologies serve multiple neuroscience disciplines.

We have called on a group of respected and internationally recognized researchers, each with in-depth expertise in implementing specific neuromethod models and techniques, to contribute to this volume. The chapters are geared to provide technical information as well as discussions on the requirements, advantages, and limitations of these neuromethods. The chapters should serve students, experienced researchers, as well as those endowed in making risk decisions. We hope that the readers will find this volume an important complement to their current repertoire of techniques and methods in their research on various aspects of neurosciences.

Bronx, NY, USA *Michael Aschner*
Seattle, WA, USA *Lucio Costa*

Contents

Contributors

GABRIELA AGUILERA-PORTILLO • *Laboratorio de Aminoácidos Excitadores, Instituto Nacional de Neurología y Neurocirugía, Mexico City, Mexico*

KRISTEN T. ASHOURIAN • *Department of Environmental Health Sciences, Robert Stempel College of Public Health & Social Work, Florida International University, Miami, FL, USA*

WILLIAM D. ATCHISON • *Department of Pharmacology and Toxicology, Michigan State University, East Lansing, MI, USA*

ANNA BAL-PRICE • *European Commission Joint Research Centre, Ispra, Italy*

MARTA BARENYS • *INSA·UB and Department of Pharmacology, Toxicology and Therapeutic Chemistry, Faculty of Pharmacy and Food Sciences, University of Barcelona, Barcelona, Spain*

AARON B. BOWMAN • *Vanderbilt University Medical Center, Department of Pediatrics, Nashville, TN, USA; Vanderbilt University Medical Center, Vanderbilt Kennedy Center, Nashville, TN, USA; Vanderbilt University Medical Center, Vanderbilt Brain Institute, Nashville, TN, USA; Vanderbilt University Medical Center, Vanderbilt Center for Stem Cell Biology, Nashville, TN, USA; Vanderbilt University Medical Center, Department of Neurology, Nashville, TN, USA; Vanderbilt University Medical Center, Department of Biochemistry, Nashville, TN, USA; Vanderbilt University Medical Center, Vanderbilt Center for Molecular Toxicology, Nashville, TN, USA*

JEREMY W. CHAMBERS • *Department of Environmental Health Sciences, Robert Stempel College of Public Health & Social Work, Florida International University, Miami, FL, USA; Biomolecular Sciences Institute, Florida International University, Miami, FL, USA*

ALINE COLONNELLO-MONTERO • *Laboratorio de Aminoácidos Excitadores, Instituto Nacional de Neurología y Neurocirugía, Mexico City, Mexico*

LI CUI • *Department of Microbiology and Immunology, University of Arkansas for Medical Sciences, Little Rock, AR, USA*

AART DE GROOT • *Neurotoxicology Research Group, Toxicology and Pharmacology Division, Institute for Risk Assessment Sciences (IRAS), Faculty of Veterinary Medicine, Utrecht University, Utrecht, The Netherlands*

JENNIFER L. FREEMAN • *School of Health Sciences, Purdue University, West Lafayette, IN, USA*

ELLEN FRITSCHE • *IUF – Leibniz Research Institute for Environmental Medicine, Duesseldorf, Germany; Heinrich-Heine-University, Duesseldorf, Germany*

DAVID P. GAVIN • *Jesse Brown Veterans Affairs Medical Center, Chicago, IL, USA; Department of Psychiatry, Center for Alcohol Research in Epigenetics, University of Illinois at Chicago, Chicago, IL, USA*

MARINA GUIZZETTI • *Oregon Health & Science University, Portland, OR, USA; VA Portland Health Care System, Portland, OR, USA*

ZHEN HE • *Division of Neurotoxicology, National Center for Toxicological Research FDA, Jefferson, AR, USA*
SYED Z. IMAM • *Division of Neurotoxicology, National Center for Toxicological Research FDA, Jefferson, AR, USA*
KETURAH G. KIPER • *School of Health Sciences, Purdue University, West Lafayette, IN, USA*
JÖRDIS KLOSE • *IUF – Leibniz Research Institute for Environmental Medicine, Duesseldorf, Germany*
TETYANA KONAK • *Division of Neurotoxicology, National Center for Toxicological Research FDA, Jefferson, AR, USA*
SERGUEI LIACHENKO • *Division of Neurotoxicology, National Center for Toxicological Research FDA, Jefferson, AR, USA*
ETHAN S. LIPPMANN • *Department of Chemical and Biomolecular Engineering, Vanderbilt University, Nashville, TN, USA; Department of Biomedical Engineering, Vanderbilt University, Nashville, TN, USA*
ALEXANDRA MAERTENS • *Center for Alternatives to Animal Testing, Bloomberg School of Public Health, Johns Hopkins University, Baltimore, MD, USA*
S. MANCINO • *Cellular and Systems Neurobiology laboratory, CEDOC, Chronic Diseases Research Centre, NOVA Medical School, Faculdade de Ciências Médicas Universidade NOVA de Lisboa, Lisbon, Portugal*
STEFAN MASJOSTHUSMANN • *IUF – Leibniz Research Institute for Environmental Medicine, Duesseldorf, Germany*
MARISOL MAYA-LÓPEZ • *Laboratorio de Aminoácidos Excitadores, Instituto Nacional de Neurología y Neurocirugía, Mexico City, Mexico*
LINDSAY T. MICHALOVICZ • *Health Effects Laboratory Division, Centers for Disease Control and Prevention – National Institute for Occupational Safety and Health, Morgantown, WV, USA*
DIANE B. MILLER • *CDC/NIOSH, Morgantown, WV, USA*
SOMSHUVRA MUKHOPADHYAY • *Division of Pharmacology & Toxicology, College of Pharmacy, Austin, TX, USA; Institute for Cellular & Molecular Biology, Austin, TX, USA; Institute for Neuroscience, The University of Texas at Austin, Austin, TX, USA*
EMMA H. NEAL • *Department of Chemical and Biomolecular Engineering, Vanderbilt University, Nashville, TN, USA*
M. DIANA NEELY • *Vanderbilt University Medical Center, Department of Pediatrics, Nashville, TN, USA; Vanderbilt University Medical Center, Vanderbilt Kennedy Center, Nashville, TN, USA; Vanderbilt University Medical Center, Vanderbilt Brain Institute, Nashville, TN, USA; Vanderbilt University Medical Center, Vanderbilt Center for Stem Cell Biology, Nashville, TN, USA*
LAURA NIMTZ • *IUF – Leibniz Research Institute for Environmental Medicine, Duesseldorf, Germany*
JAMES P. O'CALLAGHAN • *Health Effects Laboratory Division, Centers for Disease Control and Prevention – National Institute for Occupational Safety and Health, Morgantown, WV, USA; CDC/NIOSH, Morgantown, WV, USA*
JOHN PANOS • *Division of Neurotoxicology, National Center for Toxicological Research FDA, Jefferson, AR, USA*
MERLE G. PAULE • *Division of Neurotoxicology, National Center for Toxicological Research FDA, Jefferson, AR, USA*

FRANCESCA PISTOLLATO • *European Commission Joint Research Centre, Ispra, Italy*
EDGAR RANGEL-LÓPEZ • *Laboratorio de Aminoácidos Excitadores, Instituto Nacional de Neurología y Neurocirugía, Mexico City, Mexico*
JAMES RAYMICK • *Division of Neurotoxicology, National Center for Toxicological Research FDA, Jefferson, AR, USA*
MONICA RODRIGUEZ-SILVA • *Department of Environmental Health Sciences, Robert Stempel College of Public Health & Social Work, Florida International University, Miami, FL, USA*
ABEL SANTAMARÍA • *Laboratorio de Aminoácidos Excitadores, Instituto Nacional de Neurología y Neurocirugía, Mexico City, Mexico*
M. M. SERAFINI • *Department of Pharmacological and Biomolecular Sciences, University of Milan, Milan, Italy*
YAJUAN SHI • *Department of Chemical and Biomolecular Engineering, Vanderbilt University, Nashville, TN, USA*
LENA SMIRNOVA • *Center for Alternatives to Animal Testing, Bloomberg School of Public Health, Johns Hopkins University, Baltimore, MD, USA*
ANTHONY D. SMITH • *Department of Environmental Health Sciences, Robert Stempel College of Public Health & Social Work, Florida International University, Miami, FL, USA*
CHERISH A. TAYLOR • *Division of Pharmacology & Toxicology, College of Pharmacy, Austin, TX, USA; Institute for Cellular & Molecular Biology, Austin, TX, USA; Institute for Neuroscience, The University of Texas at Austin, Austin, TX, USA*
ANKE M. TUKKER • *Neurotoxicology Research Group, Toxicology and Pharmacology Division, Institute for Risk Assessment Sciences (IRAS), Faculty of Veterinary Medicine, Utrecht University, Utrecht, The Netherlands*
BARBARA VIVIANI • *Department of Pharmacological and Biomolecular Sciences, University of Milan, Milan, Italy*
REMCO H. S. WESTERINK • *Neurotoxicology Research Group, Toxicology and Pharmacology Division, Institute for Risk Assessment Sciences (IRAS), Faculty of Veterinary Medicine, Utrecht University, Utrecht, The Netherlands*
FIONA M. J. WIJNOLTS • *Neurotoxicology Research Group, Toxicology and Pharmacology Division, Institute for Risk Assessment Sciences (IRAS), Faculty of Veterinary Medicine, Utrecht University, Utrecht, The Netherlands*
RICHARD W. WUBBOLTS • *Centre for Cell Imaging (CCI), Department of Biochemistry and Cell Biology, Faculty of Veterinary Medicine, Utrecht University, Utrecht, The Netherlands*
YUKUN YUAN • *Department of Pharmacology and Toxicology, Michigan State University, East Lansing, MI, USA*
XIAOLU ZHANG • *Oregon Health & Science University, Portland, OR, USA; VA Portland Health Care System, Portland, OR, USA*

Chapter 1

In Vitro Blood-Brain Barrier Functional Assays in a Human iPSC-Based Model

Emma H. Neal, Yajuan Shi, and Ethan S. Lippmann

Abstract

The blood-brain barrier (BBB) is a key biological interface that controls trafficking between the bloodstream and brain to maintain neural homeostasis. To carry out this role, the BBB exhibits several specialized properties, including limited permeability and active transporter function. These properties are often evaluated within in vitro BBB models, which can be utilized for high-throughput screening applications such as drug discovery. Here, we detail several common methods used to qualify in vitro BBB models, including measurement of transendothelial electrical resistance (TEER), determination of permeability coefficients (P_e) for small molecules, and assessment of efflux transporter activity. We describe these methods in the context of BBB endothelial cells derived from human-induced pluripotent stem cells (iPSCs), a model commonly employed in our research group.

Key words In vitro blood-brain barrier model, Human-induced pluripotent stem cells, Brain microvascular endothelial cell, Efflux transporter, Transendothelial electrical resistance, Permeability coefficient

1 Introduction

The blood-brain barrier (BBB) consists of the endothelial cells that line brain capillaries. These brain microvascular endothelial cells (BMECs) form an incredibly restrictive interface due to the expression of complex intercellular tight junction proteins that limit free diffusion of hydrophilic molecules. Lipophilic molecules are also restricted from freely diffusing through the endothelial lipid bilayer owing to the expression of efflux transport proteins [1]. To compensate for a lack of passive transport, the BBB expresses a cohort of molecular transporters, both uni- and bi-directional, which import nutrients and export waste products between the bloodstream and brain interstitial space [2]. These properties, which also include suppressed transcytosis and lack of fenestrae [3], are highly specialized compared to most peripheral capillary beds and tightly regulated by the microenvironment of the neurovascular unit

Michael Aschner and Lucio Costa (eds.), *Cell Culture Techniques*, Neuromethods, vol. 145,
https://doi.org/10.1007/978-1-4939-9228-7_1, © Springer Science+Business Media, LLC, part of Springer Nature 2019

(NVU), which consists of pericytes, astrocytes, neurons, and microglia [4].

Due to its restrictive nature, the BBB represents a major obstacle for delivering drugs to the brain. In addition, dysfunction of the BBB is implicated in a number of neurodegenerative diseases, including but not limited to Alzheimer's disease (AD), Parkinson's disease (PD), Huntington's disease (HD), amyotrophic lateral sclerosis (ALS), and multiple sclerosis (MS). For the sake of brevity, only AD and PD are described herein. In the two-hit vascular hypothesis of AD, initial blood vessel damage contributes to BBB dysfunction and diminished brain perfusion, which can subsequently influence amyloid-β (Aβ) accumulation and neuronal injury [5]. Several neuroimaging results support this notion of BBB dysfunction in AD. Higher gadolinium leakage, observed by dynamic contrast-enhanced magnetic resonance imaging (DCE-MRI), has confirmed increased BBB permeability in several gray and white matter regions of early AD patients [6]. Also, lobar cerebral microbleeds, which indicate BBB damage, are also found in most AD patients [7, 8]. Whereas glucose transporter member 1 (GLUT-1) is selectively expressed on BBB endothelium, ^{18}F-fluorodeoxyglucose positron emission tomography (FDG-PET) studies show diminished glucose uptake and reduced GLUT-1 levels [9, 10] in humans [11] and transgenic mouse models [12] of AD, which indicate BBB disruption preceding neurodegeneration. Meanwhile, impaired p-glycoprotein, which clears xenobiotic compounds across the BBB, may also be involved in the pathogenesis of AD. On the basis of verapamil-PET results, diminished p-glycoprotein function is observed in the patients with mild AD [13] and AD [14]. Meanwhile, PD, the second most common neurodegenerative disorder in the elderly that is characterized by aggregation of α-synuclein and degeneration of dopaminergic neurons, also displays elements of BBB dysfunction. Decreased p-glycoprotein activity in the midbrain, which indicates a loss of BBB integrity, has been shown in patients with PD [15].

These issues of drug delivery and possible involvement of the microvasculature in neurodegeneration have led to the rise of in vitro BBB models, which are a useful complement to protracted in vivo studies. Cultured BMECs can be used to assess the relative permeability of drug candidates and transport rates of biologics with much higher throughput than similar measurements conducted in animals [16]. Meanwhile, NVU cell-cell interactions and their respective influence on BBB properties, as well as disease-relevant perturbations of BMEC function, can also be studied more easily in an in vitro setting. Unfortunately, in vitro models have several drawbacks, such as (1) they often do not properly mimic the phenotype of the BBB in vivo and (2) they can be difficult to construct owing to a lack of quality cells. Most in vitro BBB models have historically been constructed from primary animal BMECs, including bovine [17], porcine [18, 19], rat [20, 21], and mouse [22]. For comparison, the BBB in vivo has a theoretical maximum transendothelial electrical resistance (TEER, a measure of ion permeability) of 8000 $\Omega \times cm^2$ [23], and primary BMECs typically exhibit ranges of TEER from ~200 to 1800 $\Omega \times cm^2$ [24]. Small molecule permeability also indicates reasonable tightness in these models, and most BMEC sources exhibit active transport function. However, primary BMECs can only be isolated in low yield (because the vasculature represents only a small fraction of total brain volume [25]), and in many cases the isolation procedures are tedious and difficult to carry out. Moreover, especially with respect to drug screens and disease studies, there are concerns that species differences may prohibit extrapolation of outcomes to human biology [26]. These concerns have led to examination of primary human BMECs for constructing more representative in vitro models, but because they are only available from surgically resected tissue or cadavers, such BMECs are not suitable for high-throughput studies. Immortalized human BMECs have also been explored to overcome these issues, but the immortalization process has a negative effect on the passive barrier phenotype, yielding TEER values of 30–40 $\Omega \times cm^2$ that are more similar to peripheral endothelium than BBB [27].

Recently, endothelial cells possessing a BBB-like phenotype were derived from human-induced pluripotent stem cells (iPSCs) [28]. These so-called iPSC-derived BMECs represent an attractive alternative to primary human and immortalized BMECs. iPSCs can self-renew indefinitely, thus providing an unlimited source of BMECs similar to the immortalization procedure but with improved passive barrier properties. First-generation iPSC-derived BMECs yielded an average TEER of ~850 $\Omega \times cm^2$ when co-cultured with rat astrocytes and also demonstrated other hallmarks of the BBB, including efflux transporter activity and relative permeability to a cohort of small molecules that correlated well with in vivo uptake in rodents [28]. In subsequent studies, the addition of retinoic acid (RA) was shown to drastically enhance the passive barrier phenotype, yielding TEER of ~3000 $\Omega \times cm^2$ in monoculture and >5000 $\Omega \times cm^2$ in co-culture with human pericytes, astrocytes, and neurons [29]. Further iterations of the differentiation process have resulted in defined seeding densities and medium, which accelerate the differentiation process without compromising barrier function [30, 31]. In this book chapter, we detail the most recent iteration of the iPSC-to-BMEC differentiation process and outline the primary methods used to characterize barrier function, including TEER measurements, permeability measurements, and assessments of efflux transporter activity.

2 Materials

2.1 iPSC Maintenance and Differentiation

1. Any quality iPSC line that is properly karyotyped and exhibits minimal spontaneous differentiation.
2. mTeSR1 (STEMCELL Technologies, Cat#85850) or E8 medium (Thermo Fisher Scientific, Cat#A1517001). We routinely use E8 medium but have also used mTeSR1 medium in past publications.
3. Matrigel (Fisher Scientific, Cat#CB-40230), reconstituted to 1 mg aliquots and frozen at −80 °C. These aliquots are used to coat plates for both iPSC maintenance and differentiation. One aliquot of Matrigel is sufficient to coat two full 6-well plates.
4. Versene (Thermo Fisher Scientific, Cat#15040-066).
5. 6-well tissue culture polystyrene dishes.
6. ROCK inhibitor Y27632 (10 μM; R&D Systems, Cat#1254). Stock solutions of 10 mM are prepared in sterile ddH_2O and stored at −80 °C.
7. E6 medium (Thermo Fisher Scientific, Cat#A1516401).
8. Human endothelial serum-free medium (hESFM) (Thermo Fisher Scientific, Cat#11111-044).
9. Platelet-poor plasma-derived serum (PDS) (Fisher Scientific, Cat#AAJ64483AE). 100 mL of PDS is centrifuged at 30,000 g for 30 min to remove particulates, then sterile-filtered using 0.22 μm pore vacuum bottle-top filters (Fisher Scientific, Cat#S2GPT05RE) and stored in 5 mL aliquots at −80 °C.
10. Retinoic acid (RA) (10 μM; Sigma-Aldrich, Cat#R2625). Stock solutions of 10 mM are prepared in DMSO (Sigma-Aldrich, Cat#D8418) and stored at −80 °C.
11. Basic fibroblast growth factor (bFGF) (20 ng/mL; Peprotech, Cat#100-18b). Stock aliquots are prepared by adding 1 mg of bFGF to 10 mL of 5 mM Tris buffer (filter-sterilized, pH ~7.6) and 20 μL of human albumin solution (Sigma-Aldrich, Cat#A7223). Five hundred micro liter aliquots are then at −80 °C.
12. Extracellular matrix (ECM) coating comprised of collagen IV (Sigma-Aldrich, Cat#C5533) and fibronectin (Sigma-Aldrich, Cat#F1141). One milligram per milliliter collagen IV stock solution is prepared by dissolving 5 mg into 5 mL of 0.5 mg/mL sterile acetic acid and is stored at 4 °C. The final coating is prepared fresh by mixing 5 parts ddH_2O, 4 parts collagen IV stock solution, and 1 part fibronectin.
13. Endothelial cell culture medium: EM++ consists of hESFM, 1% PDS, 20 ng/mL bFGF, and 10 μM RA, while EM−−

consists of hESFM and 1% PDS. RA is added to EM++ immediately before use. All other components can be added to EM++ and EM−−, and each medium is stable for at least 2 weeks.

14. Accutase (Thermo Fisher Scientific; Cat#A1110501). Accutase can be aliquoted into 15 mL conicals and stored at −20 °C.

2.2 TEER Measurements

1. Transwell filters (12-well, 1.12 cm^2, 0.4 μm pore size, PET) (Fisher Scientific, Cat#07-200-161).
2. EVOM voltohmmeter (World Precision Instruments, Model#EVOM2).
3. Electrodes or Endohm cell culture cup chamber (World Precision Instruments, #STX2 or ENDOHM-12).

2.3 Permeability Measurements

1. Dilute sodium fluorescein (Sigma-Aldrich, Cat#F6377) to 10 mM in distilled water to create stock solutions. Sterile filter and store in an opaque box at 4 °C.
2. Dilute fluorescently labeled dextrans to 1 mg/mL in DPBS (Thermo Fisher Scientific, Cat#14190-144) to create stock solutions. We frequently use a 3 kDa dextran labeled with Alexa Fluor 680 (Thermo Fisher Scientific, Cat#D34681) but other sized dextrans can be employed. Store in an opaque box at −20 °C.
3. Prepare transport buffer: distilled water with 0.12 M NaCl, 25 mM $NaHCO_3$, 3 mM KCl, 2 mM $MgSO_4$, 2 mM $CaCl_2$, 0.4 mM K_2HPO_4, 1 mM HEPES, and 0.1% bovine serum albumin. All salts and albumin can be ordered in powdered form from any reputable source (Fisher Scientific or Sigma-Aldrich).
4. Dilute chosen radiolabeled compounds to 0.4 μCi in transport buffer. Radiolabeled compounds can be ordered from American Radiolabeled Compounds or PerkinElmer.
5. Scintillation counter.

2.4 Efflux Activity Measurements

1. Dilute rhodamine 123 (R123) (Thermo Fisher Scientific; Cat#R302) to a 10 mM stock solution using sterile distilled water and store at −20 °C.
2. Dilute cyclosporin A (Fisher Scientific; Cat#11-011-00) to a 10 mM stock solution using sterile DMSO and store at −20 °C.
3. Dilute H_2DCFDA (Fisher Scientific, Cat#D399) to a 10 mM stock solution using sterile DMSO and store at −20 °C.
4. Dilute MK-571 (Sigma-Aldrich; Cat#M7571) to a 10 mM stock solution using sterile distilled water and store at −20 °C.

5. Prepare lysis buffer (DPBS + 5% TX-100): Triton X-100 (TX-100) (Fisher Scientific, Cat#9002-93-1). Store at room temperature.
6. DAPI (Thermo Fisher Scientific, Cat#D1306).

3 Methods

3.1 Differentiation of iPSCs to BBB Endothelium

1. iPSCs are maintained in E8 medium on Matrigel-coated plates. Medium is changed every day, and iPSCs are passaged at a 1:6 to 1:12 ratio into freshly coated plates every 3–4 days when 60–80% confluence is reached. Passage ratios are empirically determined based on the growth rate of the iPSC line.
2. Upon reaching 60–80% confluence, cells are ready to be seeded for differentiation.
3. Collect 1 mL of spent medium per well of iPSCs that will be utilized for seeding. For example, if passaging 3 wells for seeding, collect 3 mL of spent medium. Place this spent medium in a plastic conical. Aspirate the remaining medium from each well.
4. Wash each well with 2 mL of DPBS. Aspirate DPBS.
5. Add 1 mL of Accutase to each well. Incubate cells at 37 °C for 3–5 min.
6. Using a P1000 pipet, collect the iPSCs from each well by gently spraying across the surface two to three times to detach the cells. Transfer all cells to the collected spent medium to neutralize the Accutase.
7. Pellet cells via centrifugation at 1000 RPM for 4 min. Aspirate supernatant.
8. Resuspend cells in 1 mL of E8 medium.
9. Determine cell density using a hemocytometer or automated cell counter, such as a Countess II.
10. Plate cells at a density of 150,000 live cells per well of a 6-well plate in E8 medium supplemented with 10 μM Y27632.
11. Twenty-four hours after plating, initiate differentiation by changing medium to E6 medium.
12. Change E6 medium every 24 h for 4 days.
13. After 4 days in E6 medium, change medium to EM++. Do not change medium for 48 h.
14. Twenty-four hours after changing medium to EM++ (day 5 of differentiation), prepare Transwell filters and/or plates for BMEC purification by coating with 200 μL ECM/Transwell filter or plates at the volumes listed in Table 1.
    (a) If coating plates the day before subculture (e.g., overnight), we recommend adding an equal volume of water

to each ECM-coated plate to prevent the wells from drying out during the coating process.

(b) Transwell filters must be coated for a minimum of 4 h at 37 °C. Plates must be coated for a minimum of 1 h at 37 °C.

15. Forty-eight hours after changing medium to EM++ (day 6 of differentiation), cells are ready for subculture.
16. Aspirate ECM solution from coated plates and Transwell filters. Leave lids slightly ajar and move to the back of the laminar flow hood. Transwell filters must dry for a minimum of 20 min. Plates must dry for a minimum of 5 min but no more than 30 min to avoid overdrying the wells.
17. Retrieve cells from the incubator and collect 1 mL of spent medium from each well to be subcultured. Aspirate remaining medium.
18. Wash each well with 2 mL of DPBS. Aspirate.
19. Add 1 mL of Accutase to each well and return cells to incubator.
20. Incubate cells in Accutase for 20–45 min (time varies by iPSC line used and must be determined empirically) until cells have lifted off the plate in a single cell suspension.
21. After filters have dried for a minimum of 20 min, rewet filters with 0.5 mL hESFM/filter.
22. Gently collect cells by spraying across each well two to three times using a P1000 pipet and transfer the cell suspension to the collected spent medium.
23. Collect cells via centrifugation at 1000 RPM for 4 min.
24. While cells are centrifuging, aspirate the hESFM from each filter. Add 1.5 mL of EM++ to each basolateral chamber. Omit this step if only using plates.
25. Aspirate supernatant from the cell pellet and resuspend cells in an appropriate volume of EM++ (0.5 mL/Transwell filter; see Table 1.1 for recommended well plate volumes). For 6- and 12-wells and 12-well Transwell filter inserts, cells are seeded based on a split ratio:
    (a) 1 well of a 6-well plate is split to 1 well of a 6-well plate (1:1).
    (b) 1 well of a 6-well plate is split to 3 wells of a 12-well plate (1:3) or 3 Transwell filters (1:3).
    (c) For smaller plates (24-, 48-, or 96-wells), seed 1 million cells/cm^2.
    (d) Multiply split ratio by the working volume found in Table 1 to arrive at total volume of EM++ in which to resuspend cells.

Table 1
Volumes of ECM coating and cell culture media required for the subculture phase, based on plate size

Plate type for subculture	Volume of ECM solution for coating	Working volume of EC media for cell culture
6-well	800 μL	2 mL
12-well	250 μL	1 mL
24-well	200 μL	500 μL
48-well	100 μL	400 μL
96-well	50 μL	200 μL

26. Twenty-four hours after subculture, induce barrier by changing medium from EM++ to EM−−. Begin measuring TEER at this time, which we typically refer to as day 0.
27. Once the medium is changed from EM++ to EM−−, which removes the bFGF and RA, the user should wait 24 h for barrier properties to spike. TEER will peak at 24 h and often dip drastically over 2–3 days, then slowly recover over the course of a week to reach or exceed the reading from 24 h. Thus, TEER can be measured longitudinally, but we typically measure permeability and efflux activity 24 h after removal of bFGF and RA during the first maximum peak.

3.2 TEER Measurements

1. Sterilize Chopstix using 70% ethanol. Shake Chopstix to dry ethanol from the surface of the electrodes.
2. Let Chopstix equilibrate in hESFM for 2 min.
3. To measure TEER, position the electrodes such that the shorter end is above the Transwell filter and the longer end is below. The short electrode must be fully submerged in medium. Take care not to poke the filter. Measure the resistance until the reading has fully stabilized. Repeat the measurement at the two other positions on the filter.
4. Subtract each resistance measurement from the resistance measurement of an empty filter to yield the TEER value associated with the BMEC monolayer. In our experience, there is no difference in TEER for empty filters that are coated or uncoated with collagen IV and fibronectin. Then, multiply each TEER value by the surface area of the filter to yield the final value in $\Omega \times cm^2$.
5. TEER can be measured every 24 h to noninvasively track the stability of the BMECs over time.

3.3 Permeability Measurements

3.3.1 Sodium Fluorescein Permeability

1. Prepare a working solution of 10 μM sodium fluorescein by diluting the stock solution. Then, further dilute working solution 1:80 in a separate conical by adding 62.5 μL of working solution to 4.94 mL of EM−−.
2. Using the dilute solution prepared above, construct a standard curve according to the specifications in Table 2.
3. Place each dilution into a well of a 96-well plate.
4. Upon collection of all permeability samples, measure the fluorescence of each well using a plate reader. Construct a standard curve by plotting fluorescence versus moles.
5. To prepare BMECs, aspirate medium from the apical and basolateral sides of each Transwell filter and replace with fresh EM−−. Let the cells equilibrate for 1 h to allow TEER to stabilize after the media change.
   (a) Three Transwell filters seeded with BMECs are required to calculate an average permeability.
   (b) One Transwell filter coated with ECM but not seeded with cells should be used per experiment to account for mass transfer resistance from the filter alone.
6. After allowing the filters to equilibrate, measure TEER to ensure the monolayers are still intact.
7. Aspirate medium from the apical chamber of each filter and add 0.5 mL of sodium fluorescein solution. Immediately remove 200 μL from the basolateral chamber of each well being tested and transfer it to the 96-well plate containing the wells prepared to construct the standard curve. Add 200 μL of fresh EM−− to each basolateral chamber and return all plates to the incubator.
8. Every 30 min, remove 200 μL of media from each basolateral chamber, transfer the medium to the 96-well plate, and add 200 μL of fresh media to each basolateral chamber to replace the volume removed. Media replenishment in the basolateral chamber is necessary to maintain hydrostatic pressure on the top and bottom of the filter, which could otherwise influence mass transfer rates.
9. Continue the assay for a total of 2 h (five total time points measured, including $t = 0$).
10. At the conclusion of the assay, measure the fluorescence in each well of the 96-well plate using a plate reader.
    (a) For sodium fluorescein, $\lambda_{excitation} = 460$ nm and $\lambda_{emission} = 515$ nm.
11. Calculate the average fluorescence at each time point.

Table 2
Dilution ratios and expected fluorescein compositions used to construct a calibration curve for permeability experiments

EM-- (μL)	Fluorescein (μL)	Fluorescein (μmol)	Fluorescein (μM)
0	200	0.000025	0.125022616
50	150	0.00001875	0.093766962
100	100	0.0000125	0.062511308
150	50	0.00000625	0.031255654
200	0	0	0

12. Convert fluorescence to moles for each time point using the calibration curve.
    (a) Remember that fluorescence from each sample is indicative of the 1.5 mL volume in the basolateral chamber where the sample was collected, while the calibration curve was constructed using volumes of 200 μL. Thus, the fluorescence of each sample needs to be scaled appropriately based on these volumetric differences to properly determine the number of moles.
    (b) Remember to account for the fact that you removed fluorescein from the basolateral chamber each time you took a measurement, which would result in a decrease in fluorescence. To correct each time point, add the moles from the previous time point multiplied by (0.2/1.5) to account for the fraction removed.
13. Plot moles versus time for each condition as well as the empty filter. Calculate the slope of the linear line of best fit for each condition and the empty filter. Slope will be in units of moles per time.
14. Using the assumption that the concentration of sodium fluorescein remains relatively constant in the apical chamber due to the restricted flux across the monolayer and the short time frame of the assay, the permeability coefficient (*P*) can be calculated as follows:

$$\frac{dQ}{dT} = (P)(S)(\mathrm{Cd})$$

where dQ/dT is the slope calculated in the previous step, *S* is the area of the Transwell filter, and Cd is the concentration of fluorescein in the apical chamber.

(a) Ensure the use of consistent units when calculating *P*. In our experience, inconsistent units are the most common route for misinterpreting the results in this assay.

15. To obtain the permeability of the BMEC monolayer, calculate P_e using the following equation:

$$\frac{1}{P_e} = \frac{1}{P_t} - \frac{1}{P_f}$$

where P_e is the permeability of the monolayer, P_f is the permeability of the empty filter, and P_t is the collective permeability of cells cultured on top of filters. If the iPSC-derived BMEC monolayer is high quality, expected permeability for fluorescein is in the range of 10^{-7} cm/s.

3.3.2 Fluorescently Labeled Dextran Permeability

1. The permeability of fluorescently labeled dextrans (such as Thermo Fisher Scientific #D34682) can be determined similar to sodium fluorescein described in Sect. 3.3.1. We provide details on how to calculate permeability for 3 kDa dextran below. For other molecular weight dextrans conjugated with different fluorophores, working concentrations that produce appropriate fluorescent signal for permeability assays must be empirically determined. This can be accomplished by assessing the permeation of the fluorescent molecule through a monolayer of BMECs at various concentrations. If no fluorescent signal is observed in the basolateral chamber after 1 h, the concentration of starting material in the apical chamber needs to be increased such that transport of the dextran can be reliably quantified.
2. If using the 3 kDa labeled dextran listed above, dilute the 1 mg/mL dextran stock solution to 4.11 μM by adding 125 μL of stock solution to 10 mL EM−−. This working solution should be prepared fresh for each experiment.
3. Further dilute the dextran working solution by adding 152 μL of working solution to 4.85 mL of EM−−.
4. Use this diluted working solution to prepare serial dilutions as described in Sect. 3.3.1. These serial dilutions will be used to construct a calibration curve as described in Table 2.
5. Using the working solution prepared in Sect. 3.3.2., conduct the permeability assay and subsequent permeability calculations in the same manner as the sodium fluorescein experiments in Sect. 3.3.1. Remember that the initial concentration of dextran will be different compared to the sodium fluorescein calculations.

3.3.3 Radiolabeled Compound Permeability

1. Twenty-four hours after barrier induction, aspirate medium from apical and basolateral chambers of Transwell filters to be assayed.
2. Wash the apical chamber twice with 0.5 mL of transport buffer pre-warmed to 37 °C. Add 0.5 mL to each apical chamber and 1.5 mL of transport buffer to each basolateral chamber.

3. Incubate cells for 1 h at 37 °C to allow TEER to stabilize.
4. Immediately prior to beginning the assay, measure TEER as previously described.
5. Dilute chosen radiolabeled compounds to 0.4 μCi in transport buffer. Total volume will depend on the number of filters being assayed. Take 200 μL of this stock solution and place in a scintillation vial. It represents the initial concentration of radiolabeled compound that will be used to calculate the permeability coefficient.
6. Aspirate transport buffer from apical chamber and replace with 500 μL of the radiolabeled compound solution.
7. Immediately remove 200 μL of transport buffer from the basolateral chamber, transfer to a scintillation vial, and replace with fresh transport buffer. Return cells to incubator.
8. Remove 200 μL of transport buffer from the basolateral chamber every 15 min for 1 h. Transfer each aliquot to a scintillation vial.
9. At the end of the experiment (typically 1 h), dilute each scintillation vial with 800 μL of transport buffer and measure radioactivity on a scintillation counter.
10. Calculate P_e as described in Sect. 3.3.1. A calibration curve does not need to be constructed. Plot counts per minute (CPM) as a function of time and calculate the linear slope and then divide by the filter area and CPM of the initial solution (adjusted to a per mL basis) to yield the combined permeability coefficient for BMECs and the filter. The permeability coefficient of the BMECs can then be calculated by removing the mass transfer resistance of an empty filter, as described earlier.

3.4 Efflux Transporter Activity

3.4.1 Substrate Accumulation

1. As described earlier, iPSC-derived BMECs are seeded onto 24-well plates in EM++ for 24 h and changed to EM−− for 24 h to obtain maximum barrier induction. Four wells are required for each experimental condition.
2. Stock solutions of efflux inhibitors (cyclosporin A and MK-571) are diluted in EM−− to concentrations of 10 μM. BMECs are then incubated with each inhibitor for 1 h at 37 °C. Control BMECs are left untouched.
3. While the BMECs are incubating with inhibitor, prepare two solutions, one containing EM−− and 10 μM of efflux transporter substrate and one containing EM−−, 10 μM substrate and 10 μM inhibitor. Make sure the substrate and inhibitor are appropriately matched for each efflux transporter. Three milliliters of media is required per experimental condition.
4. Media is aspirated from each well, and BMECs are then incubated with 10 μM R123 or 10 μM H2DCFDA with or without

their respective inhibitors for 1 h at 37 °C. The volume in each well is 0.5 mL. Media containing inhibitors should be added to wells that were preincubated with inhibitors.

5. At the culmination of the experiment, three wells of BMECs per condition are washed three times with DPBS, then lysed with 200 μL of lysis buffer for 10 min, and transferred to a 96-well plate. Fluorescence is measured using a plate reader.
6. The remaining wells (one per condition) are fixed with 100% ice-cold methanol for 10 min, washed three times with PBS, stained with DAPI, washed three times again with DPBS, and imaged. Six images should be acquired per well in varying locations for each condition. Calculate the number of cells per well based on the surface area, and normalize fluorescent compound uptake on a per cell basis. If efflux transporters are expressed and functionally active, R123 and H2DCFDA uptake should be increased in the presence of each respective transporter inhibitor.

3.4.2 Directional Transport

1. As described earlier, iPSC-derived BMECs are seeded onto Transwell filters in EM++ for 24 h and changed to EM−− for 24 h to obtain maximum barrier induction.
2. Efflux inhibitors (cyclosporin A and MK-571) are diluted to concentrations of 10 μM by removing 200 μL of EM−− from the side of the filter where transport is being assessed, adding an appropriate volume of efflux inhibitor to produce a 10 μM concentration for the total volume on the desired side of the filter, and returning the medium to its original location. BMECs are then incubated with each inhibitor for 1 h at 37 °C. Inhibitors are only added on the side of the filter where directional transport is being assessed. For example, if the user is examining polarized transport in the apical-to-basolateral direction, inhibitor should only be added on top of the filter. Control BMECs are left untouched.
3. While the BMECs are incubating with inhibitor, prepare two solutions, one containing EM−− with 10 μM of efflux transporter substrate and one containing EM−− with 10 μM substrate and 10 μM inhibitor. Make sure the substrate and inhibitor are appropriately matched for each efflux transporter. If performing apical-to-basolateral transport studies, 2 mL of media will be required per condition. If performing basolateral-to-apical transport studies, 5 mL of media will be required per condition.
4. Media is aspirated from the desired compartment and replaced with media containing 10 μM R123 or 10 μM H2DCFDA with or without their respective inhibitors and incubated for 1 h at 37 °C. Media containing inhibitors should be matched to the compartments that were preincubated with inhibitors.

5. After 1 h, 200 μL of media is extracted from the opposite compartment of each filter and transferred to a 96-well plate. TEER should be measured across each filter to ensure that the BMEC monolayer remains intact. Fluorescence is then measured on a plate reader as described earlier. If efflux transporters are expressed at the apical or basolateral membrane and functionally active, R123 and H2DCFDA directional transport should be increased in the presence of each respective inhibitor.

4 Notes

1. We have qualitatively noticed that maximum TEER values become lower as BMECs are differentiated from higher passage iPSCs, possibly due to the accumulation of mutations or epigenetic modifications. We typically assign a hard cutoff at passage 40 for most of our iPSC work.
2. When measuring TEER, resistance will become artificially inflated as the plates cool down from 37 °C. To prevent bias, we recommend measuring resistances on filters in a random order or working as quickly as possible.
3. The quality of PDS has a substantial influence on the differentiation process. Side-by-side comparisons of different lot numbers, tested on cultures differentiated from identical iPSC populations, have yielded BMECs with 1000–2000 $\Omega \times cm^2$ variations. A new lot of PDS should always be qualified with pilot differentiations before moving forward with scaled-up experiments.

Acknowledgments

Our research efforts in this area are supported by a NARSAD Young Investigator Award from the Brain and Behavior Research Foundation (ESL) and grant A20170945 from the Alzheimer's Disease Research Program through the BrightFocus Foundation (ESL). EHN is supported by a National Science Foundation Graduate Research Fellowship.

References

1. Abbott NJ, Rönnbäck L, Hansson E (2006) Astrocyte–endothelial interactions at the blood–brain barrier. Nat Rev Neurosci 7:41–53
2. Daneman R, Prat A (2015) The blood–brain barrier. Cold Spring Harb Perspect Biol 7:a020412
3. Obermeier B, Daneman R, Ransohoff RM (2013) Development, maintenance and disruption of the blood-brain barrier. Nat Med 19:1584–1596
4. Zlokovic BV (2008) The blood-brain barrier in health and chronic neurodegenerative disorders. Neuron 57:178–201

5. Montagne A, Nation DA, Pa J et al (2016) Brain imaging of neurovascular dysfunction in Alzheimer's disease. Acta Neuropathol (Berl) 131:687–707
6. van de Haar HJ, Burgmans S, Jansen JFA et al (2016) Blood-brain barrier leakage in patients with early Alzheimer disease. Radiology 281:527–535
7. Yates PA, Desmond PM, Phal PM et al (2014) Incidence of cerebral microbleeds in preclinical Alzheimer disease. Neurology 82:1266–1273
8. Uetani H, Hirai T, Hashimoto M et al (2013) Prevalence and topography of small hypointense foci suggesting microbleeds on 3T susceptibility-weighted imaging in various types of dementia. Am J Neuroradiol 34:984–989
9. Simpson IA, Chundu KR, Davies-Hill T et al (1994) Decreased concentrations of GLUT1 and GLUT3 glucose transporters in the brains of patients with Alzheimer's disease. Ann Neurol 35:546–551
10. Jagust WJ, Seab JP, Huesman RH et al (1991) Diminished glucose transport in Alzheimer's disease: dynamic PET studies. J Cereb Blood Flow Metab 11:323–330
11. Hunt A, Schönknecht P, Henze M et al (2007) Reduced cerebral glucose metabolism in patients at risk for Alzheimer's disease. Psychiatry Res Neuroimaging 155:147–154
12. Niwa K, Kazama K, Younkin SG et al (2002) Alterations in cerebral blood flow and glucose utilization in mice overexpressing the amyloid precursor protein. Neurobiol Dis 9:61–68
13. Deo AK, Borson S, Link JM et al (2014) Activity of P-glycoprotein, a β-amyloid transporter at the blood-brain barrier, is compromised in patients with mild Alzheimer disease. J Nucl Med 55:1106–1111
14. van Assema DM, Lubberink M et al (2012) Blood–brain barrier P-glycoprotein function in Alzheimer's disease. Brain 135:181–189
15. Kortekaas R, Leenders KL, van Oostrom JCH et al (2005) Blood–brain barrier dysfunction in parkinsonian midbrain in vivo. Ann Neurol 57:176–179
16. Naik P, Cucullo L (2012) In vitro blood–brain barrier models: current and perspective technologies. J Pharm Sci 101:1337–1354
17. Dehouck M-P, Méresse S, Delorme P et al (1990) An easier, reproducible, and mass-production method to study the blood–brain barrier in vitro. J Neurochem 54:1798–1801
18. Franke H, Galla H-J, Beuckmann CT (1999) An improved low-permeability in vitro-model of the blood–brain barrier: transport studies on retinoids, sucrose, haloperidol, caffeine and mannitol. Brain Res 818:65–71
19. Patabendige A, Skinner RA, Abbott NJ (2013) Establishment of a simplified in vitro porcine blood–brain barrier model with high transendothelial electrical resistance. Brain Res 1521:1–15
20. Al Ahmad A, Taboada CB, Gassmann M et al (2011) Astrocytes and pericytes differentially modulate blood–brain barrier characteristics during development and hypoxic insult. J Cereb Blood Flow Metab 31:693–705
21. Al Ahmad A, Gassmann M, Ogunshola OO (2009) Maintaining blood–brain barrier integrity: pericytes perform better than astrocytes during prolonged oxygen deprivation. J Cell Physiol 218:612–622
22. Song L, Pachter JS (2003) Culture of murine brain microvascular endothelial cells that maintain expression and cytoskeletal association of tight junction-associated proteins. In Vitro Cell Dev Biol Anim 39:313–320
23. Smith QR, Rapoport SI (1986) Cerebrovascular permeability coefficients to sodium, potassium, and chloride. J Neurochem 46:1732–1742
24. Helms HC, Abbott NJ, Burek M et al (2016) In vitro models of the blood–brain barrier: an overview of commonly used brain endothelial cell culture models and guidelines for their use. J Cereb Blood Flow Metab 36:862–890
25. Pardridge WM (1998) CNS drug design based on principles of blood-brain barrier transport. J Neurochem 70:1781–1792
26. Syvänen S, Lindhe Ö, Palner M et al (2009) Species differences in blood-brain barrier transport of three positron emission tomography radioligands with emphasis on P-glycoprotein transport. Drug Metab Dispos 37:635–643
27. Weksler BB, Subileau EA, Perrière N et al (2005) Blood-brain barrier-specific properties of a human adult brain endothelial cell line. FASEB J 19:1872–1874
28. Lippmann ES, Azarin SM, Kay JE et al (2012) Derivation of blood-brain barrier endothelial cells from human pluripotent stem cells. Nat Biotechnol 30:783–791
29. Lippmann ES, Al-Ahmad A, Azarin SM et al (2014) A retinoic acid-enhanced, multicellular human blood-brain barrier model derived from stem cell sources. Sci Rep 4:4160
30. Wilson HK, Canfield SG, Hjortness MK et al (2015) Exploring the effects of cell seeding density on the differentiation of human pluripotent stem cells to brain microvascular endothelial cells. Fluids Barriers CNS 12:13
31. Hollmann EK, Bailey AK, Potharazu AV et al (2017) Accelerated differentiation of human induced pluripotent stem cells to blood–brain barrier endothelial cells. Fluids Barriers CNS 14:9

Chapter 2

In Vitro Techniques for Assessing Neurotoxicity Using Human iPSC-Derived Neuronal Models

Anke M. Tukker, Fiona M. J. Wijnolts, Aart de Groot, Richard W. Wubbolts, and Remco H. S. Westerink

Abstract

The central nervous system consists of a multitude of different neurons and supporting cells that form networks for transmitting neuronal signals. Proper function of the nervous system depends critically on a wide range of highly regulated processes including intracellular calcium homeostasis, neurotransmitter release, and electrical activity. Due to the diversity of cell types and complexity of signaling processes, the (central) nervous system is very vulnerable to toxic insults.

Nowadays, a broad range of approaches and cell models is available to study neurotoxicity. In this chapter we show the applicability of human induced pluripotent stem cell (hiPSC)-derived neuronal co-cultures for in vitro neurotoxicity testing. We demonstrate that immunocytochemistry can be used to visualize networks of cultured cells and to differentiate between different cell types. Live cell imaging and electrophysiology techniques demonstrate that the neuronal networks develop spontaneous activity, including synchronized calcium oscillations that coincide with spontaneous changes in membrane potential as well as spontaneous electrical activity with defined (network) bursting. Importantly, as shown in this chapter, spontaneously active human iPSC-derived neuronal co-cultures are suitable for in vitro neurotoxicity assessment. Future application of live imaging and electrophysiological techniques on hiPSC from different donors and/or patients differentiated in different cell types holds great promise for personalized neurotoxicity assessment and safety screening.

Key words In vitro neurotoxicity screening, Human induced pluripotent stem cell-derived neuronal models, Single-cell fluorescent microscopy, Calcium homeostasis, Membrane potential, Spontaneous neuronal activity, Immunocytochemistry, Multi-well microelectrode array

1 Introduction: Neuronal Network Communication and Neurotoxicity

The (central) nervous system consists of sophisticated neuronal networks that control body function, either via direct control or indirect via input in glands. The main function of neurons that make up the nervous system is to send and receive signals, a process called neurotransmission. To that aim, neurons have a typical structure with dendrites bringing the signal toward the cell body and an axon transmitting the signal away from the cell

Michael Aschner and Lucio Costa (eds.), *Cell Culture Techniques*, Neuromethods, vol. 145, https://doi.org/10.1007/978-1-4939-9228-7_2, © Springer Science+Business Media, LLC, part of Springer Nature 2019

body. As soon as dendrites receive an excitatory chemical signal, the neuron becomes activated and translates this chemical input signal in an electrical signal, an action potential (AP). Via opening of voltage-gated sodium and potassium channels, the AP induces a change in membrane potential that travels via the cell body along the axon to the synapse at the axon terminal. There, the electrical signal (AP) is converted into a chemical signal that will be transferred to (an)other neuron(s). The first step in conversion from the electrical to the chemical signal involves opening of voltage-gated calcium channels (VGCC), resulting in a strong influx of calcium ions (Ca^{2+}). The resulting changes in the intracellular Ca^{2+} concentration ($[Ca^{2+}]_i$) are involved in a variety of cellular processes such as excitability, plasticity, motility, and viability [1, 2]. $[Ca^{2+}]_i$ is crucial for the regulation of neurotransmission as Ca^{2+} influx through VGCCs triggers the release of neurotransmitters from the presynaptic cell into the synaptic cleft [2–4]. Neurotransmitters are chemical signaling molecules that are stored in vesicles in the presynaptic neuron. There are different types of excitatory and inhibitory neurotransmitters, such as acetylcholine, dopamine, serotonin, glutamate, and gamma-aminobutyric acid (GABA). After release into the synaptic cleft by fusion of the vesicles with the presynaptic plasma membrane, neurotransmitters can bind to receptors on the postsynaptic membrane. In the receiving cell, the signal can then be converted in a new AP, or it can activate intracellular signaling pathways. The chemical signal is terminated by degradation or reuptake of the neurotransmitters from the synaptic cleft (for review see [2, 4]).

Communication in neuronal networks thus critically depends on the structure of neurons, intact neuronal membranes, and regulation of cellular and molecular mechanisms underlying neurotransmission. Additionally, proper neuronal communication also depends on supporting cells such as oligodendrocytes, astrocytes, and microglia. The multicellular nature of the networks can be confirmed with techniques as immunocytochemistry, whereas proper intracellular signaling can be studied with imaging techniques focusing on intracellular calcium levels and the membrane potential. Finally, the resulting network activity can be assessed with the use of multi-well microelectrode arrays (mwMEA).

Its complexity and poor regenerative capacity make the nervous system vulnerable to toxic insults caused by chemical, physiological, and biological agents that are present in the surrounding environment. Neurotoxicity is thus defined as an adverse effect caused by any of these agents on the structure and/or function of the nervous system. Nowadays, there is a broad range of approaches and cell models to study neurotoxicity in vitro.

1.1 Methods to Study Neuronal Network Communication

In vitro cell models should mimic the in vivo situation as closely as possible. For neurotoxicity testing, this means that the in vitro model must form functional neuronal networks with both inhibitory and excitatory neurons as well as supporting cells. In order to circumvent interspecies translation, cells from human origin are the preferred option. Recently, human induced pluripotent stem cell (hiPSC)-derived neurons became commercially available. A benefit of these cells is that they do not require long differentiation into neural progenitor cells and ultimately into functional neurons, which can take several weeks [5, 6] till months [7, 8]. It has been shown that these hiPSC-derived neurons exhibit the behavior and function of mature neurons [7, 8]. Therefore, we chose to use mixed hiPSC-derived neuronal models for the techniques described in this chapter.

First of all, it is important to determine whether the cultured cells form neuronal networks. This can be studied using fluorescent antibodies and confocal microscopy to detect specific target proteins in the cell or on the cell membrane. Besides studying network formation and complexity, immunofluorescent stainings can be used to differentiate between (neuronal) cell types in co-cultures.

Once the cells formed mixed neuronal networks, spontaneous network activity and the effect of toxic insults can be studied by looking at intracellular calcium homeostasis. Changes in $[Ca^{2+}]_i$ can be analyzed by loading the cultured cells with a high-affinity Ca^{2+}-responsive fluorescent dye. Similarly, fluorescent voltage-sensitive dyes can be used to study changes in membrane potential as an indication for the occurrence of electrical activity.

Another way to look at spontaneous neuronal network activity and (network) bursting and toxic effects hereon is by the use of electrophysiological methods. The introduction of mwMEAs provided a way to grow cells on a culture surface with an integrated array of microelectrodes. This allows for simultaneous and noninvasive recording of extracellular local field potentials at a millisecond time scale at different locations in the network grown in vitro (for review see [9]). Mammalian neuronal networks grown in vitro on mwMEA display many characteristics of in vivo neurons, including the development of spontaneous neuronal activity [10] and synchronized bursting [11]. It has also been shown that these networks are responsive to neurotransmitters [12], indicating the presence of a wide range of common neurotransmitter receptors. This technique offers consistent reproducibility across different laboratories [13, 14] and a high sensitivity and specificity [15, 16]. For these reasons mwMEAs are seen as a suitable and consistent in vitro neurotoxicity screening method. Because measurements take place in a sterile environment, this technique allows for both acute [17–19] and chronic toxicity screening [20]. Most mwMEA research was done with rat primary cortical cultures [21–24], but recently it has been shown that also hiPSC co-cultures grow on mwMEA plates, develop spontaneous activity and bursting behavior, and are suitable for neurotoxicity screening [25–27].

2 Materials

2.1 Cell Culture

For all techniques described in this chapter, we used commercially available hiPSC-derived neurons (iCell® Neurons and iCell® Glutaneurons) and astrocytes (iCell® Astrocytes). All cells were obtained from Cellular Dynamics International (Madison, WI, USA). Cells were cultured at 37 °C in a humidified 5% CO_2 incubator. From previous experiments, we know that these cells grow better on polyethylenimine (PEI)-coated surfaces compared to poly-L-lysine (PLL)- or poly-L-ornithine (PLO)-coated materials. We therefore pre-coated all our cell culture surfaces with 0.1% PEI solution diluted in borate buffer (24 mM sodium borate/50 mM boric acid in Milli-Q, pH adjusted to 8.4) unless stated otherwise. Co-cultures were grown in BrainPhys™ medium supplemented with 2% iCell Neuron supplement, 1% nervous system supplement, 1% penicillin-streptomycin, 1% N2 supplement, and 0.1% laminin (L2020 Sigma-Aldrich, Zwijndrecht, The Netherlands). Astrocytes (iCell®) were cultured in astrocyte medium (DMEM with high glucose and 10% FBS, 1% N2 supplement, and 1% penicillin-streptomycin).

Young astrocytes can proliferate rapidly and potentially overgrow neuronal cultures. Astrocytes were therefore passaged two to three times and stored in liquid nitrogen until use in co-culture with the hiPSC-derived neurons. Young astrocytes were thawed by gently swirling them for 2–3 min in a 37 °C water bath. Then, the content of the vial was transferred to a sterile 50 mL tube. The vial was rinsed three times with astrocyte medium. Total volume in the 50 mL tube was brought to 10 mL, and cells were centrifuged for 5 min at 1300 rpm. The cell pellet was dissolved in 6 mL astrocyte medium, and cells were transferred to a 25 cm^3 Geltrex™-coated culture flask (Geltrex™ was added to cover the bottom of the flask and incubated for 45–60 min at 37 °C in a humidified incubator, after which the Geltrex™ was removed). These flasks were used for a total of 1–1.4 × 10^6 cells. Astrocyte medium was replaced every 3–4 days. Cells were passaged by removing the medium, rinsing with PBS and adding 1 mL 0.0125% trypsin for ~5 min to the flask. During trypsin incubation the flask was placed at 37 °C in a humidified 5% CO_2 incubator. It is important to carefully check that cells are detached since astrocytes adhere strongly to the flask and may require more than 5 min incubation. Once the cells were detached, 9 mL of medium was added to the cells. Cells were counted and centrifuged for 5 min at 1300 rpm. For passaging, 3–4.2 × 10^6 cells were transferred to a 75 cm^3 Geltrex™-coated culture flask, and the volume was brought to 20 mL. In case cells were used for co-culturing and not for passaging, the cell pellet was dissolved in Complete iCell Neurons Maintenance Medium supplemented with 1% penicillin-streptomycin, 2% iCell Neuron medium supplement, and 1% laminin (L2020, Sigma-Aldrich, Zwijndrecht, The Netherlands) into a 14,000 cells/μL solution.

Procedures for thawing of iCell® Neurons and iCell® Glutaneurons are comparable. The vial with neurons was thawed by gently swirling it for 2–3 min in a 37 °C water bath. The cell suspension was transferred to a 50 mL tube, and the vial was rinsed three times with Complete iCell Neurons Maintenance Medium. The total volume in the 50 mL tube was brought to 10 mL, and the tube was turned upside down twice. A sample for cell counting was taken, and cells were centrifuged for 5 min at 1300 rpm. The pellet of iCell® Neurons was dissolved in dotting medium (i.e., supplemented BrainPhys™ medium with 10% laminin) to a solution of 14,000 cells/μL. In parallel, iCell® Glutaneurons were dissolved in dotting medium to a solution of 13,000 cells/μL.

When all cells were thawed or detached, we created two types of co-culture models that differ in the ratio of inhibitory neurons by addition or absence of iCell® Neurons to create different profiles of neuronal activity (see [25] for details). First, a mixture was made of ~13% iCell® Astrocytes, ~17% iCell® Neurons, and ~70% iCell® Glutaneurons (culture model A) according to Table 1. Cells were plated in 10 μL droplets (Table 2). In order to get the total

Table 1
Composition of cell models and plating density

Culture	Cell type and %	N/well	Solution	μL/well
A	~13% astrocytes	10,000	14,000 cells/μL	0.71 μL
	~17% iCell® Neurons	13,000	14,000 cells/μL	0.93 μL
	~70% iCell® Glutaneurons	52,000	13,000 cells/μL	4 μL
B	~15% astrocytes	11,250	14,000 cells/μL	0.80 μL
	~85% iCell® Glutaneurons	63,750	13,000 cells/μL	4.55 μL

Table 2
Cell culture surfaces and medium volume

Culture surface	Experiment	Total volume
48-well MEA plate (Axion Biosystems Inc., Atlanta, GA, USA)	MEA recording	300 μL/well
μ-slide 8-well chambered coverslip (ibidi GmbH, Planegg, Germany)	Immunocytochemistry	200 μL/well
Glass-bottom dishes (MatTek, Ashland MA, USA)	Live fluorescence imaging	2 mL/dish

volume to 10 μL, 4.36 μL dotting medium was added. For culture model B, a mixture was made of ~15% iCell® Astrocytes and ~85% iCell® Glutaneurons (according to Table 1). In this case, 4.65 μL dotting medium was added to bring the total plating volume to 10 μL. Following plating, droplets were allowed to adhere for 1 h after which medium was added (Table 2). On DIV1, 50% of the medium was refreshed with room temperature (RT) supplemented BrainPhys™ medium. Hereafter, 50% medium changes took place three times a week up till DIV23.

Since both co-culture models require several days to develop functional neuronal networks, MEA and imaging experiments should not be performed before DIV11. In our experience, the optimum window for performing MEA and live imaging experiments ranges from DIV14 to DIV23. It should be noted that the optimum window for measurements differs between the various available commercial models as well as culture conditions (e.g., cell density and % astrocytes).

It should be noted that using a high ratio of astrocytes may cause the cells to cluster, complicating imaging and MEA experiments. Developmental curves can be measured from DIV4 onward and can be used to determine the optimum window for assessing acute neurotoxicity. We strongly recommend to always measure a developmental curve before performing neurotoxicity assessment when starting to work with new cells and/or new culture protocols.

2.2 Confocal Microscopy for Immunocytochemistry

Immunofluorescent images of chemically fixed samples were captured with a Leica DMI4000 TCS SPEII confocal microscope. To capture images, a 20× oil immersion objective (ACS APO IMM NA 0.6) was used. The 20× objective allows for visualization of multiple neurons in one frame with the connecting dendrites and axons being clearly visible. When a 10× objective is used, too many details and sensitivity are lost, whereas a 40× objective does not capture the complexity of the network structure. It is recommended to visually scan the complete chamber to make sure the selected area is representative for the culture. We noticed that the neuronal co-cultures tend to grow differently at the periphery of the culture area compared to the center. In order to capture an area that best matches the region where MEA measurements take place, a region in the middle of the dish is chosen. Once a representative area is found, it's recommended to define an upper and lower limit in order to make a picture based on *z-stacks*. We recommend imaging a *z-stack* series, since the elaborate extensions of the cells are poorly captured in a single plane. An axially extended view provided by a maximal intensity projection visualizes the structures better. Images were captured as .lif files using Leica Application Suite Advanced Fluorescence software (LAS AF version 2.6.0; Leica Microsystems GmbH, Wetzlar, Germany).

2.3 Live Cell $[Ca^{2+}]_i$ Changes Using a Temperature Controlled Microscopy Unit

Live changes in $[Ca^{2+}]_i$ were monitored with the fluorescent dye Fura-2AM (Ex 340 and 380/Em 510, Life Technologies, Bleiswijk, The Netherlands) using an Axiovert 35 M inverted microscope (Zeiss, Göttingen, Germany) as described previously [28]. We used a 40× oil immersion objective (Plan-NeoFluar NA 1.30) to capture images. Light with an excitation wavelength of 340 and 380 nm evoked by a monochromator (TILL Photonics Polychrome IV; TILL Photonics GmBH, Gräfelfing, Germany) was directed to the sample via a 290 nm long-pass filter and beam splitter. From there, the emitted light with a wavelength of 510 nm was directed to a 440 nm long-pass filter and was collected every 0.5 s, i.e., at a sample frequency of 2 Hz for each excitation wavelength, with an Image SensiCam digital camera (TILL photonics GmBH). The degree of fluorescence in Fura-2-loaded neuronal co-cultures allowed us to sample with binning 2 × 2. TILLvisION (version 4.01) software was used to trigger the light source, camera, and data acquisition.

For the cells to exhibit spontaneous network activity, experiments must take place at 37 °C. Below this temperature no spontaneous calcium oscillations are visible (data not shown). In order to keep our samples at a stable temperature, we equipped the microscope with a custom-built heating system and bipolar temperature control unit (TC-202, Medical Systems Corp, Greenvale, New York, USA).

Changes in the ratio F340/F380 of selected regions of interest reflecting the changes in $[Ca^{2+}]_i$ were further analyzed using custom-made MS Excel macros. Since the cells form networks, spontaneous synchronous calcium oscillations are seen in all regions.

2.4 Simultaneous Live Cell Imaging of Calcium Transients and Membrane Depolarization

To monitor calcium levels, X-Rhod-1 (Ex 580/Em 602, AM-ester derivative, Life Technologies, Bleiswijk, The Netherlands) was used and for membrane depolarization the dye FluoVolt (Ex 488/Em 515, Life Technologies, Bleiswijk, The Netherlands).

Live cell microscopic fluorescence recordings were performed on a commercial NIKON Ti system equipped with an EMCCD camera (iXon Ultra897, ANDOR) using wide-field illumination. A 10× CFI Plan Fluor air objective (NA 0.3, WD 16 mm) was applied to record the data. A LED light source (Lumencor Spectra X) illuminated the samples to deliver 470/24 nm and 575/25 nm light bandwidths sequentially. Emitted fluorescence light was collected through a quadband emission filter (Chroma, DAPI/FITC/TRITC/Cy5 Quad). Triggered synchronization of the LEDs and the camera was directed with the NIS-Elements software ND acquisition module (NIKON, version 4.6) via a NIDAQ communication board (National Instruments). Halogen illumination light path was regulated by a fast shutter (Lambda SC, SUTTER Instruments) to obtain differential interference contrast (DIC) images without the delay of slow on and off glowing rates of these lamps. Fast focus control is performed by a piezo stepping stage device (Mad City Labs, Madison, WI, USA).

In the Spectra X module of the NIS-Elements software, the 470 nm and 575 nm light sources were controlled. Via the triggered acquisition module, light levels and exposure times were set (<3%, ND4 and eight filters were available for careful illumination modulation). The EMCCD camera capture area was cropped to 256 × 256, and 2 × 2 binning was applied at 20 ms exposures (sampling in frame transfer mode at 17 Mhz horizontal pixel shift readout frequency with 300 V amplification gain).

The system is equipped with temperature and CO_2 control in a stage top setup with a small water container to provide a humidified culture environment (TOKAI HIT, INUBG2EH-TiZB). The lens heating option was omitted for the air objectives employed.

Both the automated focus control system of NIKON (perfect focus system or pfs) and the compressor that provides air for the vibration isolation table can create vibrations and were thus switched off for the recordings.

Following data acquisition, recorded data was exported to MS Excel. For further frequency data analysis, one representative trace from a dish was chosen as all cells display synchronized oscillations in calcium levels and membrane potential. To calculate changes in peak amplitude, data was loaded in a custom-made MATLAB program, and averages of multiple responsive cells in the dish were calculated. Data are reported as mean ± SEM form n cells.

2.5 Multi-well Microelectrode Array Recordings

Cells were cultured on 48-well MEA plates (Fig. 1) as described previously [12, 17]. Each well contains an electrode array of 16 nanotextured gold microelectrodes that are ~40–50 μm in diameter with a 350 μm center-to-center spacing. Each well contains four integrated ground electrodes. In total this yields 768 channels in one plate that can be recorded simultaneously. A Maestro 768-channel amplifier with an integrated heating system, temperature controller, and data acquisition interface (Axion Biosystems Inc., Atlanta, GA, USA) was used for recordings of neuronal activity. Data acquisition was managed with Axion's Integrated Studio (AxIS version 2.4.2.13) and recorded as .RAW files. Files were obtained by sampling all channels simultaneously with a gain of 1200× and a sampling frequency of 12.5 kHz/channel using a band-pass filter (200–5000 Hz). Notably, the large number of electrodes, sampled at high frequency, will yield a large amount of data (~1 GB/min recording). Consequently, sufficient storage space should be available.

For data analysis .RAW files were re-recorded to obtain alpha map files. During this re-recording, spikes were detected using the AxIS spike detector (adaptive threshold crossing, Ada BandFlt v2). A variable threshold spike detector was used with a threshold set at

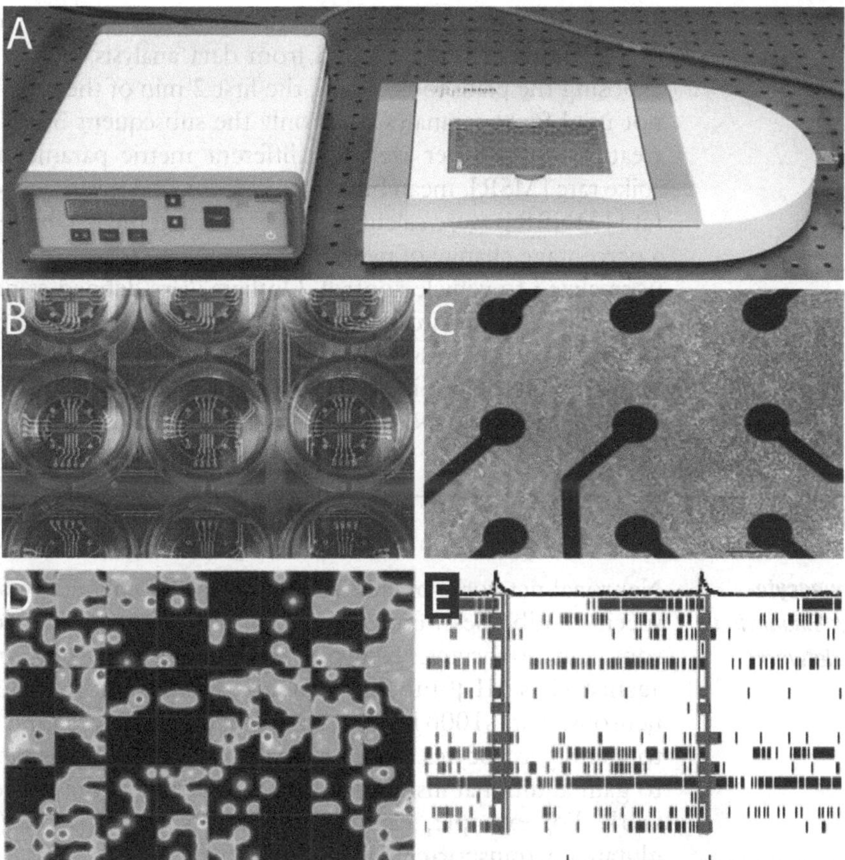

Fig. 1 Experimental setup for measurements of spontaneous neuronal network activity. Axion's Maestro platform (**a**) was used to record neuronal activity of co-cultures grown in 48-well MEA plates (**b**). Each well contains an electrode grid with 16 electrodes/well (**c**) for noninvasive extracellular field recordings. Live heat maps are shown in AxIS software during recordings (**d**). Files are loaded in neural metric to determine (**e**) spikes (black), bursts (blue), and network bursts (pink squares)

7× standard deviation of the internal noise level (rms) on each electrode. The obtained spike files were loaded in Neural Metric (version 2.04, Axion Biosystems). In this program, bursts were detected with the Poisson surprise method (minimal surprise $S = 10$ [29]). Network bursts were extracted with the adaptive threshold method (min # of spikes 10; min % of electrodes 25).

The Neural Metric output files (.csv files) were loaded in a custom-made MS Excel macro. We only used active electrodes (MSR ≥ 6 spikes/min) from active wells (≥1 active electrode) for further analysis. Electrodes were seen as bursting electrodes when the minimum burst rate was ≥0.001 bursts/s. Only wells with ≥2 bursting electrodes were included for network bursting and synchronicity analysis. Effects of test compounds were determined by comparing the activity of the baseline of a well to the activity in that well following exposure. During data analysis it is important to correct for exposure artifacts [12]. To do so, the time it took to expose all wells was excluded from data analysis. For example, if exposing the plate took 2 min, the first 2 min of the recording was not used for data analysis but only the subsequent 30 min. Then, treatment ratios per well for different metric parameters (mean spike rate [MSR], mean burst rate [MBR], and mean network burst rate [MNBR]) were calculated by expressing the $parameter_{exposure}$ as a percentage change of $parameter_{baseline}$. Next, treatment ratios were normalized to vehicle control. Outliers were defined as not within average ± 2× standard deviation and excluded for further analysis. MEA data are expressed as mean ± SEM. A one-way ANOVA was performed to determine statistical significant changes ($p < 0.05$) in MSR, MBR, and MNBR compared to vehicle control.

3 Methods and Results

3.1 Immunocytochemistry: Visualizing Neuronal Networks

Neuronal networks of culture model A were chemically fixed, and specific antibodies were used to demonstrate the presence of neurons and astrocytes. Neurons were identified using antibodies against class III β-tubulin, which is found almost exclusively in neurons. Anti-s100β is a protein specific for glial cells and was used to identify astrocytes. Notably, many other antibodies can be used to gain additional insight in the composition of the neuronal network. For example, using antibodies against vGLUT (vesicular glutamate transporter), vGAT (vesicular GABA transporter), or tyrosine hydroxylase (marker for dopaminergic neurons) will provide information on the types of neurons present.

In order to stain the cultures, a staining protocol described previously [25] was used. The cultures were chemically fixed at DIV16 and 21 with 4% PFA in 0.1 M PBS (pH 7.4) for 15 min at RT. We found that longer fixation times reduced epitope recognition by the antibodies. Following fixation, chambered coverslips were quenched for PFA, permeabilized, and incubated for 20 min at RT with 20 mM NH_4Cl in blocking buffer (2% bovine serum albumin and 0.1% saponin in PBS). Hereafter, chambers were incubated overnight at 4 °C with the primary antibody. The following primary antibodies were used: mouse anti-S100β (final dilution 1:500; Ab11178, Abcam, Cambridge, United Kingdom) to stain astrocytes and rabbit anti-βIII tubulin (final dilution 1:250; Ab18207, Abcam, Cambridge United Kingdom) to visualize the iCell® Neurons and iCell® Glutaneurons. The following day, chambers were washed thrice with blocking buffer and incubated with donkey anti-rabbit Alexa Fluor® 488 (final dilution 1:100; 715-545-152, Life Technologies, Bleiswijk, The Netherlands) and donkey anti-mouse Alexa Fluor® 594 (final dilution 1:100; 715-585-151, Life Technologies, Bleiswijk, The Netherlands) for 45 min at RT in the dark. During the last 2–3 min of the incubation time, 200 nM DAPI (staining the nuclei) was added. Chambers

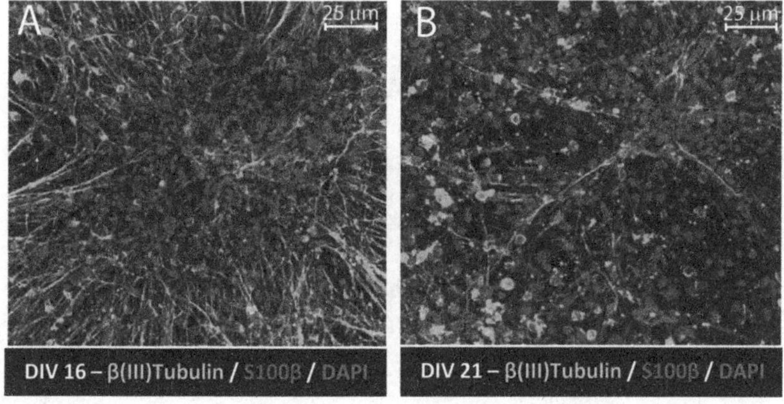

Fig. 2 Immunofluorescent stainings of co-culture model A. At DIV16 (**a**) and DIV21 (**b**), cultures were stained with β(III) tubulin (green) and S100β (red) to identify, respectively, neurons and astrocytes in the co-culture. Nuclei were stained with DAPI (blue). Scale bar depicts 25 μm

were washed again 3× with blocking buffer and sealed with two to three droplets of FluorSave (Calbiochem, San Diego, CA, USA). Now the chambers are ready for use. However, they can be stored for months at 4 °C in the dark until use.

The images of the chemically fixed and stained co-cultures of culture model A show that mixed neuronal networks with a high degree of complexity are formed (Fig. 2a, b). It is clear that the network is already strongly developed at DIV16 (Fig. 2a) and this is maintained till DIV21 (Fig. 2b). The images also show that astrocytes do not overgrow the neuronal population. The ratio astrocytes to neurons remains comparable over time and is in line with the plated ratio. Altogether, these results indicate that this co-culture consists of a mixed cell population with neurons as well as supporting astrocytes that can be used for neurotoxicity testing.

3.2 Live Cell Imaging: $[Ca^{2+}]_i$ Changes in hiPSC-Derived Co-cultures

Both dyes used for the $[Ca^{2+}]_i$ experiments described in this chapter contain an acetoxymethyl (AM) ester. The AM group allows the dye to cross the cell membrane, which allows the cells to be loaded in a noninvasive manner. Following membrane crossing, the AM group is cleaved from the dye by non-specific esterases in the cytosol. The cleaved dye is no longer able to cross the cell membrane and remains in the cytosol. The calcium-sensitive dye Fura-2 is fluorescent green (510 nm), and fluorescence increases upon Ca^{2+} binding following excitation at 340 nm, but fluorescence decreases by excitation at 380 nm. The resulting F340/380 ratio thus correlates directly with the $[Ca^{2+}]_i$.

Cells cultured in glass-bottom dishes of culture model B were loaded with 5 μM Fura-2AM for 1 h at 37 °C in a humidified 5% CO_2 incubator. After loading, cells were washed four times with 37 °C saline solution to remove excess dye. Then dishes with loaded cells were placed on the stage of the inverted microscope in the heating ring. The temperature sensor was placed in the dish,

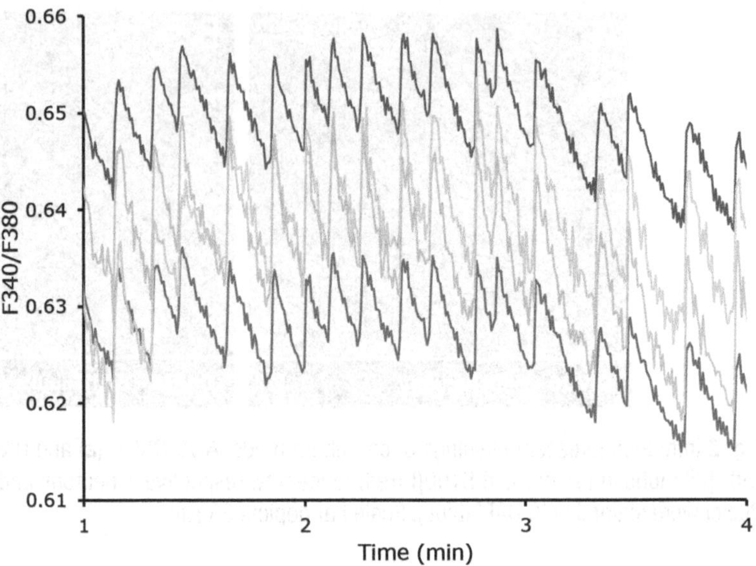

Fig. 3 Sample traces of representative neuronal cells (DIV23) showing spontaneous calcium oscillations. Each trace represents the oscillations of a single cell over a 3 min time period. Cells oscillate synchronously, indicating they are part of a single network

and cells were allowed to warm till 37 °C. As soon as this temperature was reached, a 5 min recording was started to measure spontaneous calcium oscillations.

With this method it is possible to detect spontaneous calcium oscillations in hiPSC-derived neuronal co-cultures (Fig. 3). Sample traces indicate the presence of spontaneous calcium oscillations. Cells oscillate synchronously, indicating that cells are all part of a single network.

3.3 Live Cell Imaging: Simultaneous Calcium Oscillations and Membrane Depolarization in hiPSC-Derived Co-cultures

X-rhodamine-1 (X-RhodAM) was used to study spontaneous calcium oscillations. This dye works in a comparable manner as Fura-2AM. However, the main difference is that Fura-2AM is a dual wavelength dye and X-RhodAM a single wavelength dye in the red color range (emission 595 nm). The latter has as advantage that it can be imaged at a higher sample frequency and allowing it to be used in combination with other sensor dyes. In order to study simultaneous changes in membrane depolarization, we used the dye FluoVolt. In general there are two types of probes that can be used to do this: fast- or slow-response probes. The first type reacts fast, but the magnitude of potential-dependent fluorescence change is small. The latter type reacts slower, but the magnitude of fluorescence fluctuation is high. FluoVolt is a dye that combines characteristics of fast and slow probes as it reacts fast (millisecond time scale) and yields a high magnitude of fluorescence with a small change in membrane potential (~25% change in fluorescence per 100 mV change in membrane potential).

Cells of culture model B were cultured up till DIV21 in glass-bottom dishes and used to record changes in calcium transients and membrane voltage fluctuations. One day prior to measurements, cells were transferred to an incubator close to the imaging station to minimize transport stress and temperature variation at the time of the experiment. At the day of measurements, start the microscope peripheral modules (Spectra X, Mad City Labs stage control, XY control joystick unit, halogen lamp, and TOKAI HIT control unit) before powering the microscope stand and finally turn on the PC. Distilled water was placed in the basin of the TOKAI HIT sample holder. It is important to place a dummy sample in the holder to prevent condensation on optics below the table. At this point, the climate control unit (37 °C for the basin, 40 °C for the TOP deck heating, 5% CO_2) is switched on, and the whole setup is left for acclimatization for 30 min. All experiments took place at 37 °C.

During the acclimatization time, cultures were loaded with a mixture of 5 μM X-Rhod-1, 5 μM FluoVolt, and PowerLoad solution (a solubilizing agent provided in FluoVolt kit) in life cell imaging solution (LCIS; 140 mM NaCl, 2.5 mM KCl, 1.8 mM $CaCl_2$, 1.0 mM $MgCl_2$, 20 mM HEPES, 20 mM glucose) for 15–20 min. Following incubation, cells were washed four times with 1 mL LCIS. After loading, the sample was placed in the TOKAI HIT holder (UNIVD35) to search for a representative area and select regions of interest (ROIs) to obtain live intensity preview plots (Fig. 4). Next, a 10 min baseline recording was made prior to addition of test compound and a subsequent 10 min exposure recording. For exposure recordings, a fresh stock solution of picrotoxin (PTX) in EtOH was prepared on the day of the experiment. Stock solution was further diluted in LCIS. Solvent concentration did not exceed 0.1% v/v.

Data again confirm that neuronal co-cultures form networks, since oscillations in membrane potential as well as calcium occur for all cells at the same time (Fig. 5a, b left). Exposure (dilution

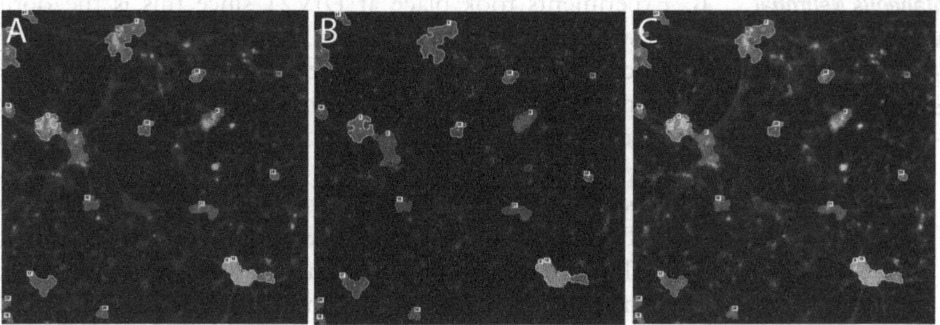

Fig. 4 Captures of neuronal cells loaded with FluoVolt (**a**), X-Rhod (**b**), and an overlay (**c**). Encircled areas are selected regions of interest for measurements as further illustrated in Fig. 5

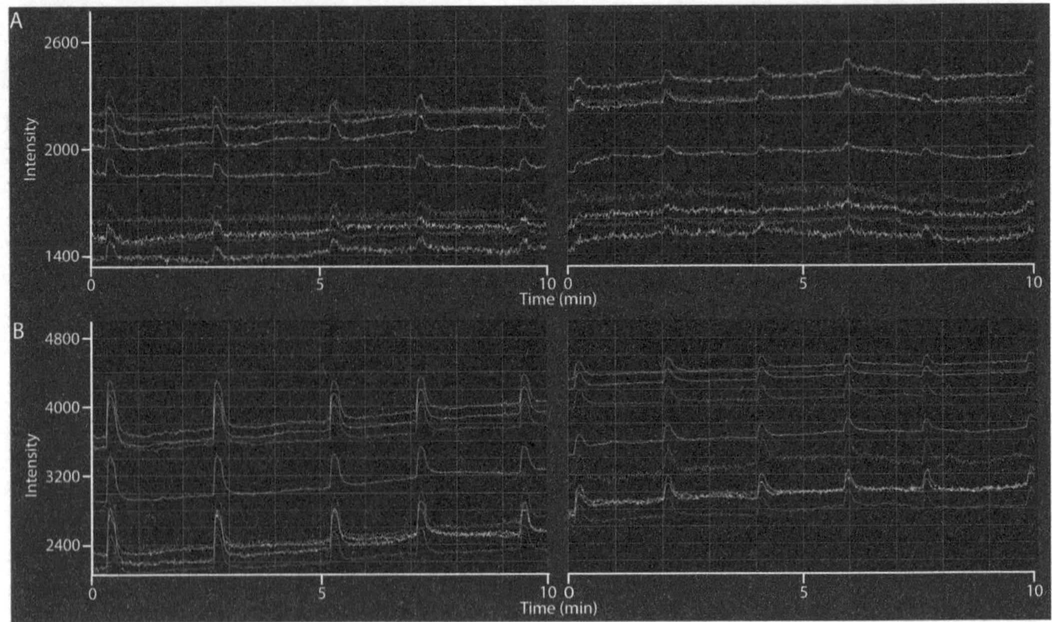

Fig. 5 Selection of traces of cells selected in Fig. 4 for assessing changes in fluorescence of FluoVolt (**a**) and X-Rhod-1 (**b**) during baseline (left) and exposure (right) to 10 μM PTX. Different cells oscillate simultaneously, indicating that changes in membrane potential (**a**) and $[Ca^{2+}]_i$ (**b**) occur at the same time and the cells are part of the same neuronal network

1:10) to 10 μM (PTX) does not affect the frequency of oscillations. However, it does result in a decrease of the amplitude of membrane potential peaks to 57.2% (±3.4, $n = 6$) of baseline (Fig. 5a). A stronger effect is seen on calcium oscillations, where amplitude decreases to 37.5% (±3.2, $n = 6$) compared to baseline (Fig. 5b). Basal fluorescence increases over time, likely as a result of ongoing de-esterification of dye that still contained the AM module. However, filter and/or subtraction methods can be used to eliminate this trend in fluorescence from data analysis.

3.4 MEA: Assessing Spontaneous Neuronal Network Activity in hiPSC-Derived Co-cultures

All MEA measurements took place at 37 °C with culture model B. Experiments took place at DIV21 since this is the optimum window in terms of activity (Table 3). It should be noted that different culture models develop differently and may therefore have a different optimum window. For this reason, it is strongly recommended to always make a developmental curve before starting toxicity experiments. In order to determine effects of test compounds on spontaneous network activity and (network) bursting of the hiPSC-derived co-culture, a 30 min baseline recording was made. Prior to the 30 min recording, plates were allowed to equilibrate in the Maestro for ~5 min. Immediately following this 30 min baseline recording, cells were exposed (dilution 1:10) to the test compounds or the solvent control, and another 30 min recording was made. Each well was exposed only once, since cumulative dosing

Table 3
Development of spontaneous neuronal activity and (network) bursting at different DIVs. Data are expressed as mean ± SEM

	MSR		MBR		MNBR	
	Frequency (Hz)	% active wells	Frequency (Hz)	% active wells	Frequency (Hz)	% active wells
DIV7	1.53 ± 0.14	93.8	0.04 ± 0.01	26.7	0.01 ± 0.01	25
DIV14	1.43 ± 0.14	93.8	0.02 ± 0	13.3	0.01 ± 0.01	100
DIV21	1.29 ± 0.18	93.8	0.05 ± 0.01	76.7	0.03 ± 0.03	65.2

may confound results due to, e.g., receptor (de)sensitization. We prepared fresh PTX stock solutions in EtOH and of strychnine in supplemented BrainPhys™ medium prior to every experiment. Stock solutions of PTX were further diluted in supplemented BrainPhys™ medium such that solvent concentration never exceeded 0.1% v/v.

Neuronal activity in culture model B can be modulated with strychnine (Fig. 6a). Exposure to the highest tested concentration strongly decreases MSR, MBR, and MNBR (Fig. 6a, right) as compared to baseline activity (Fig. 6a, left). The inhibitory effect of strychnine on MSR is concentration-dependent (Fig. 6b). This in contrast to the MBR, which increases following exposure to low concentrations of strychnine and decreases following exposure to higher concentrations. MNBR decreases by all tested concentrations of strychnine. Exposure to PTX (Fig. 6b) has little effect on the MSR, except for exposure to 1 μM, which decreases the MSR. The lowest test concentration of PTX causes a decrease in MBR; however with increasing concentrations, the MBR increases as well. On the other hand, all tested concentrations of PTX cause a decrease in MNBR.

The different effects that strychnine and PTX have on culture model B become clear from the inclusion of additional metric parameters illustrated by a heat map (Fig. 7). This heat map also indicates that for proper MEA data analysis, it is important to include more parameters than just MSR, MBR, and MNBR.

4 Conclusions

Human iPSC-derived co-cultures develop functional and spontaneously active neuronal networks consisting of mature neurons [8, 30]. Our immunocytochemistry data demonstrate the mixed nature of our co-culture models consisting of neurons and astrocytes that form complex, multicellular networks (Fig. 2).

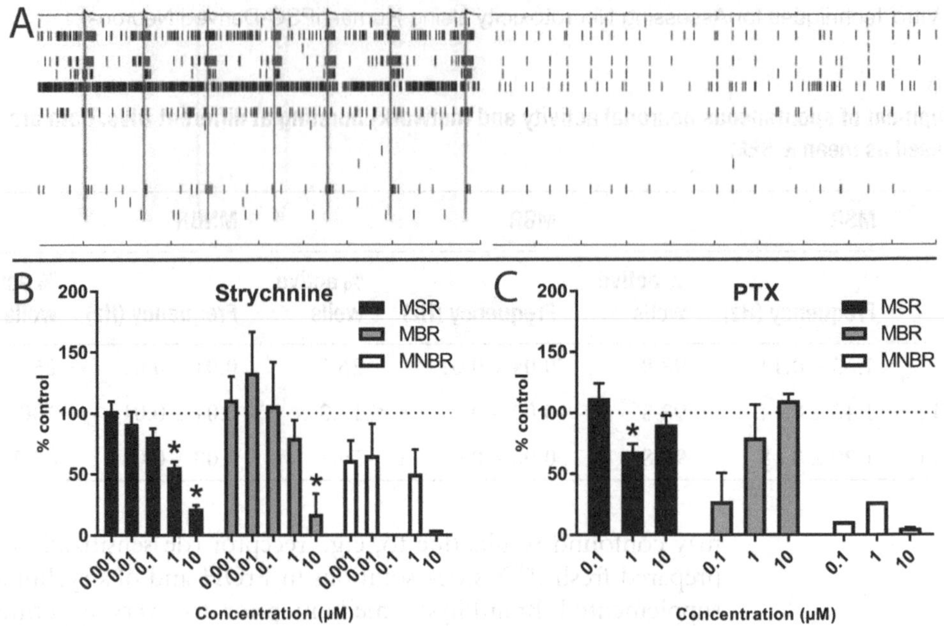

Fig. 6 Toxicological modulation of spontaneous network activity, bursting, and network bursting. Cultures were exposed at DIV21 to strychnine (**a**, **b**) or PTX (**c**). Spike raster plot depicting activity and (network) bursting before (left) and following exposure (right) to 10 µM strychnine (**a**). Each row depicts one electrode, each tick representing one spike (field potential) in a 200 s interval, bursts are depicted in blue, and network bursts are encircled in pink squares. Data are expressed as MSR, MBR, or MNBR as % change relative to vehicle control; mean ± SEM from $n = 1–12$; $^*p < 0.05$ (**b**, **c**)

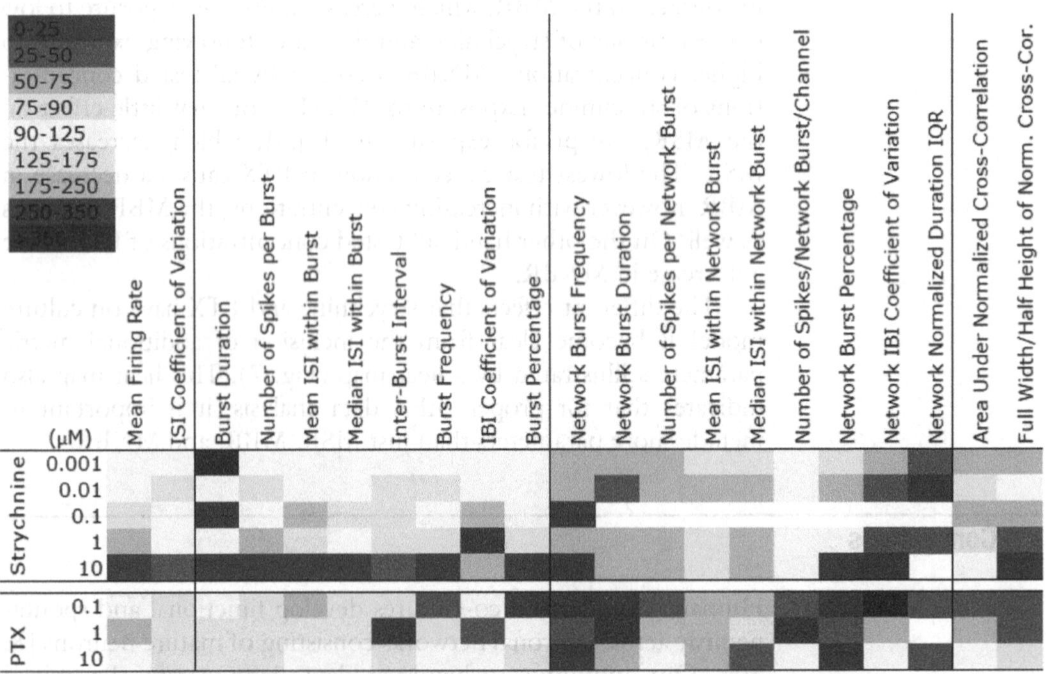

Fig. 7 Heat map of the effects of strychnine and PTX (concentration in µM) on selected metric parameters on culture model B. Color scaling is based on the magnitude of the % of change relative to the vehicle control. No average could be calculated for white cells

We have shown that hiPSC-derived neuronal co-cultures are amenable to multiple real-time recording techniques, including live cell imaging and electrophysiology. Our calcium imaging data indicate that the co-culture models develop spontaneous calcium oscillations and spontaneous changes in membrane potential (Figs. 3 and 5). Since calcium oscillations and changes in membrane potential occur in multiple cells at the same time (Figs. 3 and 5), it can be concluded that functional networks are formed. Moreover, MEA recordings demonstrate that neuronal co-cultures develop spontaneous network activity and (network) bursting (Table 3 and Fig. 6).

Spontaneously active human iPSC-derived neuronal co-cultures are suitable for (preliminary) neurotoxicity assessment [18, 25, 27, 31, 32] as is confirmed by our data (Figs. 5, 6, and 7). However, it must be noted that model composition, e.g., the ratio of GABAergic and glutamatergic neurons and the presence of astrocytes, greatly influences the model's characteristics [25]. Therefore, a careful model characterization must be performed prior to toxicity testing.

The increasing availability of hiPSC from different donors and/or patients differentiated in different cell types, e.g., GABAergic, glutamatergic, dopaminergic neurons and astrocytes as well as peripheral neurons, holds great promise for future personalized toxicity and safety screening. Using hiPSC-derived neurons in combination with the techniques described here will provide a good starting point for neurotoxicity assessment.

Acknowledgments

We gratefully acknowledge members of the Neurotoxicology Research Group for helpful discussions. This work was funded by a grant from the National Centre for the Replacement, Refinement and Reduction of Animals in Research (NC3Rs; project number 50308-372160), the Netherlands Organisation for Health Research and Development (ZonMW; InnoSysTox project number 114027001), and the Faculty of Veterinary Medicine (Utrecht University, The Netherlands).

References

1. Clapham DE (2007) Calcium signaling. Cell 131:1047–1058. https://doi.org/10.1016/J.CELL.2007.11.028
2. Westerink R (2006) Targeting exocytosis: ins and outs of the modulation of quantal dopamine release. CNS Neurol Disord Drug Targets 5: 57–77. https://doi.org/10.2174/187152706784111597
3. Barclay JW, Morgan A, Burgoyne RD (2005) Calcium-dependent regulation of exocytosis. Cell Calcium 38:343–353
4. Südhof TC (2014) The molecular machinery of neurotransmitter release (nobel lecture). Angew Chem Int Ed 53:12696–12717. https://doi.org/10.1002/anie.201406359

5. Kuijlaars J, Oyelami T, Diels A et al (2016) Sustained synchronized neuronal network activity in a human astrocyte co-culture system. Sci Rep 6:36529. https://doi.org/10.1038/srep36529
6. Görtz P, Fleischer W, Rosenbaum C et al (2004) Neuronal network properties of human teratocarcinoma cell line-derived neurons. Brain Res 1018:18–25. https://doi.org/10.1016/j.brainres.2004.05.076
7. Odawara A, Katoh H, Matsuda N, Suzuki I (2016) Physiological maturation and drug responses of human induced pluripotent stem cell-derived cortical neuronal networks in long-term culture. Sci Rep 6:26181. https://doi.org/10.1038/srep26181
8. Paavilainen T, Pelkonen A, Mäkinen ME-L et al (2018) Effect of prolonged differentiation on functional maturation of human pluripotent stem cell-derived neuronal cultures. Stem Cell Res 27:151–161. https://doi.org/10.1016/j.scr.2018.01.018
9. Johnstone AFM, Gross GW, Weiss DG et al (2010) Microelectrode arrays: a physiologically based neurotoxicity testing platform for the 21st century. Neurotoxicology 31:331–350
10. Robinette BL, Harrill JA, Mundy WR, Shafer TJ (2011) In vitro assessment of developmental neurotoxicity: use of microelectrode arrays to measure functional changes in neuronal network ontogenyl. Front Neuroeng 4:1. https://doi.org/10.3389/fneng.2011.00001
11. Cotterill E, Hall D, Wallace K et al (2016) Characterization of early cortical neural network development in multiwell microelectrode array plates. J Biomol Screen 21:510–519. https://doi.org/10.1177/1087057116640520
12. Hondebrink L, Verboven AHA, Drega WS et al (2016) Neurotoxicity screening of (illicit) drugs using novel methods for analysis of microelectrode array (MEA) recordings. Neurotoxicology 55:1–9. https://doi.org/10.1016/j.neuro.2016.04.020
13. Novellino A, Scelfo B, Palosaari T et al (2011) Development of micro-electrode array based tests for neurotoxicity: assessment of interlaboratory reproducibility with neuroactive chemicals. Front Neuroeng 4:4. https://doi.org/10.3389/fneng.2011.00004
14. Vassallo A, Chiappalone M, De Camargos Lopes R et al (2017) A multi-laboratory evaluation of microelectrode array-based measurements of neural network activity for acute neurotoxicity testing. Neurotoxicology 60:280–292. https://doi.org/10.1016/j.neuro.2016.03.019
15. McConnell ER, McClain MA, Ross J et al (2012) Evaluation of multi-well microelectrode arrays for neurotoxicity screening using a chemical training set. Neurotoxicology 33:1048–1057. https://doi.org/10.1016/j.neuro.2012.05.001
16. Valdivia P, Martin M, LeFew WR et al (2014) Multi-well microelectrode array recordings detect neuroactivity of ToxCast compounds. Neurotoxicology 44:204–217. https://doi.org/10.1016/j.neuro.2014.06.012
17. Nicolas J, Hendriksen PJM, van Kleef RGDM et al (2014) Detection of marine neurotoxins in food safety testing using a multielectrode array. Mol Nutr Food Res 58:2369–2378. https://doi.org/10.1002/mnfr.201400479
18. Hondebrink L, Kasteel EEJ, Tukker AM et al (2017) Neuropharmacological characterization of the new psychoactive substance methoxetamine. Neuropharmacology 123:1–9. https://doi.org/10.1016/j.neuropharm.2017.04.035
19. Bradley JA, Luithardt HH, Metea MR, Strock CJ (2018) In vitro screening for seizure liability using microelectrode array technology. Toxicol Sci 163:240–253. https://doi.org/10.1093/toxsci/kfy029
20. Dingemans MML, Schütte MG, Wiersma DMM et al (2016) Chronic 14-day exposure to insecticides or methylmercury modulates neuronal activity in primary rat cortical cultures. Neurotoxicology 57:194–202. https://doi.org/10.1016/j.neuro.2016.10.002
21. Hogberg HT, Sobanski T, Novellino A et al (2011) Application of micro-electrode arrays (MEAs) as an emerging technology for developmental neurotoxicity: evaluation of domoic acid-induced effects in primary cultures of rat cortical neurons. Neurotoxicology 32:158–168. https://doi.org/10.1016/j.neuro.2010.10.007
22. Alloisio S, Nobile M, Novellino A (2015) Multiparametric characterisation of neuronal network activity for in vitro agrochemical neurotoxicity assessment. Neurotoxicology 48:152–165. https://doi.org/10.1016/j.neuro.2015.03.013
23. Zwartsen A, Hondebrink L, Westerink RH (2018) Neurotoxicity screening of new psychoactive substances (NPS): effects on neuronal activity in rat cortical cultures using microelectrode arrays (MEA). Neurotoxicology 66:87–97. https://doi.org/10.1016/j.neuro.2018.03.007
24. Frank CL, Brown JP, Wallace K et al (2017) From the cover: developmental neurotoxicants disrupt activity in cortical networks on microelectrode arrays: results of screening 86 com-

pounds during neural network formation. Toxicol Sci 160:121–135. https://doi.org/10.1093/toxsci/kfx169

25. Tukker AM, Wijnolts FMJ, de Groot A, Westerink RHS (2018) Human iPSC-derived neuronal models for in vitro neurotoxicity assessment. Neurotoxicology 67:215–225. https://doi.org/10.1016/J.NEURO.2018.06.007
26. Odawara A, Matsuda N, Ishibashi Y et al (2018) Toxicological evaluation of convulsant and anticonvulsant drugs in human induced pluripotent stem cell-derived cortical neuronal networks using an MEA system. Sci Rep 8:10416. https://doi.org/10.1038/s41598-018-28835-7
27. Tukker AM, De Groot MWGDM, Wijnolts FMJ et al (2016) Is the time right for in vitro neurotoxicity testing using human iPSC-derived neurons? ALTEX 33:261–271. https://doi.org/10.14573/altex.1510091
28. Heusinkveld HJ, Thomas GO, Lamot I et al (2010) Dual actions of lindane (γ-hexachlorocyclohexane) on calcium homeostasis and exocytosis in rat PC12 cells. Toxicol Appl Pharmacol 248:12–19. https://doi.org/10.1016/j.taap.2010.06.013
29. Legéndy CR, Salcman M (1985) Bursts and recurrences of bursts in the spike trains of spontaneously active striate cortex neurons. J Neurophysiol 53:926–939. https://doi.org/10.1152/jn.1985.53.4.926
30. Hyysalo A, Ristola M, Mäkinen MEL et al (2017) Laminin α5 substrates promote survival, network formation and functional development of human pluripotent stem cell-derived neurons in vitro. Stem Cell Res 24:118–127. https://doi.org/10.1016/j.scr.2017.09.002
31. Kasteel EEJ, Westerink RHS (2017) Comparison of the acute inhibitory effects of Tetrodotoxin (TTX) in rat and human neuronal networks for risk assessment purposes. Toxicol Lett 270:12–16. https://doi.org/10.1016/j.toxlet.2017.02.014
32. Ishii MN, Yamamoto K, Shoji M et al (2017) Human induced pluripotent stem cell (hiPSC)-derived neurons respond to convulsant drugs when co-cultured with hiPSC-derived astrocytes. Toxicology 389:130–138. https://doi.org/10.1016/j.tox.2017.06.010

Chapter 3

Oxidative Stress Signatures in Human Stem Cell-Derived Neurons

M. Diana Neely and Aaron B. Bowman

Abstract

Compelling evidence suggests that oxidative stress plays a significant role in the pathogenesis of many neurodegenerative diseases as well as the neuronal/glial demise resulting from exposure to environmental stressors. The cellular redox balance is maintained by a host of cellular redox systems with set points that are regulated at a subcellular level. Overwhelming deviations of these redox system balances result in oxidative stress and ultimately deficient functioning of cellular organelles and biomolecules such as protein, lipids, and nucleic acids. The analysis of cellular and subcellular redox states is challenging due to the spatiotemporal compartmentation of redox systems and the cell type, stressor, and exposure paradigm specificity of the responses to a particular insult. Due to this complexity, multiple approaches to examine the presence and nature of oxidative stress in biological systems can be used to enhance rigor and may include the analysis of the level of reactive oxygen/nitrogen species (RONS) present, the evaluation of cellular redox systems, and the modification of biomolecules. We describe here three such methods applied to stem cell-derived neurons: (1) the chloromethyl 2′, 7′-dichlorodihydrofluorescein diacetate (DCF) assay to assess cellular RONS levels, (2) a method to determine the state of the cellular GSH redox system, and (3) a procedure to assess oxidative stress-induced lipid modification. This multifold approach to assess the cellular redox state can establish an "oxidative stress signature" specific for a stressor, a cell type, and exposure paradigm. This threefold approach allows for a better comparison of how different biological systems react to a particular stressor or how different stressors (or exposure paradigms) affect a particular biological model system.

Key words Oxidative stress, Human neurons, Human-induced pluripotent stem cells, Reactive oxygen/nitrogen species (RONS), Glutathione (GSH), Lipid peroxidation, F_2-isoprostanes

1 Introduction

Cellular oxidative stress is believed to play a major role not only in the pathogenesis of many neurodegenerative diseases but also in the environmental neuro-stressor-induced loss of neuronal functionality [1–10]. Cellular redox homeostasis is central to life, and redox processes play a role in the large majority of biological processes and are likely as important to cellular homeostasis as other signaling pathways such as phosphorylation/dephosphorylation and the

Michael Aschner and Lucio Costa (eds.), *Cell Culture Techniques*, Neuromethods, vol. 145, https://doi.org/10.1007/978-1-4939-9228-7_3, © Springer Science+Business Media, LLC, part of Springer Nature 2019

ubiquitin cascade [11–13]. An optimal balance of prooxidants and antioxidants exists for any given physiological condition, and the set point of the redox potential of a given reaction may vary in different subcellular locations [11]. The cellular redox balance is maintained by several cellular redox systems, the most prominent ones include the nicotinamide nucleotide (NAD and NADP) and disulfide reductase systems (glutathione and thioredoxin) [11, 14, 15]. Deviations from these redox set points result in redox signaling through the action of sophisticated mechanisms. More pronounced deviations result in oxidative stress that may cause damage of biomolecules and ultimately lead to disease [11–13]. Analysis of cellular redox states is challenging due to the spatiotemporal compartmentation resulting from substantial differences in redox potential in different subcellular compartments and among different cell types [11–13]. In addition, different oxidative stressors affect the cellular redox balance in different ways [16].

Many different methods to measure cellular oxidative stress have been developed. Oxidative stress is an imbalance in the prooxidant-antioxidant homeostasis in which the prooxidant is dominant resulting, in an aerobic system, in an increase in the so-called reactive oxygen and nitrogen species (RONS). RONS are generated exogenously via environmental stressors and endogenously through multiple cellular mechanisms [13]. This large class of molecules can appear in the form of free radicals (e.g., hydroxyl radical, nitric oxide) or non-radicals (such as H_2O_2, peroxynitrite) with reactivities that differ by up to 11 orders of magnitude. Thus, some assays are based on the quantification of RONS directly. Other assays are based on the assessment of changes in the systems that regulate the cellular redox state, such as the nicotinamide, nucleotide, and disulfide reductase systems. Finally, the extent of oxidative stress can also be assessed by quantifying oxidative stress-induced modifications of cellular macromolecules including DNA, RNA, proteins, and lipid (Fig. 1). The outcomes of all these measures depend on the prooxidant (oxidative stressor) examined, the exposure paradigm (acute, chronic), the time point when the oxidative stress is assessed, and the cell type assessed [16–19].

Here we will describe three different methods to measure oxidative stress levels in neurons derived from human-induced pluripotent stem cells (hiPSC) sampling the cellular redox state at three different levels: (1) the DCFH-DA assay, an indicator of levels of intracellular RONS directly; (2) determination of intracellular GSH levels, an indicator of the state of one of the cellular redox systems; and (3) quantification of isoprostanes, a lipid derivative resulting from the action of oxidative stress on lipids (Fig. 1). Examining cellular oxidative stress using different approaches provides a more in-depth view of the neuronal redox state and reveals a combination of stressor and exposure paradigm-specific oxidative stress phenotypes that we have termed the “oxidative stress signature” of a neuronal population [16].

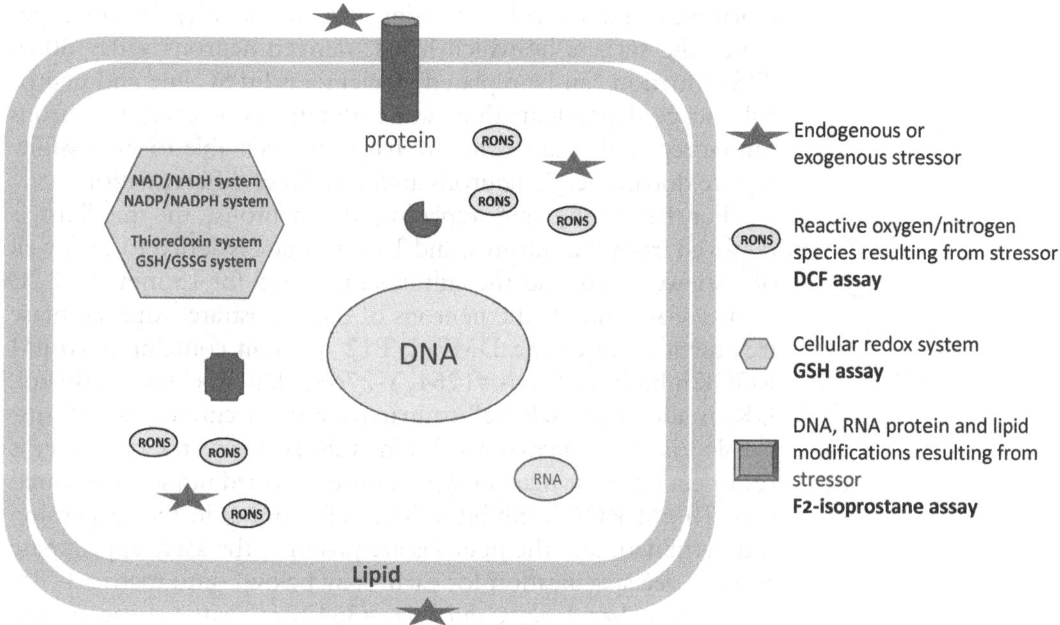

Fig. 1 Oxidative stress induced by a stressor ( ) (symbol incorrect, should be a red star as in figure) results in reactive oxygen/nitrogen species RONS ( ) the level of which can be determined to assess the level of oxidative stress. Here we describe the DCF assay to measure RONS. Cellular redox systems play important roles in maintaining the cellular redox balance, and changes in any of the redox systems resulting from an oxidative stressor can be measured. Here we describe an assay designed to measure cellular GSH levels. Oxidative stress ultimately results in the modification of cellular macromolecules ( ) such as DNA, RNA, proteins, and lipids, all of which can be assessed. Here we describe the quantification of F_2-isoprostanes that result from oxidative modification of the lipid arachidonic acid

2 Materials and Methods

2.1 Preparing hiPSC-Derived Neurons for Oxidative Stress Measures

Protocols for the differentiation of different neuronal lineages from hiPSC will not be covered in this review but rather we refer here to the abundant literature available [16, 20–23]. In order to assess oxidative stress elicited by a toxicant(s), the differentiated neurons need to be replated at a cell density and into a culture vessel compatible with the assay to be applied. Typically, we differentiate and maintain neurons in six-well plates for approximately 25 days. At this time, the cultures are dense and need to be replated for further differentiation or toxicant exposure, which has to be performed at a known cell density and moles of toxicant/cell (see section below "Toxicant exposure"). Around day 25, human neuronal cultures can be replated without significant loss of neurons; harvesting and replating of neurons can be performed at earlier or later stages of differentiation as desired; however, the process becomes more difficult at later stages as the loss of viability due to the stress of replating increases with increasing neuron maturity. In our experience the success by which hiPSC-derived neurons > day 30 of differentiation can be replated efficiently is hiPSC line and neuronal lineage dependent; thus, while glutamatergic cortical neurons can be replated at any time, we have not been able to successfully replate dopaminergic neurons after day 30 of differentiation.

For harvesting and replating the neurons, the medium is removed from the cultures, and 1 ml accutase is added to each well of a six-well plate and the cultures incubated for 15 min at 37 °C to dissociate and lift the neurons of their substrate. After accutase treatment 3 ml of the DMEM/F12 medium containing 10 μM ROCK inhibitor (Tocris #1254, Y-27632 dihydrochloride, Bristol, UK) is added to each well to inactivate the accutase, the cell suspension is centrifugated for 3 min at 200× g, and the cell pellet is resuspended in the neuron-type appropriate medium supplemented with 10 μM ROCK inhibitor. The cell number in the suspension is determined, and the neurons are plated at the assay-appropriate density (see specification for each assay below) onto matrigel (BD Biosciences, San Jose, California, #354277)-coated tissue culture plates. The next day the ROCK inhibitor-containing medium is replaced with the same medium lacking ROCK inhibitor; with neurons of > day 30 of differentiation, applying ROCK inhibitor for additional 1 or 2 days can help with survival of the replated neurons. Should a comparison of responses between younger and more mature neurons be desired, the duration of ROCK inhibitor presence before and/or during the oxidant exposure and oxidative stress assay should be kept the same for all cultures so data can be compared.

Since the efficiency of neural differentiation of hiPSC can show significant variability between hiPSC lines as well as from experiment to experiment of the same cell line, it is crucial that the neuronal cultures are validated for each differentiation so oxidative stress-induced phenotypes can be compared between experiments for which the neuronal populations are comparable. This can be achieved by immune staining and/or quantifying gene expression by reverse transcriptase quantitative polymerase chain reaction (QRT-PCR) for the appropriate markers [16, 20–23].

2.2 Toxicant (Stressor) Exposure

Accumulation of toxicants and thus sensitivity of cultured neurons not only depends on the extracellular concentration of the toxicant present in the medium but is influenced by the mass of exposure, i.e., the ratio of moles of toxicant/cell, and, thus, is dependent on the cell density and the actual volume (not just the concentration) of medium containing a certain stressor [24]. Therefore, for comparisons of the effect of a certain oxidative stressor using different assays, the cell density and volume of medium/cell should be kept constant.

Assessing oxidative stress should be performed in the absence of cell death; thus, conditions of stressor exposure (stressor concentration and time of exposure) should be chosen such that at the time of the measurement of oxidative stress, there is no loss of cell viability. Dead cells, especially human neurons, often display a significant autofluorescence which can interfere with measurements that involve measurements of fluorescence signals [25]. In addition, in order to obtain pertinent information of the role oxidative stress plays for the development of a certain neuronal phenotype, or any biological process, the presence/development/degree/and type of oxidative stress present should be assessed as early as possible, rather than when the neurons are already dead. Thus, the absence of neuronal death after an oxidative stress exposure paradigm should be confirmed initially by a viability assay such as for example, a Cell Titer Blue Assay (Promega, kit G8081). This assay uses the indicator dye resazurin that is reduced in viable cells to the highly fluorescent product resorufin. Fluorescence is measured using a plate reader equipped with a filter set for excitation/emission maxima of 560/590 nm. To follow good laboratory practice for assays of this kind, it is important to confirm that the signal intensity measured lies within the dynamic (linear) range of the assay, i.e., the signal correlates linearly with the cell number. The range of this linear relationship depends on cell type (ability to reduce resazurin), cell density, the ratio of the volume of Cell Titer Blue reagent/volume culture medium, and dye incubation time. We typically plate 100 μl of postmitotic neurons at 0.5×10^6 cells/ml per well of a black 96-well plate, use 20 μl of dye/100 μl of medium, and read the signal at 1 h, 1.5 h, and 2 h after adding the Cell Titer Blue dye. The cell density at the time of plating will have to be adjusted if mitotically active NPCs are used. Cell viability can of course be determined with a wide variety of other available assays.

2.3 Measuring Cellular RONS: Chloromethyl 2′, 7′-Dichlorodi hydrofluorescein Diacetate (DCF) Assay

2.3.1 Principal of the Assay

Chloromethyl 2′, 7′-dichlorodihydrofluorescein diacetate (DCFH-DA) is a widely used small-molecule fluorescent probe which is membrane permeable. Once in the cell, it is hydrolyzed to the DCFH carboxylate anion which is then retained within the cell. A two-electron oxidation of DCFH results in the formation of the fluorescent product dichlorofluorescein (DCF) which can be monitored and quantified by fluorescence-based methods such as microscopy, flow cytometry, or plate reader. This probe has been claimed to measure cellular peroxides directly; however, it is not specific to peroxides but rather detects the radicals to which these peroxides and other RONS are converted [13, 26]. While we describe a kinetic measurement approach here in that we measure the DCF signal changes during toxicant/stressor exposure, the DCF assay can also be used for endpoint determinations.

2.3.2 Method

The method described here involves measurement of the fluorescent signal using a plate reader; when analysis by flow cytometry or microscopy is desired, the plating of the neurons for the assay and the oxidant exposure paradigm need to be adjusted accordingly. Whatever method is to be chosen, the instrument needs to be equipped with a filter set appropriate for the dye's excitation/emission maxima of 492–495/517–527 nm.

Typically, we perform the assay in sextuplicate for each condition. For an assay using a plate reader, we replate at a density of 150,000 cells/cm^2 (i.e., 100 μl/well of a 0.5 × 10^6 cells/ml suspension of differentiated neurons) into black 96-well plates (Corning, Corning, New York No. 3603) coated with matrigel. The next day the neurons are washed twice with Hanks Balanced Salt Solution (HBSS) containing Ca[2+] and Mg[2+] ions to ensure that all the phenol red indicator dye from the medium is removed. If the assay needs to be performed in medium, phenol red-free medium should be used. Before the neurons are loaded with the dye, the signal of neurons in the absence of dye (background) is read using a plate reader equipped with filters appropriate for ex/em 492–495/517–527. After this initial read, the neurons are loaded with 2 μM CM-H2DCFDA (Thermo Fisher No. C6827) in HBSS containing 25 mM glucose for 25 min at room temperature in the dark, and then the cells are washed two times with HBSS before toxicant/stressor exposure. All chemicals to be applied to the neurons are diluted in HBSS containing 25 mM glucose. We typically use 1 mM H_2O_2 (Sigma) as a positive control in the DCF assay. Fluorescence signals are measured immediately after adding the toxicants (time 0 min) and then every 20 min for 4 h (or as desired).

2.3.3 Data Analysis and Interpretation

Due to slight differences in cell densities, dye loading efficiencies, and experiment-to-experiment technical and biological variation, the data are expressed (after subtracting the background fluorescence, which is the signal before dye loading; see above) in relative terms as a fold increase relative to vehicle-treated cells. Due to the fact that the DCF assay is an indirect measure of RONS, the interpretation of the DCF assay can be somewhat ambiguous. Thus, a lack of signal could be a manifestation of a lack of catalyst for the conversion of RONS to the radicals that are ultimately detected by the DCF assay, rather than an actual absence of intracellular RONS; while an increase in signal could be a manifestation of an increased local availability of such a catalyst unrelated to the presence of RONS [26]. These observations support the conclusion that the use of multiple different assays that are designed to assess oxidative stress outcomes at different levels (i.e., measuring RONS directly, assessing cellular redox systems, and modification of cellular macromolecules) gives an advantage to assessing the potential induction of cellular oxidative stress by a compound (Fig. 1).

2.4 Assessment of the GSH-Dependent Disulfide Reductase System

2.4.1 Principal of the Assay

The glutathione (GSH/GSSG) system is one of the two major cellular disulfide reductase systems (the other being the thioredoxin system) which play an important role in the maintenance of the cellular redox balance [6, 15, 27]. GSH is the most abundant small molecular thiol in the brain present at concentrations of ~1–3 mM [6]. A cellular stressor can cause oxidation of GSH or react covalently with the thiol group, both of which can result in a decrease of cellular GSH levels and thus the cellular GSH/GSSG ratio, a determinant of the cellular redox potential [28, 29]. In order to assess if an oxidative stressor affects the thiol system, either the cellular GSH content or the ratio of GSH/GSSG is typically determined. In hiPSC-derived neurons, we found GSSG levels to be very low (<5% of total GSH; our own unpublished observations) and therefore difficult to accurately quantify. Thus, we routinely limit our measurements to cellular GSH (reduced form only) using the GSH-Glo kit available from Promega (Promega, V6911). The assay is based on a two-step reaction: (1) the conversion of a luciferin derivative into luciferin in the presence of GSH, a reaction catalyzed by glutathione-S-transferase and (2) generation of a luminescent signal proportional to the cellular GSH content from luciferin in a reaction catalyzed by firefly luciferase.

2.4.2 Method

As a first step, the cell density that results in a GSH signal within the linear range of the assay should be determined. If the signal from the cell cultures in question exceeds the signals within the linear range of the GSH standard curve, the cell density needs to be titrated so that the cellular GSH signals lie within the linear range of the standard curve. Since GSH content can vary from neuron type and between differentiation stages, appropriate cell density needs to be determined for each neuronal lineage and developmental stage separately. Neurons are plated at the predetermined density (for us typically this is ≤75,000 cells/cm^2; i.e., 25,000 cells/well) into black matrigel-coated 96-well plates. Fifteen to 18 matrigel-coated wells need to be kept free of cells (but with medium); these cell-free wells will be used for the GSH assay standard curve and background determination. The neurons are then exposed with oxidative stressors for the required time. In order to confirm specificity of the assay, a few cell cultures are incubated with 1 mM buthionine sulfoximine (BSO, Sigma, #B2515), a specific and irreversible inhibitor of γ-glutamylcysteine synthetase, an enzyme crucial for GSH biosynthesis [30, 31] for 24 h; this should reduce GSH content by ≥90% [32, 33].

We perform the GSH-Glo assay according to the manufacturer's protocol (Promega, GSH-Glo Glutathione Assay, TB369-revised 8/13), except that we use only 80% of the reagent volumes, to assure that a full 100 assays can be performed with each kit. In brief, all reagents are brought to room temperature and kept in the dark. Immediately before the assay is started, the GSH standard curve solutions and the 1X GSH-Glo reagent are prepared. For the GSH standard curve, we dilute the 5 mM GSH stock solution provided in the kit by a factor of 100 in deionized water to a stock solution of 50 μM and then make a twofold serial dilution of this 50 μM solution including a 0 μM blank. Five microliters in triplicate of each GSH dilution are pipetted into the cell-free wells (see above) from which all the medium has been removed to prepare the standard curve. The culture medium is then carefully and completely removed from the cell containing wells and 80 μl of a 1X GSH-Glo solution added to each well containing cells, the GSH standards and the blank (0 μM GSH). The plate is gently tapped for mixing and incubated at room temperature in the dark for 30 min; then 80 μl of luciferin detection reagent is added to each well and the plate tapped gently again for mixing and incubated at room temperature in the dark for 15 min. Thereafter the GSH luminescence signals are read in a plate reader equipped to detect luminescence. The luminescence signal integration time should lie within 0.25–1 s/well.

2.4.3 Data Analysis and Notes

As for the DCF assay (see above), the GSH assay is based on reading signals in a plate reader, and thus there are experiment-to-experiment variabilities resulting from the machine performance, temperature, and cell line- and differentiation stage-dependent differences. In order to present the data in a cohesive way, and to control for the biological and technical variation, we normalize all our data, after subtraction of the background (0 μM GSH) to the vehicle-treated control values. Actual GSH concentrations/well can be derived from the GSH standard curve; however, normalization to protein concentration or cell number is challenging, due to the fact that the typical protein assays cannot be performed from the same wells used for the GSH assay and the cell number at the time the assay is performed can be difficult to determine due to potential loss of cells resulting from the stress of replating and/or the potential increase due to proliferation of still dividing neural precursor cells between the time of replating and time of the GSH assay. Thus, GSH levels are best reported as a percentage of the control (vehicle-treated cells).

2.5 Measuring Oxidative: Stress-Induced Lipid Modification – F2-Isoprostane Quantification

2.5.1 Principal of the Assay

Ultimately, oxidative stress leads to damage of biomolecules, including DNA, RNA, proteins, and lipids [9, 27, 34, 35] (Fig. 1). Due to the high abundance of polyunsaturated fatty acids in the brain compared to other organs, lipid peroxidation is a primary outcome of RONS-induced brain injury [36]. Indeed, elevated levels of lipid peroxidation products have been used as biomarkers for neurodegenerative disease [37–39]. Thus, the third assay we describe here measures oxidative damage to lipids in human neurons by the quantification of F_2-isoprostanes, a major class of lipid oxidation products. Isoprostanes are initially formed from the unsaturated lipid arachidonic acid in situ on the lipids and can be released from the lipid backbone as free acids by phospholipase [38, 40–42]. They are a unique series of prostaglandin-like molecules formed in a nonenzymatic free radical-initiated peroxidation of arachidonic acid and are ubiquitously present reliable markers of lipid peroxidation. Their quantification is considered by many to be a "gold standard" measure of oxidative stress [39, 43, 44].

2.5.2 Method

Mass spectrometric (MS)-based assays are widely accepted as the most accurate means for quantification of isoprostanes [43]. In order to ensure enough cellular material for F_2-isoprostane analysis, we plate at least one well of a six-well plate with 1×10^6 neurons (1×10^6 neurons/10 cm^2) for each experimental condition. After the desired oxidative stress treatment, the neurons are scraped off the plates with a cell scraper (Sarstedt, Newton, NC, #83.1830), centrifugated at 200× g for 3 min, the supernatant is removed, and cell pellets are immediately flash frozen in liquid nitrogen and stored at −80 °C. Arachidonic acid, from which the F_2-isoprostanes are derived, is susceptible to ex vivo (ex-cell culture) oxidation, and thus flash freezing of the neurons is crucial for accurate determination of F_2-isoprostane levels. Quantifying F_2-isoprostanes with MS-based methods are involved and require expensive instrumentation and are thus best performed by a laboratory specializing in these methods. Our F_2-isoprostane determinations are carried out by Vanderbilt University Eicosanoid Core Laboratory. F_2-isoprostanes are quantified using gas chromatography/mass spectrometry with selective ion monitoring (GC/MS) as described [50, 51]. This method, in brief, includes the following steps Cell pellets are resuspended in 3 ml of Folch solution, sonicated, lipids extracted, and dried. The dried lipids are resuspended in 0.5 ml of methanol containing 0.005% butylated hydroxytoluene, vortexed, and then subjected to chemical saponification using 15% KOH to hydrolyze bound F_2-isoprostanes. The cell lysates are adjusted to pH 3, and 1 ng of 4H2-labeled 15-F2a-isoprostane d4-15-F2t-isoprostane (8-iso-PGF2a) internal standard (Cayman Chemical; Cat. No. 316351) is added as internal standard. F_2-isoprostanes are then purified by sequential C18 and silica Sep-Pak extraction, derivatized to the corresponding pentafluorobenzyl esters, and the derivatives separated by thin-layer chromatography. The purified pentafluorobenzyl ester derivatives are further derivatized to trimethylsilyl ether derivatives and then quantified via GC/MS. Quantification of F_2-isoprostanes is achieved with a precision of ±6% and an accuracy of 96% with a lower limit of detection of 0.002 ng/ml.

2.5.3 Data Analysis and Notes

Since the cell pellet is resuspended in Folch solution for the F_2-isoprostane quantification, a protein assay for normalization of the F_2-isoprostane level per mg cellular protein is not possible.

F_2-isoprostane levels can be reported on a per cell basis, however. To express the data as ng F_2-isoprostane/cell, the number of cells at the time of harvest has to be known. Thus, the survival rate after replating and/or the proliferation rate between replating and harvest of the cells for control and treated cells should be known, or the cells have to be counted during the harvest; cell counting, however, often requires the harvesting of cells using a protease and then spinning, resuspending, and counting the cells and then another spinning before the cell pellet can be flash frozen and stored. These procedures can be quite stressful to the cells and might lead to additional oxidative stress; thus, we prefer to use the much faster method of harvesting the cells by scraping them off the plate with a cell scraper, centrifugating them, and then immediately flash freezing the cell pellets. For proliferating cells or cells that show appreciable cell death due to replating, this method requires previous knowledge of the proliferation rate or death rate, respectively, so actual preharvest cell numbers can be adequately estimated. F_2-isoprostane levels can then be reported as ng F_2-isoprostane/million cells.

3 Conclusion

Cells have developed many different stress response systems by which they homeostatically regulate oxidative stress and the resulting RONS [12, 13, 45]. The cellular manifestation of oxidative stress depends on the stressor itself, exposure paradigms, and cell types. We have recently demonstrated that two chemically very different prooxidants, the metal ion manganese and the lipophilic pesticide rotenone, both well-established inhibitors of oxidative phosphorylation [46–49], result in very different neuronal phenotypes, depending on which oxidative stress assay was used [16]. We thus postulate that the assessment of oxidative stress in neurons requires the analysis of oxidative stress at different levels and might include the determination of cellular or subcellular RONS levels, the assessment of a redox-balancing system such as the nicotinamide or sulfhydryl systems, as well as the oxidative modifications of cellular macromolecules such as lipids, protein, or nucleic acids (Fig. 1). Such a multiplex analysis of oxidative stress outcomes allows the establishment of what we have termed the "oxidative stress signature," a combination of oxidative stress phenotypes that is characteristic for the prooxidant applied, the exposure paradigm, and the cell type assessed. Comparisons of such "oxidative stress signatures" provide us with the possibility of establishing a more detailed comparison of the response of different neuronal lineages or developmental stages to the same prooxidant and/or different prooxidants on the same neuronal cell population.

Acknowledgments

This work was supported in part by NIH NIEHS RO1 ES016931 (ABB) and ES010563 (ABB).

References

1. Jenner P (2003) Oxidative stress in Parkinson's disease. Ann Neurol 53(Suppl 3):S26–S36; discussion S36–28. https://doi.org/10.1002/ana.10483
2. Multhaup G, Ruppert T, Schlicksupp A, Hesse L, Beher D, Masters CL, Beyreuther K (1997) Reactive oxygen species and Alzheimer's disease. Biochem Pharmacol 54(5):533–539
3. Uttara B, Singh AV, Zamboni P, Mahajan RT (2009) Oxidative stress and neurodegenerative diseases: a review of upstream and downstream antioxidant therapeutic options. Curr Neuropharmacol 7(1):65–74. https://doi.org/10.2174/157015909787602823
4. Guo JD, Zhao X, Li Y, Li GR, Liu XL (2018) Damage to dopaminergic neurons by oxidative stress in Parkinson's disease (review). Int J Mol Med 41(4):1817–1825. https://doi.org/10.3892/ijmm.2018.3406
5. Puspita L, Chung SY, Shim JW (2017) Oxidative stress and cellular pathologies in Parkinson's disease. Mol Brain 10(1):53. https://doi.org/10.1186/s13041-017-0340-9
6. Ren X, Zou L, Zhang X, Branco V, Wang J, Carvalho C, Holmgren A, Lu J (2017) Redox signaling mediated by thioredoxin and glutathione systems in the central nervous system. Antioxid Redox Signal 27(13):989–1010. https://doi.org/10.1089/ars.2016.6925
7. Angelova PR, Abramov AY (2018) Role of mitochondrial ROS in the brain: from physiology to neurodegeneration. FEBS Lett 592(5):692–702. https://doi.org/10.1002/1873-3468.12964
8. Gandhi S, Abramov AY (2012) Mechanism of oxidative stress in neurodegeneration. Oxidative Med Cell Longev 2012:428010. https://doi.org/10.1155/2012/428010
9. Grimm S, Hoehn A, Davies KJ, Grune T (2011) Protein oxidative modifications in the ageing brain: consequence for the onset of neurodegenerative disease. Free Radic Res 45(1):73–88. https://doi.org/10.3109/10715762.2010.512040
10. Browne SE, Ferrante RJ, Beal MF (1999) Oxidative stress in Huntington's disease. Brain Pathol 9(1):147–163
11. Jones DP, Sies H (2015) The redox code. Antioxid Redox Signal 23(9):734–746. https://doi.org/10.1089/ars.2015.6247
12. Sies H (2015) Oxidative stress: a concept in redox biology and medicine. Redox Biol 4:180–183. https://doi.org/10.1016/j.redox.2015.01.002
13. Sies H, Berndt C, Jones DP (2017) Oxidative stress. Annu Rev Biochem 86:715–748. https://doi.org/10.1146/annurev-biochem-061516-045037
14. Herrmann JM, Dick TP (2012) Redox biology on the rise. Biol Chem 393(9):999–1004. https://doi.org/10.1515/hsz-2012-0111
15. Comini MA (2016) Measurement and meaning of cellular thiol: disulfide redox status. Free Radic Res 50(2):246–271. https://doi.org/10.3109/10715762.2015.1110241
16. Neely MD, Davison CA, Aschner M, Bowman AB (2017) From the cover: manganese and rotenone-induced oxidative stress signatures differ in iPSC-derived human dopamine neurons. Toxicol Sci 159(2):366–379. https://doi.org/10.1093/toxsci/kfx145
17. Bornhorst J, Meyer S, Weber T, Boker C, Marschall T, Mangerich A, Beneke S, Burkle A, Schwerdtle T (2013) Molecular mechanisms of Mn induced neurotoxicity: RONS generation, genotoxicity, and DNA-damage response. Mol Nutr Food Res 57(7):1255–1269. https://doi.org/10.1002/mnfr.201200758
18. Kumar KK, Lowe EW Jr, Aboud AA, Neely MD, Redha R, Bauer JA, Odak M, Weaver CD, Meiler J, Aschner M, Bowman AB (2014) Cellular manganese content is developmentally regulated in human dopaminergic neurons. Sci Rep 4:6801. doi:srep06801 [pii]. https://doi.org/10.1038/srep06801
19. Desole MS, Esposito G, Migheli R, Sircana S, Delogu MR, Fresu L, Miele M, de Natale G, Miele E (1997) Glutathione deficiency potentiates manganese toxicity in rat striatum and brainstem and in PC12 cells. Pharmacol Res 36(4):285–292. doi:S1043-6618(97)90197-3 [pii]. https://doi.org/10.1006/phrs.1997.0197
20. Neely MD, Litt MJ, Tidball AM, Li GG, Aboud AA, Hopkins CR, Chamberlin R, Hong

CC, Ess KC, Bowman AB (2012) DMH1, a highly selective small molecule BMP inhibitor promotes neurogenesis of hiPSCs: comparison of PAX6 and SOX1 expression during neural induction. ACS Chem Neurosci 3(6):482–491. (PMC888888). https://doi.org/10.1021/cn300029t

21. Chambers SM, Fasano CA, Papapetrou EP, Tomishima M, Sadelain M, Studer L (2009) Highly efficient neural conversion of human ES and iPS cells by dual inhibition of SMAD signaling. Nat Biotechnol 27(3):275–280
22. Kriks S, Shim JW, Piao J, Ganat YM, Wakeman DR, Xie Z, Carrillo-Reid L, Auyeung G, Antonacci C, Buch A, Yang L, Beal MF, Surmeier DJ, Kordower JH, Tabar V, Studer L (2011) Dopamine neurons derived from human ES cells efficiently engraft in animal models of Parkinson's disease. Nature 480(7378):547–551. doi:nature10648 [pii]. https://doi.org/10.1038/nature10648
23. Shi Y, Kirwan P, Livesey FJ (2012) Directed differentiation of human pluripotent stem cells to cerebral cortex neurons and neural networks. Nat Protocol 7(10):1836–1846. https://doi.org/10.1038/nprot.2012.116
24. Meacham CA, Freudenrich TM, Anderson WL, Sui L, Lyons-Darden T, Barone S Jr, Gilbert ME, Mundy WR, Shafer TJ (2005) Accumulation of methylmercury or polychlorinated biphenyls in in vitro models of rat neuronal tissue. Toxicol Appl Pharmacol 205(2):177–187. https://doi.org/10.1016/j.taap.2004.08.024
25. Hennings L, Kaufmann Y, Griffin R, Siegel E, Novak P, Corry P, Moros EG, Shafirstein G (2009) Dead or alive? Autofluorescence distinguishes heat-fixed from viable cells. Int J Hyperthermia Off J Eur Soc Hyperthermic Oncol N Am Hyperthermia Group 25(5):355–363. https://doi.org/10.1080/02656730902964357
26. Winterbourn CC (2014) The challenges of using fluorescent probes to detect and quantify specific reactive oxygen species in living cells. Biochim Biophys Acta 1840(2):730–738. https://doi.org/10.1016/j.bbagen.2013.05.004
27. Pacifici RE, Davies KJ (1991) Protein, lipid and DNA repair systems in oxidative stress: the free-radical theory of aging revisited. Gerontology 37(1–3):166–180. https://doi.org/10.1159/000213257
28. Cotgreave IA, Gerdes RG (1998) Recent trends in glutathione biochemistry – glutathione-protein interactions: a molecular link between oxidative stress and cell proliferation? Biochem Biophys Res Commun 242(1):1–9. https://doi.org/10.1006/bbrc.1997.7812
29. Dringen R, Gutterer JM, Hirrlinger J (2000) Glutathione metabolism in brain metabolic interaction between astrocytes and neurons in the defense against reactive oxygen species. Eur J Biochem 267(16):4912–4916
30. Griffith OW (1982) Mechanism of action, metabolism, and toxicity of buthionine sulfoximine and its higher homologs, potent inhibitors of glutathione synthesis. J Biol Chem 257(22):13704–13712
31. Griffith OW, Meister A (1979) Potent and specific inhibition of glutathione synthesis by buthionine sulfoximine (S-n-butyl homocysteine sulfoximine). J Biol Chem 254(16):7558–7560
32. Neely MD, Boutte A, Milatovic D, Montine TJ (2005) Mechanisms of 4-hydroxynonenal-induced neuronal microtubule dysfunction. Brain Res 1037(1–2):90–98. doi:S0006-8993(04)01950-X [pii]. https://doi.org/10.1016/j.brainres.2004.12.027
33. Aquilano K, Baldelli S, Cardaci S, Rotilio G, Ciriolo MR (2011) Nitric oxide is the primary mediator of cytotoxicity induced by GSH depletion in neuronal cells. J Cell Sci 124(Pt 7):1043–1054. https://doi.org/10.1242/jcs.077149
34. Cadet J, Davies KJA, Medeiros MH, Di Mascio P, Wagner JR (2017) Formation and repair of oxidatively generated damage in cellular DNA. Free Radic Biol Med 107:13–34. https://doi.org/10.1016/j.freeradbiomed.2016.12.049
35. Davies KJ (2001) Degradation of oxidized proteins by the 20S proteasome. Biochimie 83(3–4):301–310
36. Reed TT (2011) Lipid peroxidation and neurodegenerative disease. Free Radic Biol Med 51(7):1302–1319. https://doi.org/10.1016/j.freeradbiomed.2011.06.027
37. Montine TJ, Neely MD, Quinn JF, Beal MF, Markesbery WR, Roberts LJ, Morrow JD (2002) Lipid peroxidation in aging brain and Alzheimer's disease. Free Radic Biol Med 33(5):620–626. doi:S0891584902008079 [pii].
38. Pratico D (2010) The neurobiology of isoprostanes and Alzheimer's disease. Biochim Biophys Acta 1801(8):930–933. https://doi.org/10.1016/j.bbalip.2010.01.009
39. Miller E, Morel A, Saso L, Saluk J (2014) Isoprostanes and neuroprostanes as biomarkers of oxidative stress in neurodegenerative diseases. Oxidative Med Cell Longev 2014:572491. https://doi.org/10.1155/2014/572491

40. Morrow JD, Hill KE, Burk RF, Nammour TM, Badr KF, Roberts LJ 2nd (1990) A series of prostaglandin F2-like compounds are produced in vivo in humans by a non-cyclooxygenase, free radical-catalyzed mechanism. Proc Natl Acad Sci U S A 87(23):9383–9387
41. Kayganich-Harrison KA, Rose DM, Murphy RC, Morrow JD, Roberts LJ 2nd (1993) Collision-induced dissociation of F2-isoprostane-containing phospholipids. J Lipid Res 34(7):1229–1235
42. Stafforini DM, Sheller JR, Blackwell TS, Sapirstein A, Yull FE, McIntyre TM, Bonventre JV, Prescott SM, Roberts LJ 2nd (2006) Release of free F2-isoprostanes from esterified phospholipids is catalyzed by intracellular and plasma platelet-activating factor acetylhydrolases. J Biol Chem 281(8):4616–4623. https://doi.org/10.1074/jbc.M507340200
43. Milne GL, Dai Q, Roberts LJ 2nd (2015) The isoprostanes – 25 years later. Biochim Biophys Acta 1851(4):433–445. doi:S1388-1981(14) 00216-9 [pii]. https://doi.org/10.1016/j.bbalip.2014.10.007
44. Milne GL, Yin H, Hardy KD, Davies SS, Roberts LJ 2nd (2011) Isoprostane generation and function. Chem Rev 111(10):5973–5996. https://doi.org/10.1021/cr200160h
45. D'Autreaux B, Toledano MB (2007) ROS as signalling molecules: mechanisms that generate specificity in ROS homeostasis. Nat Rev Mol Cell Biol 8(10):813–824. https://doi.org/10.1038/nrm2256
46. Galvani P, Fumagalli P, Santagostino A (1995) Vulnerability of mitochondrial complex I in PC12 cells exposed to manganese. Eur J Pharmacol 293(4):377–383
47. Xiong N, Huang J, Chen C, Zhao Y, Zhang Z, Jia M, Hou L, Yang H, Cao X, Liang Z, Zhang Y, Sun S, Lin Z, Wang T (2012) Dl-3-n-butylphthalide, a natural antioxidant, protects dopamine neurons in rotenone models for Parkinson's disease. Neurobiol Aging 33(8):1777–1791. doi:S0197-4580(11) 00079-0 [pii]. https://doi.org/10.1016/j.neurobiolaging.2011.03.007
48. Zhang S, Fu J, Zhou Z (2004) In vitro effect of manganese chloride exposure on reactive oxygen species generation and respiratory chain complexes activities of mitochondria isolated from rat brain. Toxicol In Vitro 18(1):71–77. doi:S0887233303001632 [pii].
49. Degli Esposti M (1998) Inhibitors of NADH-ubiquinone reductase: an overview. Biochim Biophys Acta 1364(2):222–235
50. Milatovic D, Montine TJ, Aschner, M (2011) Measurement of isoprostanes as markers of oxidative stress. Methods Mol Biol 758:195–204
51. Milne GL, Gao B, Terry ES, Zackert WE, Sanchez, SC (2013) Measurement of F2-isoprostanes and isofurans using gas chromatography-mass spectrometry. Free Radic Biol Med 59:36–44

Chapter 4

Glial Reactivity in Response to Neurotoxins: Relevance and Methods

Lindsay T. Michalovicz and James P. O'Callaghan

Abstract

Microglia and astrocytes become activated in response to diverse toxic exposures, regardless of the cellular or molecular targets affected; biomarkers of these responses, therefore, can be used to detect and localize damage to any area of the CNS. A variety of cellular and molecular markers of reactive microglia and astrocytes have been implemented to reveal all types of neural injuries, including those caused by chemical insults of the CNS. Recent advances in approaches to evaluate the cell-specific transcriptome in the CNS allow for an expansion of the existing repertoire of glial activation biomarkers. Here, we show how the approach we used to validate assays of glial fibrillary acidic protein (GFAP) as a biomarker of astrogliosis can be extended to a cell signaling-based assay via phosphorylation of signal transducer and activator of transcription 3 (STAT3). We also introduce new methods to assess cell type-specific gene expression, glial-specific pharmacological inhibition, and genetic manipulation that can be used to evaluate glial reactivity, with the overall goal of defining the microglial and astroglial activation phenotype that results from exposures to broad classes of neurotoxicants.

Key words Astrocytes, Microglia, Hypertrophy, Gene expression, Protein expression, ELISA, Neuroinflammation

1 Introduction

The history of neuroscience in large measure has been dominated by a focus on the structure and function of neurons with relatively little attention devoted to the other nervous system cell type: glia. More recently, however, microglia, oligodendroglia, and astrocytes, the principal glial subtypes, have been subject to renewed attention at the molecular to cellular levels. Microglia and astrocytes, in particular, have received attention for their role in neuroinflammation and their potential contribution to disease (for examples, see More et al. [1], Cai et al. [2], Phillips et al. [3], Crotti and Glass [4], and Hooten et al. [5]). Our focus here is on the propensity of microglia and astrocytes to become "activated" in response to neurotoxic insults regardless of the brain area

Michael Aschner and Lucio Costa (eds.), *Cell Culture Techniques*, Neuromethods, vol. 145,
https://doi.org/10.1007/978-1-4939-9228-7_4, © Springer Science+Business Media, LLC, part of Springer Nature 2019

affected or the particular neurotoxicant or neurotoxic mixture involved. It is this "reactive" property of microglia and astrocytes that make them a useful source for broadly applicable biomarkers of neurotoxicity. We review, herein, some of the approaches we have taken to validate existing biomarkers of glial activation and present contemporary methods for discovering and characterizing novel glial biomarkers of neurotoxicity.

1.1 Glial Reactivity and CNS Damage

As the "microsensors" of the CNS, the activation of glia serves as a biological response/sensor, driven by the brain's reaction to diverse sets of insults [6, 7]. Since the glial activation response is triggered regardless of affected brain area or neuron type, it can serve as a sensitive and specific measure of neurotoxicity that is not specific to the compound or mixture of compounds that induced it. Microglia and astrocytes are the main "reactive" glial cells within the CNS, with each responding to various stimuli, both centrally and peripherally generated. With their similarity to macrophages, microglia are typically viewed as the "resident immune cells" of the CNS, secreting proinflammatory mediators like cytokines and chemokines and demonstrating phagocytic functions once activated. Astrocytes are also immune-like cells capable of secreting cytokines and chemokines and creating "glial scars" to isolate areas of significant damage but additionally provide metabolic support to neurons and their synapses and are crucial to the formation of the blood-brain barrier. While oligodendrocytes are also major constituents of the glial population of the CNS and can be affected by toxic exposures, their role in neurotoxicity has been less well studied; most focus has been on developmental effects on oligodendroglia, such as those caused by lead, alcohol, and anesthetics [8–10].

Glia can be activated by a variety of stimuli, including infection, disease (e.g., Alzheimer's and Parkinson's), and cell damage (e.g., toxicity- or traumatic injury-induced) (for review, see Burda and Sofroniew [7]). Generally, glial activation results in a combination of morphological changes to the cell, as well as the presentation of a different molecular and biochemical pattern. The morphological changes to microglia and astrocytes are the most straightforward indication of glial activation; however, this response cannot be generalized across both cell types. Microglia transition from a ramified state to a more amoeboid state, losing their processes and, consequently, resulting in a larger cell body when activated. While reactive microglia have historically been classified into categories (M1 and M2), their response is more graded, and their morphology can represent a continuum between ramified, "resting" cells to activated, amoeboid phagocytic cells. Astrocytes also experience hypertrophy (reactive gliosis, astrogliosis) but also tend to increase the number and/or size of their processes, as opposed to losing them. Additionally, reactive astrocytes, under severe injury conditions such as trauma, can undergo proliferation, cell migration, and conglomeration to form "scars." This response can result from signaling molecules that are released by neighboring cells or carried to the brain by circulation.

At the molecular level, each cell presents a unique response upon activation that is largely dependent upon the type and severity of the stimuli. The responses in microglia and astrocytes are regulated via both autocrine and paracrine signaling mechanisms. As immune-like cells, microglia act similarly to macrophages largely mounting an inflammatory response (neuroinflammation) upon detection of pathogen-associated or damage-associated molecular patterns. The signaling instigated through ligation of microglial pattern recognition receptors results in the altered expression of pro- and anti-inflammatory signaling molecules, which can stimulate surrounding microglia or other neural cells. The hallmark of astrogliosis is the increased expression of glial fibrillary acidic protein (GFAP), which reflects the hypertrophy and associated accumulation of glial intermediate filaments in reactive astrocytes. Like microglia, astrocytes also secrete inflammatory molecules, as well as respond to external signals received from microglia and other cells.

It is important to note that while neurotoxicity often results in neuronal cell damage, glial reactivity is not exclusive to neurodegeneration. This is crucial to the efficacy of glial reactivity as a means to detect neurotoxicity, because many toxic compounds may have more subtle effects on neurons or result in neuroinflammation as opposed to degeneration. While the study of disease states like traumatic brain injury, Alzheimer's disease, Parkinson's disease, multiple schlerosis, etc. has made it difficult to separate neural damage from neuroinflammation [11–14], it is possible for neuroinflammation (in the absence of damage) to create adverse symptomology. A striking example of this is observed in Gulf War Illness, where veterans of the 1991 Persian Gulf War have presented with a persistent, likely toxicity-driven, multi-symptom illness [15, 16] that seems to be the result of underlying chronic neuroinflammation [17–19]. While these individuals present with various cognitive impairments [16, 20–22], the only structural brain pathology that has been noted constitutes some small changes in white matter volume [23, 24], further highlighting the importance of non-neuronal indices of neurotoxicity.

1.2 Methods for Detecting and Measuring Glial Reactivity

The utility of glial reactivity as a broadly applicable index of neurotoxicity is that it overcomes the unpredictable nature of a given toxic compound's target by taking advantage of a cellular response that occurs without regard for the location of the damage or for the agent(s) causing the damage. The glial response to neurotoxicity, by definition, is an indirect response of glial cell types to the damaged target, one that reflects cellular interconnectivity and

Table 1
"Toolbox" neurotoxicants and their targets

Toxicant	Regional target	Cellular target
Trimethyltin	Limbic structures	Neurons
Triethyltin	Limbic structures	Neurons
Kainic acid	Limbic structures	Neurons
Domoic acid	Limbic structures	Neurons
Ibotenic acid	Limbic structures	Neurons
MPTP	Neostriatum	Dopaminergic neurons
Amphetamine	Neostriatum	Dopaminergic neurons
Methamphetamine	Neostriatum	Dopaminergic neurons
MDA	Neostriatum	Dopaminergic neurons
MDMA	Neostriatum	Dopaminergic neurons
6-Hydroxydopamine	Neostriatum	Dopaminergic neurons
Cadmium	Striatum	Neurons, glia
Methylmercury	Cortex, hippocampus	Neurons
Methylazoxymethanol	Cortex, hippocampus	Neurons
Bilirubin	Cerebellum	Purkinje neurons
Colchicine	Hippocampus	Dentate neurons
3-Acetyl pyridine	Inferior olive	Neurons
Iminodipropionitrile	Cortex, brain stem, olfactory bulb	Neurons
MK-801 (Dizocilpine)	Cortex	Neurons
Ketamine	Cortex	Neurons
5,7-Dihydroxytryptamine	Hippocampus	Serotonergic neurons
Fenthion	Eye	Retinal neurons
2,6-Dichlorobenzonitrile	Olfactory bulb	Sensory neurons

communication of "damage signals" to activate microglia and astroglia. The requirement of cell-to-cell signaling to elicit glial activation emphasizes the importance of having a model that keeps this signaling intact. Thus, it is a requirement to evaluate these responses in an intact animal model of toxic exposure. The utility of a given method for the measurement of glial reactivity is dependent upon its ability to be used in the assessment of various toxic agents. Thus, the methods developed for this task should prove to be relatively consistent across a battery of toxic compounds targeting different cells in different brain areas (see Table 1).

As described previously, microglia and astrocytes exhibit hallmarks of reactivity, and these changes serve as the basis of the methods used to detect these cellular activation responses. The morphological changes exhibited by activated microglia and astrocytes are best evaluated by neurohistology utilizing cell-specific markers. However, due to the morphological continuum of reactive gliosis and the more qualitative evaluation involved with histology, the evaluation of salient protein biomarkers associated with cell morphological changes or the signaling cascades involved in the "activation" response by Western blotting and enzyme-linked immunosorbent assays (ELISAs) techniques is more desirable.

As noted previously, a large component of microglial and astroglial reactivity is the expression and secretion of inflammatory mediators. While cytokines and chemokines are typically measured, particularly clinically, via protein-based analysis, the measurement of these proteins in the brain has proven difficult and inconsistent without explanation. Thus, evaluation of gene expression can serve as a means to detect glial reactivity when protein-based measurements fail. Large-scale gene expression analysis, such as RNA sequencing, can be applied in conjunction with novel genetically modified mouse strains to capture a more extensive response profile to a neurotoxic insult.

2 Materials

2.1 Animal Tissue Preparation

In general, all experiments use either C57BL/6J mice purchased from the Jackson Laboratory or Long Evans rats purchased from Charles River Laboratories, depending on the neurotoxic model used. For instance, trimethyltin instigates a much more robust response in rats than in mice [25], whereas MPTP (1-methyl-4-phenyl-1,2,3,6-tetrahydropyridine) is very effective at damaging dopaminergic nerve terminals in mice, but not so in rats [26]. Conscious decapitation is the preferred method of euthanasia when evaluating brain targets as the use of anesthetics can alter the expression of various genes and proteins [27, 28]. The brain is rapidly and carefully removed from the skull and moved to a thermoelectric cold plate for dissection of discrete brain areas, typically the cortex, hippocampus, and striatum (Fig. 1). First, fine curved tweezers are used to pinch and separate the left and right hemispheres along the midline of the brain (along the corpus callosum) to allow folding back of the cortex. The cortex is then gently teased away from the underlying tissue exposing the striatum and hippocampus (Fig. 1a). The hippocampus is toward the posterior of the brain, having an elongated and curved shape, and can be rolled out from the underlying brain tissue (Fig. 1b). The striatum is toward the anterior of the brain and is removed by "cupping out" the visibly striated tissue (Fig. 1c, d). At this point, the entire cortex

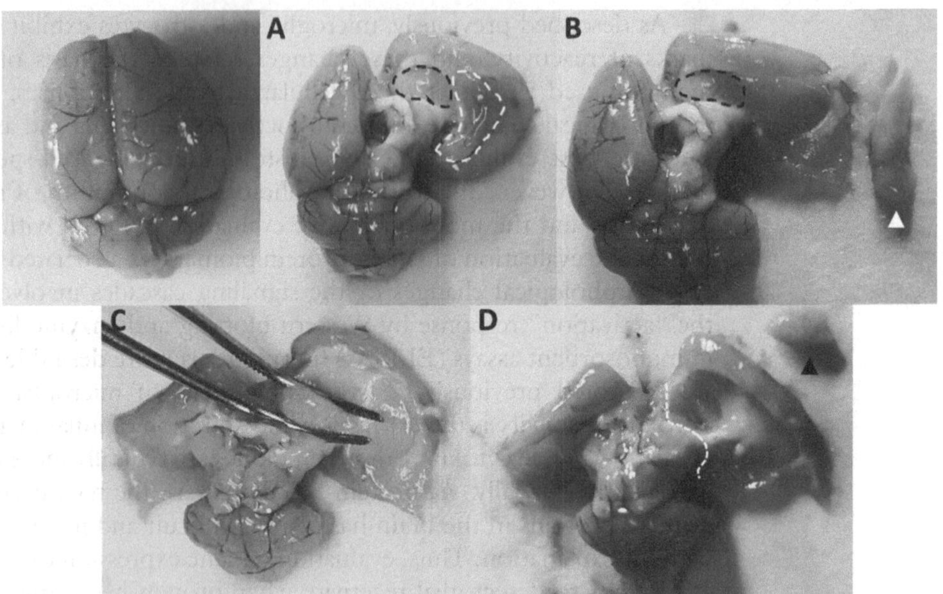

Fig. 1 Dissection of discreet brain regions. Representative images are shown for mouse brain. (a) After breaking the white matter connections of corpus callosum along the midline, the cortex can be folded back to reveal the striatum (within black dashes) and the hippocampus (within white dashes). (b) The hippocampus (arrowhead) can be rolled away from the cortex for isolation, leaving the striatum behind. (c) The striatum is then scooped out using fine curved forceps. (d) After removal of the striatum (arrowhead), the cortex can be removed from the rest of the brain along the dashed line

should be easily separated from the rest of the brain. Care should be taken to remove the white matter from the surface of the dissected brain regions. For the isolation of RNA, the tissues (in microcentrifuge tubes) are frozen on dry ice and moved to a –80 °C freezer. For protein evaluation, the tissue is weighed and then sonicated in ten volumes of hot (90–95 °C) 1% sodium dodecyl sulfate (SDS) buffer (in 1× phosphate buffered saline). For example, 100 μg of tissue weight is homogenized in 1000 μL of 1% SDS. The homogenization in SDS buffer is crucial for the GFAP ELISA (Sect. 3.1) to adequately release GFAP protein into the supernatant while still generating a sample that is suitable for ELISA.

2.2 Microwave Tissue Fixation

When evaluating the expression of phosphoproteins, like pSTAT-3^{Tyr705}, it is best to sacrifice the animal by focused microwave irradiation (Muromachi Microwave Fixation Applicator, model TMW-4012C, Tokyo, Japan) to preserve the phosphorylation state of the protein [29]. Using the appropriate water-jacketed adaptor, microwave fixation is achieved using 3.5 kW of applied power for 0.9 seconds for mice (approx. weight 30 g) and 1.5–1.7 seconds for rats (approx. weight 350–400 g).

2.3 ELISA Assay Preparation

Protein homogenates are diluted in 0.5% Triton X-100 in 1× phosphate buffered saline (PBS). The antibodies used in this assay are rabbit polyclonal α-GFAP (1:300; DAKO), mouse monoclonal α-GFAP (1:1500; Sigma-Aldrich), and alkaline phosphatase-conjugated α-mouse IgG (1:1500; Jackson ImmunoResearch Laboratories, Inc.), all diluted in 1× PBS. The colorimetric reaction with alkaline phosphatase is made using the alkaline phosphatase substrate kit (Bio-Rad), per manufacturer's instruction. Wash buffers are 1× PBS or 0.5% Triton X-100 in 1× PBS (PBS-T), and BLOTTO blocking buffer is 5% dry milk in 1× PBS. All steps are performed at room temperature, except for initial coating of the plate with polyclonal antibody (37 °C).

2.4 Translating Ribosome Affinity Purification (TRAP)

The isolation of actively translating mRNA bound to the cell-specific, eGFP-tagged ribosomes present in the bacTRAP transgenic mouse lines is relatively straightforward and consists of a combined procedure of standard immunoprecipitation and RNA purification [30–33]. A detailed protocol for the "TRAP-ing" procedure, as well as mouse genotyping, is publically available online at www.bactrap.org.

3 Methods

3.1 Quantitative GFAP ELISA

GFAP is ubiquitously recognized as the hallmark of astrocytes, present in both resting and reactive cells, making it a common astrocytic marker for histology. While certainly useful for morphological evaluation of astrocytes and their hypertrophy upon activation, using histology as a means of assessing neurotoxicity when the cellular targets within the brain are unpredictable would prove cumbersome, requiring multiple sections throughout the entirety of the brain to assess the response. Furthermore, a qualitative assessment of gliosis or toxicity, in general, is less desirable than a quantitative measure that could hint at the level of toxicity or damage that results from a particular dosage of exposure. As astrogliosis results in the accumulation of astrocytic filaments and, thus, the GFAP which is associated with these filaments, the increased expression of GFAP can also be used to *quantitate* astrogliosis.

Using the "toolbox" of known neurotoxicants described in Table 1, a quantitative ELISA for GFAP was developed [34, 35] and validated for its efficacy at measuring levels of astrogliosis associated with neurotoxicity [36]. As outlined by O'Callaghan and Sriram [36], the GFAP assay demonstrated that all types of neurotoxic compounds, regardless of target brain region or cell type, resulted in the increased expression of GFAP even at doses not associated with detectable histological or behavioral changes (Fig. 2a). Furthermore, this response is specific to conditions of neural damage, as the enhanced expression of GFAP is not seen under strictly neuroinflammatory conditions (Fig. 2b).

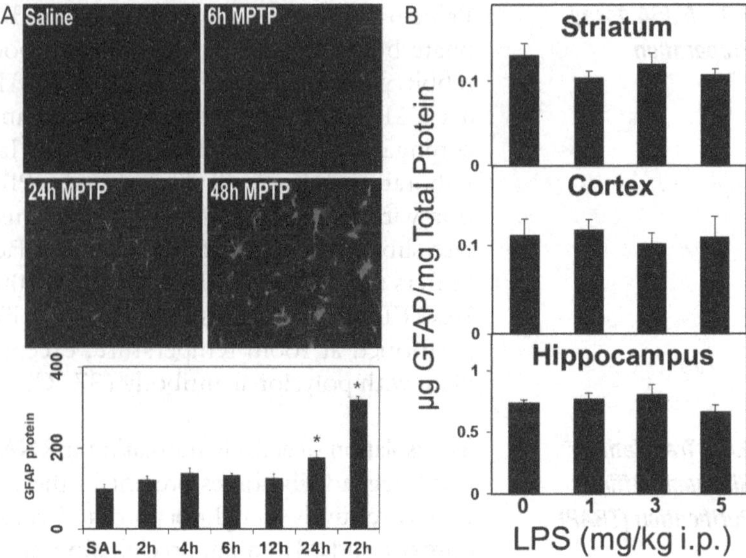

Fig. 2 Time course of GFAP expression post-MPTP and post-LPS. (a) Mice were exposed to 1-methyl-4-phenyl-1,2,3,6-tetrahydropyridine (MPTP) (12.5 mg/kg, s.c.), and the induction of GFAP was measured by immunohistochemistry and ELISA. (b) Mice were exposed to LPS at various concentrations and GFAP measured in multiple brain regions by ELISA. Bars represent mean ± S.E.M., $*p \leq 0.05$. (Modified from O'Callaghan and Sriram [36] and O'Callaghan et al. [37])

In this assay, the plate is incubated with polyclonal GFAP antibody for a minimum of 1 h, washed four times with PBS, and then blocked with BLOTTO for 1.5 h. Samples and appropriate protein standards (100 μL/well) are added to the plate, incubated for 1 h, and then washed four times with PBS-T. The protein standards constitute a set of serial dilutions ranging from 8 to 0.2 ng of GFAP for mice and 24–0.6 ng of GFAP for rats, accounting for differences in brain anatomical size across the two species. This standard can be created with either purified GFAP protein or a total protein sample with known GFAP concentration. The monoclonal GFAP and alkaline phosphate-conjugated antibodies are added simultaneously to the plate, incubated for 1 h, and then washed four times with PBS-T. Finally, the plate is incubated for 20 min with P-nitrophenyl and stopped with 0.4N NaOH prior to reading at 405 nm.

3.2 Western Blot Analysis of pSTAT3^{Tyr705}

While astrogliosis and the measurement of GFAP have been established as a quantitative measure to determine the level of neurotoxicity, astrogliosis is also associated with neuroinflammation. This correlation allows for the use of neuroinflammatory markers associated with astrogliosis as indicators of neurotoxicity, in addition to the previously described GFAP assay. As such, it has been shown that activation of the JAK2-STAT3 pathway, indicated by phosphorylation of STAT3 at Tyr705 (pSTAT3^{Tyr705}), correlates with astrogliosis [37, 38]. More specifically, neurotoxic insult results in an increased expression of inflammatory cytokines and chemokines in the brain that is detectable prior to astrogliosis, which then activates the JAK2-STAT3 pathway in reactive astrocytes (Fig. 3). Interestingly, while associated with neuroinflammation, the activation of STAT3 by neurotoxic damage is not suppressible with anti-inflammatory treatment, yet is responsive to neuroprotective drugs [37]. This response in astrocytes is distinguishable from purely neuroinflammatory activation of STAT3, through an inflammagen like lipopolysaccharide, which is suppressible with anti-inflammatory treatment and likely originates from microglia [37]. Thus, STAT3 activation can also be used as an indicator of astrogliosis.

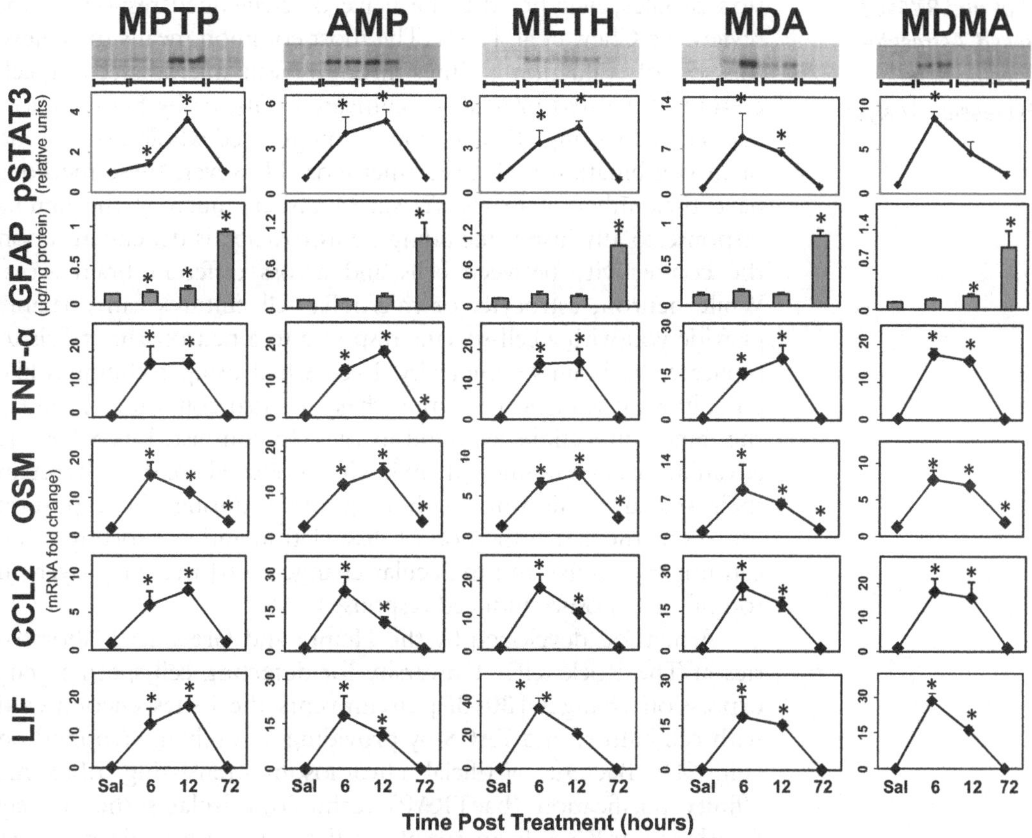

Fig. 3 Time course of STAT3 activation following neurotoxicant exposure. Mice were exposed to MPTP (12.5 mg/kg, s.c.), amphetamine (AMP) (10 mg/kg, s.c., three injections at 2 h intervals), methamphetamine (METH) (20 mg/kg, s.c., three injections at 2 h intervals), 3,4-methylenedioxyamphetamine (MDA) (10 mg/kg, s.c., three injections at 2 h intervals), or 3,4-methylenedioxy methamphetamine (MDMA) (20 mg/kg, s.c., three injections at 2 h intervals). pSTAT3^{Tyr705} and GFAP proteins were measured by immunoblot and ELISA, respectively. Inflammatory cytokine/chemokine mRNA (TNF-a, OSM, CCL2, and LIF) was measured by qRT-PCR. Points represent mean ± S.E.M. and * indicates $p \leq 0.05$. (Modified from O'Callaghan et al. [37])

While detection of pSTAT3^{Tyr705} is achievable through standard Western blotting procedures, the key to reliably detecting phosphorylated proteins is protecting against their dephosphorylation postmortem [29]. Microwave fixation of the brain tissue is a simple, straightforward method of achieving stable phosphorylation without affecting the downstream tissue processing. The optimal Western blotting conditions for detecting pSTAT3^{Tyr705} are to use a 10% SDS-PAGE (sodium dodecylsulfate-polyacrylamide gel electrophoresis) separating gel to resolve samples of 20 μg of total protein, typically with a concentration of 1 μg/μL in Laemmli/β--mercaptoethanol buffer, with overnight protein transfer to nitrocellulose membrane.

3.3 ALDH1L1 bacTRAP Transgenic Mice and RNAseq-Based, Astrocyte-Specific Gene Expression Analysis

There are many methods that have been developed for the isolation of specific cell populations in order to evaluate their unique expression profiles, whether they be naïve or challenged responses (for review see Chew et al. [39]). The most common means to achieve this sort of evaluation is either through tissue microdissection, cell culture of isolated primary or cultivated cells, or by fluorescence-assisted cell sorting (FACS) from homogenized whole tissue alone or in combination with other methods. However, these methods have their inherent issues. As stated earlier, much of the brain's response to any insult, including neurotoxicity, is dependent upon the connectivity between cells and across different brain areas. While neuron, astrocyte, or microglia cell cultures can certainly provide you with a cell-specific response to a neurotoxin, it fails to replicate the brain's complexity. Even when using techniques like co-culture, it is unclear whether these in vitro conditions produce the same connections as found in vivo. In contrast, FACS has the advantage of beginning with an in vivo model where the cells retain their spatial relationship with each other during the exposure. However, the actual method of dissociation and cell sorting often can impart cellular and molecular changes [40] that may layer on top of the exposure-induced response.

A method developed by the Heintz and Greengard laboratories at The Rockefeller University for detecting cell-specific gene expression changes [30–33] circumvents the issues encountered with cell culture and FACS by providing a purely in vivo method (Fig. 4). Bacterial artificial chromosome-translating ribosome affinity purification (bacTRAP) technology isolates the actively translating mRNA from specific cell types. Transgenic mice are generated using a BAC that expresses eGFP-tagged ribosomal RNA (L10a) expressed under the control of a cell-specific marker gene. While one may question the use of a BAC clone as opposed to a traditional transgenic construct, the bacTRAP transgenics employ more than just the promoter region of the cell-specific gene that is typically used to control traditional transgenics, using

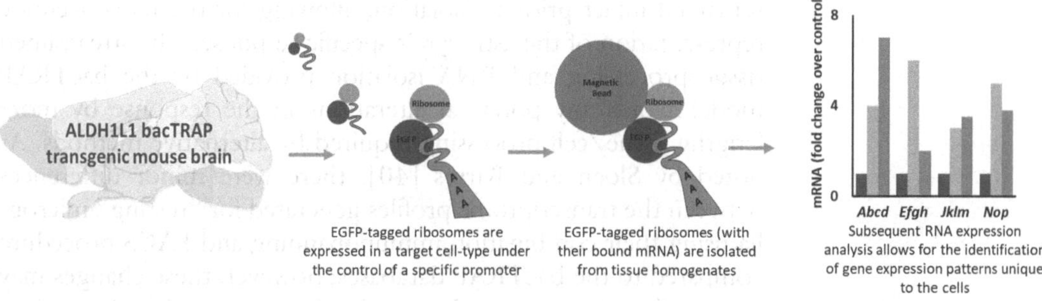
ALDH1L1 bacTRAP transgenic mouse brain
Ribosome
EGFP
Magnetic Bead
EGFP-tagged ribosomes are expressed in a target cell-type under the control of a specific promoter
EGFP-tagged ribosomes (with their bound mRNA) are isolated from tissue homogenates
mRNA (fold change over control)
8
4
0
Abcd Efgh Jklm Nop
Subsequent RNA expression analysis allows for the identification of gene expression patterns unique to the cells

Fig. 4 bacTRAP methodology. By utilizing targeted, cell-specific expression of eGFP-tagged ribosomes, the bacTRAP transgenic mice allow for the isolation of cell-specific mRNA that are actively being transcribed by and, thus, bound to ribosomes. Following isolation of the cell-specific mRNA, gene expression patterns can be evaluated by a number of methods

a large portion of the gene upstream of the 5′ start site that includes not only the promoter but also the transcriptional control elements constituting the need to use a BAC to contain the large amount of DNA. Due to their size, BAC-based transgenics also tend to have a lower number of inserts of the transgene, a condition ideal when expressing rRNA components as the goal is to isolate translating mRNAs without affecting ribosomal numbers in the cell and, by extension, transcriptional rates. By using the eGFP tag to affinity purify ribosomes from brain tissue, the translating mRNA attached to the ribosomes can be isolated and utilized to evaluate the gene expression response of the target cell type. This can then be combined with any method of gene expression analysis, including real-time PCR, microarray, and RNA sequencing, to develop a response profile to the exposures administered in vivo.

The utility of the bacTRAP technology is reliant upon the availability of high fidelity, cell type-specific markers. While a number of bacTRAP transgenic lines exist for various cell types throughout the brain, the Aldh1L1 bacTRAP mouse, relying on the control of the astrocyte-enriched, aldehyde dehydrogenase one family member L1 gene [32], is specific for isolating astrocyte expression profiles [41–45]. Historically, it has been difficult to identify a microglial marker that captures the majority of the brain's microglial population and does not overlap with macrophages. Often, the markers are specific to the different activation stages of the cells, such as CD11b or Iba1 which increase in activated microglia. However, a bacTRAP transgenic mouse was recently created for microglia using CD11b [46].

As an in vivo animal model, the Aldh1L1 bacTRAP mouse can be treated with any potential neurotoxicant and evaluated for astrocyte-specific gene expression postmortem, preserving the integrity of the astrocyte responses. Isolating the gene expression profile in this manner means that all intra- and extracellular communication remained intact prior to isolation, allowing for the most accurate representation of the astrocyte's specific response. The streamlined tissue processing and RNA isolation provided by the bacTRAP model avoids any potential alterations in the response by more lengthy tissue/cell processing required by alternative methods. As noted by Sloan and Barres [40], there were minor differences between the transcriptome profiles generated for "resting" microglia using their combination immunopanning and FACS procedure compared to the bacTRAP databases; however, these changes may be significantly greater when evaluating the rapid and transient responses to exposure. By removing the requirement to extensively process the brain tissue and/or cells prior to transcriptomic analysis, it is important to note that by isolating only the actively translating mRNA bound to the ribosomes, the RNA yield following the immunoprecipitation and purification procedures of the TRAP protocol is low compared to traditional RNA isolation. As high-throughput gene expression analysis like microarray and RNA sequencing generally requires at least 1 μg of RNA for analysis, it is necessary to pool samples from smaller brain areas, like the striatum and hippocampus, or use nearly the entire cortical region for a mouse.

Beyond capturing the cell-specific gene expression profiles to neurotoxic exposures, the bacTRAP technology also allows for the comparison of cell-specific responses to the profile of the total brain, mixed cell population. As RNA yields are higher without the restriction of isolating ribosome-bound mRNA, this can be achieved with larger brain areas, like the cortex, by setting aside a small piece of cortical tissue for "traditional" RNA isolation prior to the TRAP procedure. With smaller brain areas that require pooling for analysis, preparing separate pools by hemisphere can achieve similar results as long as the exposure effects are not expected to be bilateral or homogenous across the brain. This comparison can reveal a great deal regarding the cell-specific responses to a neurotoxic exposure, not only highlighting the astrocyte-enriched gene expression profile but also contextualizing the contribution of the astrocyte in the level of a particular gene's expression. By exposing the Aldh1L1 bacTRAP mice to the "toolbox" of known neurotoxicants and evaluating by high-throughput gene expression analysis, the large datasets generated by microarray or RNA sequencing can be cross-compared to narrow down common neurotoxicant-responsive genes that can be used as markers of agent-induced astrogliosis.

3.4 Evaluating Microglial Reactivity

As noted earlier, the study of microglia has largely been impeded by the absence of a ubiquitous microglia-specific marker, impacting the success of FACS and other technologies like bacTRAP transgenics to isolate cell-specific responses. As such, much of the study of microglial reactivity relies on morphological evaluation by histology using several microglial markers, including Iba1, CD11b, and Isolectin B4. The other methods of studying microglia reactivity involve using pharmacological inhibitors or genetically modified mice to evaluate the brain's response to a neurotoxic insult in the absence of fully functioning microglia. While several pharmacological agents have been identified to inhibit microglia [47], one of the most commonly used inhibitors is the tetracycline antibiotic, minocycline. Minocycline was found to significantly reduce neuroinflammation, leading to the hypothesis that the drug is a selective microglial inhibitor [48, 49]. Additionally, the recently developed pharmaceutical colony-stimulating factor 1 receptor (CSF1R) inhibitor, pexidartinib, has been identified to render microglia quiescent, severely reducing the microglial population [50, 51]. While much of the work using this drug has focused on protective effects related to neurodegenerative diseases and injury [52–56], the potential for pexidartinib to be used as a tool to evaluate neurotoxicity still requires investigation.

In addition to pharmacological inhibition, genetically modified mice can also be used to evaluate the microglial response to a neurotoxic exposure. While several knockout/conditional knockout mouse lines exist that can be manipulated to study microglia, the most commonly used line is the fractalkine (Cx3cr1) knockout mouse. While these mice do not display the "absence" of microglia which may be achieved with using a pharmacological inhibitor like pexidartinib, these mice exhibit a suppressed or altered microglia immune response [57–61]. As with the inhibitors, this provides a less direct method of evaluating the microglial contribution to a neurotoxic response, because the knockout response profile must be compared to the wild type in order to evaluate the role of the microglia in neurotoxicity.

4 Conclusions

In this chapter, we have highlighted the usefulness of glial reactivity as an indicator or biomarker of neurotoxicity and presented several reliable methods for evaluating the response of astrocytes and microglia to neurotoxic exposures. The significance of measuring glial reactivity to assess neurotoxicity lies in the ability of this process to highlight areas of toxicant-induced damage in the absence of understanding the chemical's mechanism of action or cellular/regional target. By shifting the focus away from the neurons themselves, the evaluation of whether a novel (or existing) chemical, drug, or other agent has detrimental effects on the brain can be reliably measured.

5 Notes

While it is highly recommended to use microwave fixation-based euthanasia for the evaluation of phosphorylated proteins, pSTAT-3^{Tyr705} in particular seems to be relatively stable and can be reliably screened in banked, fresh frozen tissue.

Acknowledgments

The findings and conclusions in this report are those of the authors and do not necessarily represent the official position of the National Institute for Occupational Safety and Health, Centers for Disease Control and Prevention.

References

1. More SV, Kumar H, Kim IS, Song SY, Choi DK (2013) Cellular and molecular mediators of neuroinflammation in the pathogenesis of Parkinson's disease. Mediat Inflamm 2013: 952375
2. Cai Z, Hussain MD, Yan LJ (2014) Microglia, neuroinflammation, and beta-amyloid protein in Alzheimer's disease. Int J Neurosci 124(5):307–321
3. Phillips EC, Croft CL, Kurbatskaya K, O'Neill MJ, Hutton ML, Hanger DP, Garwood CJ, Noble W (2014) Astrocytes and neuroinflammation in Alzheimer's disease. Biochem Soc Trans 42(5):1321–1325
4. Crotti A, Glass CK (2015) The choreography of neuroinflammation in Huntington's disease. Trends Immunol 36(6):364–373
5. Hooten KG, Beers DR, Zhao W, Appel SH (2015) Protective and toxic neuroinflammation in amyotrophic lateral sclerosis. Neurotherapeutics 12(2):364–375
6. Kreutzberg GW (1996) Microglia: a sensor for pathological events in the CNS. Trends Neurosci 19(8):312–318
7. Burda JE, Sofroniew MV (2014) Reactive gliosis and the multicellular response to CNS damage and disease. Neuron 81(2):229–248
8. Deng W, Poretz RD (2003) Oligodendroglia in developmental neurotoxicity. Neurotoxicology 24(2):161–178
9. de la Monte SM, Kril JJ (2014) Human alcohol-related neuropathology. Acta Neuropathol 127(1):71–90
10. Bosnjak ZJ, Logan S, Liu Y, Bai X (2016) Recent insights into molecular mechanisms of propofol-induced developmental neurotoxicity: implication for the protective strategies. Anesth Analg 123(5):1286–1296
11. Ellwardt E, Zipp F (2014) Molecular mechanisms linking neuroinflammation and neurodegeneration in MS. Exp Neurol 262(Pt A): 8–17
12. Vivekanantham S, Shah S, Dewji R, Dewji A, Khatri C, Ologunde R (2015) Neuroinflammation in Parkinson's disease: role in neurodegeneration and tissue repair. Int J Neurosci 125(10):717–725
13. de Oliveira AC, Candelario-Jalil E, Fiebich BL, Santos Mda S, Palotás A, dos Reis HJ (2015) Neuroinflammation and neurodegeneration: pinpointing pathological and pharmacological targets. Biomed Res Int 2015:487241
14. Kempuraj D, Thangavel R, Natteru PA, Selvakumar GP, Saeed D, Zahoor H, Zaheer S, Iyer SS, Zaheer A (2016) Neuroinflammation induces neurodegeneration. J Neurol Neurosurg Spine 1(1):pii:1003
15. Smith BN, Wang JM, Vogt D, Vickers K, King DW, King LA (2013) Gulf war illness: symptomatology among veterans 10 years after deployment. J Occup Environ Med 55(1): 104–110
16. Steele L (2000) Prevalence and patterns of Gulf War illness in Kansas veterans: association of symptoms with characteristics of person, place, and time of military service. Am J Epidemiol 152(10):992–1002
17. Koo BB, Michalovicz LT, Calderazzo S, Kelly KA, Sullivan K, Killiany RJ, O'Callaghan JP (2018) Corticosterone potentiates DFP-

induced neuroinflammation and affects high-order diffusion imaging in a rat model of Gulf War Illness. Brain Behav Immun 67:2–46

18. Locker AR, Michalovicz LT, Kelly KA, Miller JV, Miller DB, O'Callaghan JP (2017) Corticosterone primes the neuroinflammatory response to Gulf War Illness-relevant organophosphates independently of acetylcholinesterase inhibition. J Neurochem 142(3):444–455
19. O'Callaghan JP, Kelly KA, Locker AR, Miller DB, Lasley SM (2015) Corticosterone primes the neuroinflammatory response to DFP in mice: potential animal model of Gulf War Illness. J Neurochem 133(5):708–721
20. Dursa EK, Barth SK, Schneiderman AI, Bossarte RM (2016) Physical and mental health status of Gulf War and Gulf Era veterans: results from a large population-based epidemiological study. J Occup Environ Med 58(1):41–46
21. Fukuda K, Nisenbaum R, Stewart G, Thompson WW, Robin L, Washko RM, Noah DL, Barrett DH, Randall B, Herwaldt BL, Mawle AC, Reeves WC (1998) Chronic multisymptom illness affecting Air Force veterans of the Gulf War. JAMA 280(11):981–988
22. White RF, Steele L, O'Callaghan JP, Sullivan K, Binns JH, Golomb BA, Bloom FE, Bunker JA, Crawford F, Graves JC, Hardie A, Klimas N, Knox M, Meggs WJ, Meling J, Philbert MA, Grashow R (2016) Recent research on Gulf War illness and other health problems in veterans of the 1991 Gulf War: effects of toxicant exposures during deployment. Cortex 74:449–475
23. Heaton KJ, Palumbo CL, Proctor SP, Killiany RJ, Yurgelun-Todd DA, White RF (2007) Quantitative magnetic resonance brain imaging in US army veterans of the 1991 Gulf War potentially exposed to sarin and cyclosarin. Neurotoxicology 28(4):761–769
24. Chao LL, Rothlind JC, Cardenas VA, Meyerhoff DJ, Weiner MW (2010) Effects of low-level exposure to sarin and cyclosarin during the 1991 Gulf War on brain function and brain structure in US veterans. Neurotoxicology 31(5):493–501
25. Perretta G, Righi FR, Gozzo S (1993) Neuropathological and behavioral toxicology of trimethyltin exposure. Ann Ist Super Sanita 29(1):167–174
26. Giovanni A, Sieber BA, Heikkila RE, Sonsalla PK (1994) Studies on species sensitivity to the dopaminergic neurotoxin 1-methyl-4-phenyl-1,2,3,6-tetrahydropyridine. Part 1: systemic administration. J Pharmacol Exp Ther 270(3):1000–1007
27. Ichiyama T, Nishikawa M, Lipton JM, Matsubara T, Takashi H, Furukawa S (2001) Thiopental inhibits NF-kappaB activation in human glioma cells and experimental brain inflammation. Brain Res 911(1):56–61
28. National Research Council (US) Committee on Recognition and Alleviation of Pain in Laboratory Animals (2009) Recognition and alleviation of pain in laboratory animals. National Academies Press (US), Washington, D.C. 4, Effective Pain Management
29. O'Callaghan JP, Sriram K (2004) Focused microwave irradiation of the brain preserves in vivo protein phosphorylation: comparison with other methods of sacrifice and analysis of multiple phosphoproteins. J Neurosci Methods 135(1–2):159–168
30. Doyle JP, Dougherty JD, Heiman M, Schmidt EF, Stevens TR, Ma G, Bupp S, Shrestha P, Shah RD, Doughty ML, Gong S, Greengard P, Heintz N (2008) Application of a translational profiling approach for the comparative analysis of CNS cell types. Cell 135:749–762
31. Heiman M, Schaefer A, Gong S, Peterson JD, Day M, Ramsey KE, Suárez-Fariñas M, Schwarz C, Stephan DA, Surmeier DJ, Greengard P, Heintz N (2008) A translational profiling approach for the molecular characterization of CNS cell types. Cell 135:738–748
32. Dougherty JD, Schmidt EF, Nakajima M, Heintz N (2010) Analytical approaches to RNA profiling data for the identification of genes enriched in specific cells. Nucleic Acids Res 38:4218–4230
33. Heiman M, Kulicke R, Fenster RJ, Greengard P, Heintz N (2014) Cell type-specific mRNA purification by translating ribosome affinity purification (TRAP). Nat Protoc 9:1282–1291
34. O'Callaghan JP (1991) Quantification of glial fibrillary acidic protein: comparison of slot-immunobinding assays with a novel sandwich ELISA. Neurtoxicol Teratol 13(3):275–281
35. O'Callaghan JP (2002) Measurement of glial fibrillary acidic protein. In: Maines MD, Costa LG, Hodgson E, Reed DJ, Sipes IG (eds) Current protocols in toxicology. Wiley, New York. Sections 12.8.1–12.8.12
36. O'Callaghan JP, Sriram K (2005) Glial fibrillary acidic protein and related glial proteins as biomarkers for neurotoxicity. Expert Opin Drug Saf 4(3):433–442
37. O'Callaghan JP, Kelly KA, VanGilder RL, Sofroniew MV, Miller DB (2014) Early activation of STAT3 regulates reactive astrogliosis induced by diverse forms of neurotoxicity. PLoS One 9(7):e102003

38. Sriram K, Benkovic SA, Hebert MA, Miller DB, O'Callaghan JP (2004) Induction of gp130-related cytokines and activation of JAK/STAT3 pathway in astrocytes precedes up-regulation of glial fibrillary acidic protein in the 1-methyl-4-phenyl-1,2,3,6-tetrahydropyridine model of neurodegeneration: key signaling pathway for astrogliosis in vivo? J Biol Chem 279: 19936–19947
39. Chew LJ, DeBoy CA, Senatorov VV Jr (2014) Finding degrees of separation: experimental approaches for astroglial and oligodendroglial cell isolation and genetic targeting. J Neurosci Methods 236:125–147
40. Sloan SA, Barres BA (2018) Assembling a cellular user manual for the brain. J Neurosci 38(13):3149–3153
41. Bellesi M, de Vivo L, Tononi G, Cirelli C (2015a) Transcriptome profiling of sleeping, waking, and sleep deprived adult heterozygous Aldh1L1-eGFP-L10a mice. Genom Data 6:114–117
42. Bellesi M, De Vivo L, Tononi G, Cirelli C (2015b) Effects of sleep and wake on astrocytes: clues from molecular and ultrastructural studies. BMC Biol 13:66
43. Boulay AC, Saubaméa B, Adam N, Chasseigneaux S, Mazaré N, Gilbert A, Bahin M, Bastianelli L, Blugeon C, Perrin S, Pouch J, Ducos B, Le Crom S, Genovesio A, Chrétien F, Declèves X, Laplanche JL, Cohen-Salmon M (2017) Translation in astrocyte distal processes sets molecular heterogeneity at the gliovascular interface. Cell Discov 3:17005
44. Morel L, Chiang MSR, Higashimori H, Shoneye T, Iyer LK, Yelick J, Tai A, Yang Y (2017) Molecular and functional properties of regional astrocytes in the adult brain. J Neurosci 37(36):8706–8717
45. Sakers K, Lake AM, Khazanchi R, Ouwenga R, Vasek MJ, Dani A, Dougherty JD (2017) Astrocytes locally translate transcripts in their peripheral processes. Proc Natl Acad Sci U S A 114(19):E3830–E3838
46. Boutej H, Rahimian R, Thammisetty SS, Béland LC, Lalancette-Hébert M, Kriz J (2017) Diverging mRNA and protein networks in activated microglia reveal SRSF3 suppresses translation of highly upregulated innate immune transcripts. Cell Rep 21(11): 3220–3233
47. Samanani S, Mishra M, Silva C, Verhaeghe B, Wang J, Tong J, Yong VW (2013) Screening for inhibitors of microglia to reduce neuroinflammation. CNS Neurol Disord Drug Targets 12(6):741–749
48. Henry CJ, Huang Y, Wynne A, Hanke M, Himler J, Bailey MT, Sheridan JF, Godbout JP (2008) Minocycline attenuates lipopolysaccharide (LPS)-induced neuroinflammation, sickness behavior, and anhedonia. J Neuroinflammation 5:15
49. Möller T, Bard F, Bhattacharya A, Biber K, Campbell B, Dale E, Eder C, Gan L, Garden GA, Hughes ZA, Pearse DD, Staal RG, Sayed FA, Wes PD, Boddeke HW (2016) Critical data-based re-evaluation of minocycline as a putative specific microglia inhibitor. Glia 64(10):1788–1794
50. Elmore MR, Lee RJ, West BL, Green KN (2015) Characterizing newly repopulated microglia in the adult mouse: impacts on animal behavior, cell morphology, and neuroinflammation. PLoS One 10(4):e0122912
51. Elmore MR, Najafi AR, Koike MA, Dagher NN, Spangenberg EE, Rice RA, Kitazawa M, Matusow B, Nguyen H, West BL, Green KN (2014) Colony-stimulating factor 1 receptor signaling is necessary for microglia viability, unmasking a microglia progenitor cell in the adult brain. Neuron 82(2):380–397
52. Rice RA, Spangenberg EE, Yamate-Morgan H, Lee RJ, Arora RP, Hernandez MX, Tenner AJ, West BL, Green KN (2015) Elimination of microglia improves functional outcomes following extensive neuronal loss in the hippocampus. J Neurosci 35(27):9977–9989
53. Dagher NN, Najafi AR, Kayala KM, Elmore MR, White TE, Medeiros R, West BL, Green KN (2015) Colony-stimulating factor 1 receptor inhibition prevents microglial plaque association and improves cognition in 3xTg-AD mice. J Neuroinflammation 12:139
54. Szalay G, Martinecz B, Lénárt N, Környei Z, Orsolits B, Judák L, Császár E, Fekete R, West BL, Katona G, Rózsa B, Dénes Á (2016) Microglia protect against brain injury and their selective elimination dysregulates neuronal network activity after stroke. Nat Commun 7:11499
55. Spangenberg EE, Lee RJ, Najafi AR, Rice RA, Elmore MR, Blurton-Jones M, West BL, Green KN (2016) Eliminating microglia in Alzheimer's mice prevents neuronal loss without modulating amyloid-β pathology. Brain 139(Pt 4):1265–1281
56. Rice RA, Pham J, Lee RJ, Najafi AR, West BL, Green KN (2017) Microglia repopulation resolves inflammation and promotes brain recover after injury. Glia 65(6):931–944
57. Staniland AA, Clark AK, Wodarski R, Sasso O, Maione F, D'Acquisto F, Malcangio M (2010) Reduced inflammatory and neuropathic pain and decreased spinal microglial response in fractalkine receptor (CX3CR1) knockout mice. J Neurochem 114(4):1143–1157

58. Cho SH, Sun B, Zhou Y, Kauppinen TM, Halabisky B, Wes P, Ransohoff RM, Gan L (2011) CX3CR1 protein signaling modulates microglial activation an protects against plaque-independent cognitive deficits in a mouse model of Alzheimer disease. J Biol Chem 286(37):32713–32722

59. Mattison HA, Nie H, Gao H, Zhou H, Hong JS, Zhang J (2013) Suppressed pro-inflammatory response of microglia in CX3CR1 knockout mice. J Neuroimmunol 257(1–2): 110–115

60. Milior G, Lecours C, Samson L, Bisht K, Poggini S, Pagani F, Deflorio C, Lauro C, Alboni S, Limatola C, Branchi I, Tremblay ME, Maggi L (2016) Fractalkine receptor deficiency impairs microglial and neuronal responsiveness to chronic stress. Brain Behav Immun 55:114–125

61. van der Maten G, Henck V, Wieloch T, Ruscher K (2017) CX_3C chemokine receptor 1 deficiency modulates microglia morphology but does not affect lesion size and short-term deficits after experimental stroke. BMC Neurosci 18(1):11

Chapter 5

Neuron-Glia Interactions Studied with In Vitro Co-Cultures

S. Mancino, M. M. Serafini, and Barbara Viviani

Abstract

The complexity of neuronal cell structures and functions requires specific methods of culture to determine how alteration in or among cells gives rise to brain dysfunction and disease. In this context, the primary culture of neuronal cells plays an important role in the study of this topic, especially related to neuronal cells survival and differentiation, nutritional requirements, but also neuronal development and spine formation. For all these investigations and applications, it is very important that primary neurons are cultured under conditions that resemble the in vivo environment as closely as possible. In this line, glia-neuron sandwich co-cultures are an extremely useful tool in vitro to evaluate cell-to-cell interaction relaying on the release of soluble factors and could be a suitable method in the study of the contribution of glia-secreted molecules to neuronal development and spine formation. To this end, this chapter describes the procedures to set up a sandwich co-culture system from primary rat glial cells and hippocampal neurons, and highlights advantages and disadvantages of this approach and its possible application in the investigation of individual glial factor impact on neuronal properties.

Key words Primary hippocampal neurons, Primary glial cells, Sandwich co-culture, Cell-to-cell interaction, In vitro cell systems, Synaptogenic factors, Neuronal development

1 Introduction

Recent evidence has provided new insights into the neuron-glia interrelation, and the role of glial cells in neuronal function and morphology has reached new horizons. Indeed, complex interdependencies among these elements have been demonstrated for development and maintenance of the nervous system [1]. The signals connecting neurons to glia and vice versa include ion fluxes, neurotransmitters, cell adhesion molecules, and specialized signaling molecules. Neurons dynamically regulate a wide range of glial activities, including their proliferation, differentiation, maturation, and myelination [2, 3] in connection with their status of activity, through the release of substances and chemical messengers at synaptic junctions and in extrasynaptic regions of neurons [2]. Meanwhile, glial cells contribute to maintain normal functioning of the nervous system by both controlling the extracellular

Michael Aschner and Lucio Costa (eds.), *Cell Culture Techniques*, Neuromethods, vol. 145,
https://doi.org/10.1007/978-1-4939-9228-7_5, © Springer Science+Business Media, LLC, part of Springer Nature 2019

environment and supplying metabolites and growth factors to neurons. Specifically, glia produces trophic and synaptogenic factors, regulates neurotransmitter and ion concentrations, and removes toxins and debris from the extracellular space of the central nervous system (CNS), maintaining an extracellular milieu that is optimally suited for neuronal function grow [4]. In this line, glia also plays a key role in synaptic plasticity by secretion of synaptogenic factors able to regulate synapse formation and synaptic strength and may participate in information processing by coordinating activity among different sets of neurons [5]. In the same way, and more specifically, astroglial cells secrete neurotransmitters (such as glutamate, ATP, and GABA), neuromodulators (such as adenosine and D-serine), cytokines (such as interleukins and tumor necrosis factor-α) [6], neurohormones (such as atrial natriuretic peptide) and other humoral factors (such as eicosanoids) that modulate synaptic networks affecting information processing [7]. In addition to neurons, glia communicate with other glial cells through intracellular waves of calcium, diffusion of chemical messengers, specialized cell contact-mediated communication via gap junctions, but also by extracellular signaling molecules that are released and propagated. Physiologically, the bidirectional communication existing between glia and other cells of the nervous system (i.e., neurons, other glial cells, and blood vessel cells) allows glia to link cells and structures that are not functionally connected, to continuously monitor and modulate their activity as a function of local needs [8]. The consequences of such interactions are the neuronal synchronization, the modulation of synaptic function, and the regulation of cerebral blood flow. The recognition of this connection led to the definition of a new function named "gliotransmission." The term gliotransmission describes the release of factors from physiologically stimulated glia able to activate a rapid response in neighboring cells [8]. All these factors have different targets and roles, but could be also involved in neuronal development, remodeling of neuronal topography, formation of synaptic connections and dendritic spine growth [2, 9]. In fact, neurons grown in the absence of astrocytes produce fewer functional synapses than neurons grown either with astrocytes or with astrocyte-conditioned medium, suggesting that factors secreted by astrocytes are necessary for synaptic development [10]. Moreover, studies using purified neuron and astroglial cultures revealed that neurons form few and weak synapses in the absence of glia [10] and mice with genetically inhibited gliogenesis display rampant neuron loss, diminished motor output [11], and altered synaptogenesis [12]. Indeed, evidence has provided a framework whereby glia-derived secreted factors promote the formation and maturation of excitatory synapses [13].

Neuron-glia communication is not only necessary in the physiology of the nervous system: aberration of normal glial functions turns to be detrimental, leading to toxicity and pathogenesis (for an extensive review, visit the Nature collection Glial cells in health and disease at the web link https://www.nature.com/collections/ypjcncrzxn).

Understanding the mechanistic details of this complex interconnection in health and disease requires a robust in vitro system that can effectively assess the impact of individual glial factors on neuronal properties.

A classical approach to study neuron-glial interactions in vitro is the co-culture system. Co-cultures of different cells of the nervous system (i.e., neurons, astrocytes, microglia, endothelial cells) represent the easiest approach to dissect the intercommunication among different cell populations in physiological, pathological, pharmacological, and toxicological conditions taking into account the molecular mechanisms involved.

Classically, co-cultures are based on the use of two different cell population representatives of the nervous system, such as different subtypes of neurons, astrocytes, microglia, pericytes or endothelial cell, differently combined. The setup most commonly used to obtain co-cultures allows the contact between the considered cell populations (mixed culture) or partially separates them. In the former setup, cells are sequentially seeded on the same surface (i.e., the bottom of a well), and in the latter, cells are grown on two different surfaces (i.e., transwell system) or are separated using microfluidic devices (Fig. 1) [14, 15].

An extension of the co-culture models is triple cultures resulting from the combination of three different populations like in blood-brain barrier (BBB) models that contain endothelial cells, astrocytes, and a third population that might be neurons or pericytes [15] (Fig. 1). The advent of the organ on the chip allowed to obtain 3D co-culture in a dynamic in vitro setup that enables exchange of oxygen, nutrients, and metabolites [16]. This system allows to combinate single-organ systems to obtain a multi-organ system, which emulates systemic interaction of different organs [16].

The highest complexity and diversity in co-cultures system has been reproduced by generating brain organoids, three-dimensional structures based on the use of human pluripotent stem cells which allows to model mature brain features [17].

The present chapter will describe an example of no-contact culture (sandwich co-culture), based on the growth of primary hippocampal neurons on a glass coverslip facing a glia monolayer grown on the bottom of a 12-well plate (Fig. 2). In our experimental model, the glass coverslip holding neurons is separated by the glia monolayer through small paraffin dots at the edges of the coverslip. The glia monolayer can be a mixed culture, containing both microglia and astrocytes, or a monoculture of purified microglia or astrocytes. This setup has been also used in our

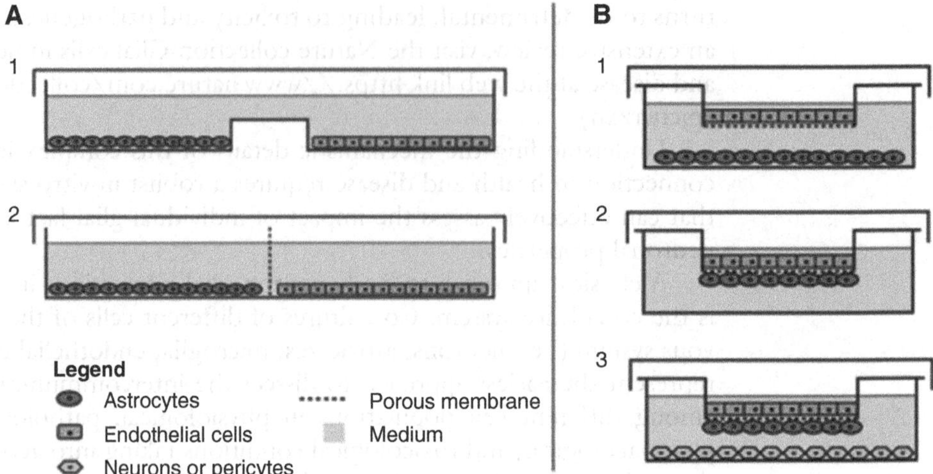

Fig. 1 Different types of co-culture systems. (**a**) Microfluidic systems with cells separately seeded, in (1) the medium is directly in communication, in (2) medium is communicating through a porous membrane. (**b**) Transwell systems with (1) noncontact co-culture with cell monolayers separated by a gap, (2) in contact co-cultured with two cell types separated by a membrane, and (3) triple-cultured primary and immortalized cell lines

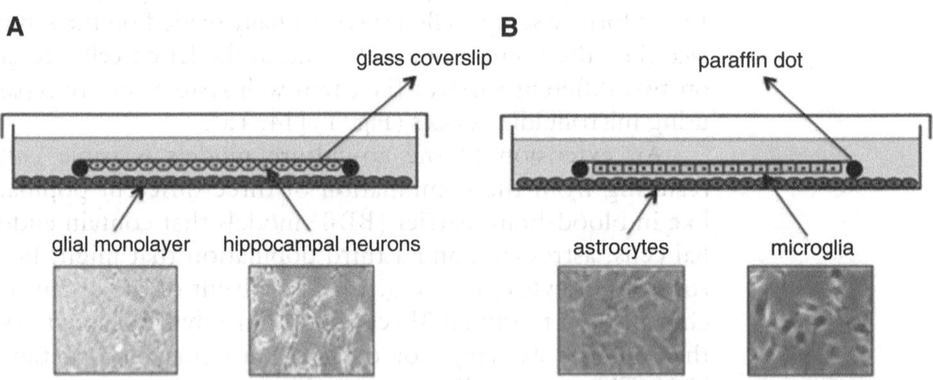

Fig. 2 Schematic representation of a typical sandwich co-culture. (**a**) Hippocampal neuron-glia co-culture and (**b**) microglia-astrocytes co-culture prepared in a 12-well plate. Glial cells or astrocytes are seeded at the bottom of Petri dishes, while neurons or microglia, respectively, on the top of glass and placed facedown in the well. Paraffin dots on the coverslips do not allow direct contact among cell populations

laboratory to study astrocyte-microglia interaction by seeding astrocytes on the glass coverslips and microglia on the bottom of the well (Fig. 2). Although conceptually similar to the transwell system, this setup allows a narrowest gap between neurons and glia that favors cell-to-cell communication and neuronal differentiation. Because there is no contact between cell populations, sandwich co-cultures allow to study how different cell populations can reciprocally influence their functions and viability through the release of soluble mediators. This tool became a valuable method for improving cultivation success needed for neurons discriminating the cell type-specific molecular mechanisms involved.

By separating populations, co-culture systems can be treated as comprising "monoculture modules." In general, the possibility to separate the two populations allows to control the existing interactions, which can be key to achieve a stable system. We got advantage of this approach in studying the role of native proinflammatory cytokine IL-1β released by gp-120 stimulated glia on neuronal death progression [18, 19]. By exposing the same set of neurons to gp-120, in the presence or not of a mixed glia monolayer, we were able to prove the need of glia population to induce neuronal death. The molecular mechanisms involved were also investigated. To do that, neurons and glia in co-culture were separated after being exposed together to the toxic challenge and were analyzed for different parameters. Neurons were subjected to (1) intracellular calcium measurement, (2) Western blot analysis to evaluate the tyrosine phosphorylation of the NMDA receptor (a pathway related to NMDA-induced intracellular calcium increase and cellular location of the receptor), and (3) confocal microscopy to analyze NMDAR distribution at the synapse. Glia was analyzed for cytokines expression [19]. Gp-120 was able to selectively trigger the production of IL-1β [19] which in turn activated the Src family of tyrosine kinases in neurons [20]. Loading neurons with an irreversible Src kinase specific inhibitor (Ca-pYEEIE) prior to be transferred to glia, and consequently treating them with gp-120, allowed to act on neurons without interfering with the Src family signal transduction in glia. Through this approach, we proved the relevance of this pathway in gp-120 and IL-1β-induced calcium increase in neuron, activation of the NMDAR, increased localization at the synaptic spine, spine density reduction, and finally neuronal death [20]. Sandwich co-cultures have thus been a useful tool also to elucidate how glia-derived secreted factors temporally affect synaptic structure and neuronal morphology while investigating molecular mechanisms in specific cell types in isolation.

To summarize, the great advantage of a sandwich co-culture system in studies over the other in vitro systems is the possibility of separating the two cell populations at any time (e.g., prior to or after a treatment) while retaining their integrity. This allows the investigator to (1) manipulate the cell types differently before they are cocultivated together, thus providing information on the involvement of specific mediators or biochemical pathways, (2) perform different biochemical measurements on the two cell populations separately at the end of the treatment, and (3) evaluate the activity on highly differentiated neurons in the presence or absence of the glial feeder layer. In the same line, co-cultures could be a useful tool to elucidate how the diverse glia-derived secreted factors are released in a regulated fashion for temporal and spatial control of neuronal morphology and structure.

The simplicity of this system, compared to organoids, helps in identifying key biochemical processes that can be further addressed in more complex co-cultured systems through a tiered approach.

The lack of contact between neurons and glia, or astrocytes and microglia, may, however, represent a disadvantage since it does not mirror physiological conditions. To overcome this issue, hippocampal neurons can be plated directly onto glial cells, as can microglia onto astrocytes, but the advantage of easily manipulating one of the two cell populations forming the co-culture would be lost.

Collectively, this evidence shows that communication between neurons and glia is critical for the formation, stability, morphology, and functionality of neurons and points to the importance of methods to investigate these issues. In this line, the neuron-glia co-culture systems described in this study could be a useful tool to address questions regarding cellular population interactions and intercommunication through secretion of soluble factors.

In this chapter, we describe procedures to set up a sandwich co-culture system and specific methods for the preparation of hippocampal neurons, glial cells, astroglial cells, and coated glass coverslips.

2 Methods

Before entering into the details of the glia-neurons sandwich co-culture method, we should provide some general comments. The aim of co-culture cells preparation is not only to study the interaction between glia and neural cells, in developing and mature neurons, but also to obtain a highly differentiated neuronal culture. At this end, this technique consists of different steps and is largely based on the method of Goslin and Banker [21]. The success of this method depends on careful planning, organization, and attention to detail since, as shown in Fig. 3, it takes around 20 days to prepare neurons for experiments. Glia culture must be prepared at least 10 days before the planned neuron dissection to allow glial trophic monolayer to reach confluency before seeding of hippocampal neurons for neuron-glia sandwich co-culture preparation.

Primary glial cells are seeded at the bottom of multiwell plates and can be a mixed culture containing both microglia and astrocytes or a monoculture of purified microglia or astrocytes. Mixed culture monolayer is viable and suitable to grow neurons for up to 1 month from its preparation.

Cultures of hippocampal neurons are prepared and plated onto glass coverslips. These coverslips are then placed above the glial monolayer (Fig. 3), separated through small paraffin dots at the edges of the coverslip.

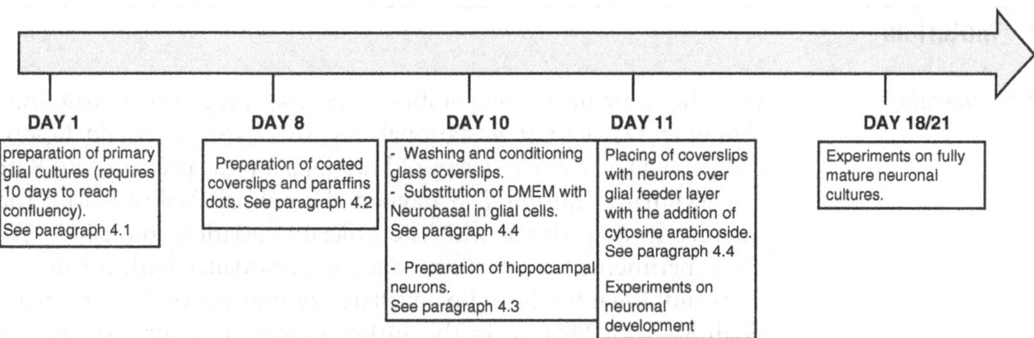

Fig. 3 Sandwich co-culture preparation timeline. This timeline describes the major steps involved in preparing the sandwich hippocampal glia-neuron co-culture system. These steps are explained in Sects. 4.1, 4.2, 4.3, and 4.4

A similar co-culture system of astrocytes and microglial cells has been also used in order to study the molecular mechanisms involved in communication between these two cell populations with some minor changes in the protocol. Following the protocol described by McCarthy and DeVellis [22] and Giulian and Baker [23] description, which takes advantage of the difference in degree of attachment of the two cell types to tissue culture plastic, indeed, microglial cells detach from astrocytes by shaking mixed glial cultures on an orbital shaker, and astrocytes are then further depleted of microglial cells by treatment with L-leucine methyl ester (L-LME) that selectively kills microglia. In this case, astrocytes are seeded at the bottom of multi-well plates while microglia on the top of glass coverslips. In this way, microglial cells are juxtaposed, without touching, to the astrocyte monolayer. This astrocyte-microglia co-culture is usually used within 48 h of its preparation.

Other smaller steps, including preparing paraffin dots, cleaning coverslips and coating, and washing and conditioning glass coverslips, must be carefully integrated into the calendar to ensure that all the materials are ready at the time of hippocampal neuron preparation for cellular co-culture. Important parameters to obtain successful co-cultures are sterility, rapidity in dissecting brain tissues and plating the obtained cells, and fresh reagents. This means that, if possible, cell isolation, culturing, and preparation of all solutions should be performed in the shortest time in a laminar flow hood and all the surgical instruments should be autoclaved for 21 min at 121 °C to be sterilized.

Here we describe, first of all, the main step to prepare primary glial cultures (Day 1, see Sect. 4.1), then the preparation of the glass coverslips (Day 8, see Sect. 4.2), and followed by the preparation and seed of hippocampal neurons (Day 10, see Sect. 4.3) for the neuron-glia cell co-culture (Day 11, see Sect. 4.4).

3 Materials

3.1 Animals

For the co-culture preparation, we use pregnant female rats (Sprague–Dawley) at gestational day 18 (E18) to obtain hippocampal neurons and 1–2-day-old rat pups for glia preparation. Rats are acclimated during 2–3 days in the animal facility before the sacrifice in accordance with the official legislation guidelines. All the experiments here performed are in accordance with the guidelines laid down by the animal welfare committees of the Università degli Studi di Milano. In the author's laboratory, after the isoflurane (3–4%) anesthetization and CO_2 exposure, animals are beheaded. For pregnant rats, the abdomen is opened, and rat embryos are taken. From each fetal rat, the whole brain is dissected and kept in cold HBSS as following described in paragraph 3.3.

3.2 Common Dissection Equipment

Isoflurane dispenser and chamber connected to a CO_2 tank Dissecting tools, sterile:

- Ninety-five percent ethanol (for tool sterilization)
- Stainless steel scissors with ~4 and ~2 cm blades
- Curved forceps (two pairs)
- Dumont forceps, no. 3c and no. 5
- 100 (two), 60 (one), and 35 mm (six) Petri dishes
- Bistoury (Aesculap)

Dissecting microscope, e.g., Zeiss Stemi DV4
Disposables:

- 1.5 mL microcentrifuge tubes, sterile
- Fifty milliliter conical polystyrene centrifuge tubes
- One milliliter and 200 μl pipet tips with filters, sterile
- Twelve and/or twentyfour-well tissue culture plates
- Coated coverslips
- Seventy-five square centimeters canted-neck flasks with screw caps

Reagents:

- HBSS (see recipe)
- 1× trypsin/EDTA (ethylenediamine tetra-acetic acid) solution (Sigma-Aldrich – Merck)
- Ten milligrams per milliliter DNase I stock solution (see recipe)
- High-glucose MEM/10% (v/v) FBS (see recipe)
- 0.04% (w/v) trypan blue

3.3 Common Equipment for Cell Culture

Laminar flow hood equipped with UV lamp to prepare cell cultures

37 °C, 95% air/5% CO_2, and 95% relative humidity incubator to grow cell cultures

100, 60, and 35 mm Petri dishes, plates, and flasks to be used during dissection and to plate and grow cell cultures

Centrifuge and water bath heater

Equipment for the determination of cell number and viability: cell counter, hemocytometer or Burker chamber, squared glass coverslips (22 × 22), and 0.04% (w/v) Trypan blue solution (see recipe)

Pipettes and pipette tips (P20, P200, P1000)

3.4 Primary Glial Cultures

Specific disposables:

- One hundred micrometer nylon cell strainer (BD Falcon)

Specific equipment:

- Stirring plate

Specific reagents:

- High-glucose MEM/20% (v/v) FBS (see recipe)

3.5 Primary Hippocampal Cultures

Specific reagents:

- Neurobasal complete medium (Gibco–Thermo Fisher Scientific) (see recipe).
- Two millimeter cytosine-1-b-d-arabino-furanoside (CyARA) stock solution in distilled water. Store up to 1 year, protected from light, at −20 °C.
- Microwave oven to sterilize glass coverslip or, if needed, tips.
- Seventy-five square centimeter tissue culture flasks.
- Hemocytometer.

3.6 Glass Coverslips for Sandwich Co-culture

Disposables:

- 5 to 10 mL sterile syringe with 0.95 × 40 mm needle
- Sterile 12 and/or 18 mm glass coverslips

Reagents:

- 1× poly-L-ornithine solution (see recipe)
- 1× PBS (Sigma-Aldrich – Merck)
- High-glucose MEM/10% FBS (see recipe)
- Neurobasal complete medium for primary hippocampal cells
- Two millimeters cytosine arabinoside solution

Equipment:

- Paraffin wax
- Microwave oven 5–10 mL sterile syringe with 0.95 × 40 mm needle germicide lamp (e.g., as equipped on a flow hood)
- Heating magnetic stirrer
- Sterile sharpened forceps

4 Methods

4.1 Primary Glial Cultures

Generally, in the author's laboratory, isolating and culturing cortical glial cells require 2 h from dissection to plating.

4.1.1 Preparation of Cortical Glial Cells

Glial cells (astrocytes, microglia and oligodendrocytes) are obtained by mechanical and enzymatic dissociation (e.g., trypsin) of cerebral tissue classically from 1- to 2-day-old rat pups (Sprague–Dawley). With this procedure, the author usually obtains ~5 × 10^6 cells from two pups. Confluent cultures of glial cells can be used to grow and differentiate neurons, to assemble the sandwich co-culture or to obtain purified cultures of astrocytes and microglia with the method described by McCarthy and DeVellis [22]. To obtain a glial monolayer for neuron-glia co-culture, cells are seeded in 24-well plates at a density of 50,000 cells/mL per well or in 12-well plates at a density of 100,000 cells/mL. The choice of the type of well plates is suggested by the endpoint to be analyzed and the number of neurons needed. Cultures reach confluence in ~10 days.

In the following paragraph, all the steps from the dissection to preparation of the glial monolayer are described.

4.1.2 Dissection

1. Sterile-dissecting tools reported in the Materials paragraph are needed. Once all the material is ready, animal is sacrificed following the officially approved procedure, and heads are placed and stored in a 35 mm Petri dish with 2 mL of HBSS. It is important to keep this Petri dish on ice along the whole procedure to reduce cellular metabolic activity and preserve tissue integrity. Carefully pry away the skin and thin skull using either sharpened scissors or curved forceps. To ease the removal of the brain, cut the left and right side of the skull in the center perpendicularly to the midline, gently open the skull with sharpened scissors, and finally remove the brain with a small spatula, and place it in a 60 mm Petri dish with 5 mL of HBSS.

4.1.3 Cortice Isolation

To separate the cortices, move one brain to another 35 mm Petri dish with 2 mL of cold HBSS. Working under the dissecting microscope, place the two tips of Dumont no. 5 forceps along the brain midline under the cortices, one at the anterior and the other at the posterior end, and with a single cut, isolate the cortices from the brainstem. Remove the meninges on the cortical surface by pulling them off gently with Dumont forceps until the surface of the cortex appears completely white. At the end, store all the cortices on ice in a 60 mm Petri dish with 5 mL of HBSS.

4.1.4 Cell Disassociation

All the cortices are placed and grouped in the center of a 35 mm Petri dish on ice. Here, with the use of bistouries, cut the tissues several times in different directions to obtain a homogenate.

Resuspend the minced cortices with a 2 mL of HBSS with a disposable sterile plastic pipette. Repeat the washing until all minced cortices are removed. Usually, three washing for a total of 6 mL of HBSS are sufficient. After mechanical dissociation, proceed to the enzymatic one by adding 750 μL of 10× trypsin and 750 μL of 10 mg/mL DNase in a 50 mL tube. Seal the tube with Parafilm to be able to vigorously agitate it in a water bath at 37 °C for 15 min. This process favors the enzymatic digestion of the tissue. At the end, allow the undissociated tissue to be collected at its bottom.

Carefully collect dissociated cells in suspension with a disposable sterile plastic pipette and transfer to 50 mL centrifuge tube with 12 mL MEM complete medium 10% FBS.

The leftover undissociated tissue will be processed once more by repeating the above procedure. To this purpose add again 6 mL of HBSS, 750 μL of 10× trypsin and 750 μL of 10 mg/mL DNase to the 50 mL tube containing undissociated tissues.

Collect all dissociate cells together and filter the suspension through a 100 mm nylon cell strainer in a new 50 mL centrifuge tube to eliminate any residual undigested tissue.

Centrifuge the pooled cells for 5 min at 200–300× *g* at room temperature, aspirate the medium, and recover the pelleted cells.

Pellet is then resuspended in ~5 mL of MEM complete medium supplemented with 10% FBS. To optimize disaggregation of the pellet, first add 2 mL of MEM while pipetting with a 2 mL disposable plastic pipette until the solution becomes homogeneous, and then dilute with additional 3 mL of MEM 10% FBS.

Count cells with the hemocytometer under an inverted phase contrast microscope (*see* **Note 1**) after having mixed 100 μL of cell suspension to 200 μL of 0.04% (w/v) trypan blue solution (1:3) (see recipe).

4.1.5 Preparation of Glia for the Sandwich Co-culture

From the 5 mL cell suspension, withdraw a volume containing a sufficient number of viable cells to prepare the desired number of 24- or 12-well plates at a ratio of 50,000 or 100,000 cells/well, respectively. Dilute the obtained cell suspension to 50,000 or 100,000 cells/mL with high-glucose MEM/20% FBS, and pipet 1 mL cell suspension into each well of a 24- or 12-well plate.

4.1.6 Growing Cultures to Confluence

Incubate plates for 24 h. At the end add fresh MEM supplemented with 20% FBS. This passage allows to eliminate all unattached dead cells.

After 5 days, replace culture medium with fresh MEM supplemented with 10% FBS. Grow to confluence replacing the medium twice a week.

4.2 Glass Coverslips for Sandwich Co-culture

At least 1 day before seeding the hippocampal neurons, prepare paraffin dots on 12 or 18 mm glass coverslip. Place glass coverslips in a multiwell plate (one coverslip per well), and sterilize under UV light for 2–3 h or in a microwave oven for 10 min at the highest power.

In the meanwhile, heat paraffin wax to ~100 °C. When coverslips are ready, take an aliquot of paraffin with an insulin syringe, and apply three small drops near the outer edges of each coverslip at roughly equal distances from each other (*see* **Note 2**).

Re-sterilize the coverslips by UV irradiation for 30 min with a germicide lamp.

In a laminar flow hood, add 1 mL of 1× poly-L-ornithine solution in each well containing the coverslips making sure that they are completely covered by polyornithine. Keep them to room temperature up to 2 days or for 2 h at 37 °C. Coating will allow to promote cellular attachment. Plates can be stored sealed with Parafilm in the dark.

Before cell seeding, remove completely polyornithine, wash with PBS, add 1 mL of neurobasal medium, and incubate at 37 °C, 5% CO_2 to equilibrate before cell seeding.

4.3 Primary Hippocampal Cultures

Hippocampal cultures were prepared from 18-day-old fetal rats. From one litter of 18-day-old embryos (i.e., 9–12 embryos), it is possible to obtain ~7 × 10^6 cells. According to the plating density experiment of the author's group, 80,000 cells/coverslip yielded a highly differentiated hippocampal culture with a sufficient number of cells for the independent measurement of several parameters of neurotoxicity. The procedures for isolating hippocampal neurons from rat embryos are an adaptation of the method initially described by Goslin and Banker [21]. In the author's laboratory, isolation and seeding of hippocampal neurons require 1.5–2 h.

4.3.1 Dissection

Arrange sterile-dissecting tools and anesthetized animal with the officially approved procedure. Set the animal on a dissecting table, with the ventral side up, and sterilize the abdomen by pouring 95% ethanol over it. Grasp the abdominal skin with forceps, and cut the abdomen completely open from the vagina to the thoracic cavity, cutting the diaphragm. Gently grasp uterine horns at one of the constrictions and lift it up. Remove the horns by cutting the attachments to the abdominal cavity, and place them in a 100 mm Petri dish filled with 10 mL of cold HBSS. Remove the fetuses from the uterine horns, and place them in a 100 mm Petri dish with cold HBSS by grasping at the upper constriction of each yolk sac and cutting along one side; the fetuses will slip out of it. Place the heads in a 60 mm Petri dish with cold HBSS. Each single step of this procedure should be made on ice to arrest metabolic activity. Place one head in a 35 mm Petri dish filled with 2 mL of HBSS, with the skullcap toward the operator. Under the dissection microscope, gently grasp the cut edges of the skin and skull with both the Dumont forceps and pull in opposite directions to expose the brain. Remove the brain by lifting upward out of the skull with curved-tip forceps, and place it in a 60 mm Petri dish with 5 mL of HBSS as previously described.

4.3.2 Isolation of Hippocampi

To remove the hippocampi, split the brain in half between the cerebral hemispheres, and separate the cortex from the diencephalon and the brainstem. Discard the brainstem, and orient one of the hemispheres with the medial surface upward. The hippocampus is found on the posterior half of the hemisphere. Remove the meninges, and dissect out the hippocampus by cutting along the boundary between the hippocampus and the adjoining cortex. Then, transfer the hippocampus to a 35 mm Petri dish with 1 mL of HBSS on ice.

4.3.3 Preparation of Hippocampi

Carefully transfer the hippocampi isolated from rat embryos in a sterile 1.5 mL microcentrifuge tube with a 1 mL pipette and centrifuge for 4 min at 100–150× g at room temperature. Remove the supernatant with a pipette, and add to the pellet 400 μL of 1× trypsin/EDTA and 80 μL of 10 mg/mL DNase I. Gently shake and place in a 37 °C incubator for 5 min (*see* **Note 3**).

Prepare three 1.5 mL Eppendorf tubes with 400 μL of MEM supplemented with 10% FBS each and one with neurobasal medium.

Gently aspirate off the trypsin/DNAse solution with a sterile 1 mL pipette (avoid vacuum), add 400 mL of MEM complete medium 10% FBS, and gently shake until the hippocampi are floating in the medium (*see* **Note 4**).

Collect the pelleted at the bottom of the Eppendorf tube with a 1 mL pipette in the smallest possible volume of medium and transfer to the first of the four tubes with 400 μL of MEM complete medium 10% FBS. Repeat for each tube to wash the hippocampi free from trypsin and DNAse I.

Dissociate the hippocampi by gently pipetting in the last tube containing neurobasal medium, and adjust the volume of the cell suspension to 1 mL with neurobasal medium.

Count cells in a hemocytometer mixing 10 μL of cell suspension with 20 μL of 0.04% trypan blue solution (1:3 dilution).

4.3.4 Preparation of Hippocampal Neurons for the Sandwich Co-culture

Pipet the volume containing the desired number of cells into each well of a 24- or 12-well plate containing a coated coverslip and incubate plates overnight.

4.4 Preparation of Sandwich Co-culture

1. The same day that hippocampal neurons are prepared, replace the medium in confluent cortical glial monolayers with 1 mL Neurobasal medium, and return them to the incubator overnight (*see* **Note 5**).
2. The next morning, place the coverslip over the confluent glial monolayer (*see* **Note 6**) with the help of a pair of sharpened forceps.
3. To each well containing neurons and glia, add cytosine arabinoside to a final concentration of 5 μM to reduce the proliferation of glial cells (*see* **Note 7**).
4. Maintain the co-cultures routinely feeding them once every 7 days by replacing one-third of Neurobasal with fresh medium (*see* **Note 8**).

5 Notes

1. Trypan blue is used to distinguish viable from dead cells. Viable cells exclude trypan blue, while dead or damaged cells are stained dark blue.
2. The temperature of the paraffin is very important; if it is too hot, it spreads too thin and wide, while if it is too cold, the dots do not adhere to the coverslip and will subsequently detach.
3. DNase is added to prevent that the DNA released by damaged cells makes the dissociation medium too viscous during digestion.
4. The serum in the MEM inhibits residual trypsin and prevents over-digestion of the cells.
5. The addition of Neurobasal to glia prior to the addition of glass coverslips with hippocampal neurons allows conditioning to favor early phases of neural maturation.
6. Neurons can be co-cultured with glia even later than the day after preparation. In this case, glial medium should be replaced with neurobasal medium in which neurons have been grown and have been differentiated (conditioned neurobasal).
7. Addition of cytosine arabinoside, toxic to dividing cells, is important to block astrocytes proliferation in the neuronal culture. Thus, its addition after the neuronal cells are attached allows to obtain a 98% pure neuronal culture on the glass coverslip.

8. During feeding, it is important not to change the culture medium completely since neurons depend upon glial cells to condition the medium for long-term survival. Under this condition, neurons reach a high degree of maturation becoming highly innervated.

6 Reagents and Solutions

Use deionized, distilled water or equivalent for all recipes and protocol steps.

Cytosine Arabinoside, 2 mM

Dissolve 2 mg cytosine-1-β-D-arabino-furanoside (cytosine arabinoside) in 5 mL water. Store up to 6 months at −20 °C.

DNase I, 10 mg/mL

Dissolve 100 mg of 536 Kunitz units/mg DNase I (Sigma-Aldrich – Merck) in 10 mL HBSS (see recipe). Store up to 6 months in 1.5 mL aliquots at −20 °C.

HBSS

To 850 mL H_2O, add:

Hanks Balanced Salts powder (Sigma-Aldrich – Merck), enough for 1 l.

Ten milliliter 10 mM HEPES (Sigma-Aldrich – Merck).

Ten milliliter penicillin/streptomycin stock solution (Sigma-Aldrich – Merck).

Adjust volume to 1 l with H_2O.

Sterilize using a 0.22 μm cellulose-acetate disposable vacuum-filtration system (Merck Millipore).

Store up to 1 month at 4 °C.

The stock solution of penicillin/streptomycin contains 10,000 U/mL penicillin and 10 mg/mL streptomycin. Store this stock solution up to reported expiration date in 5 mL aliquots at −20 °C.

High-Glucose MEM/10%, or 20% FBS

Five hundred milliliter minimal essential medium with Earle's salts (MEM; Sigma-Aldrich – Merck).

0.6% (w/v) D(+)- glucose (Sigma-Aldrich – Merck).

Five milliliter penicillin/streptomycin stock solution (Sigma-Aldrich – Merck).

Five milliliter of 200 mM L-glutamine.

Shake vigorously to dissolve glucose.

Ten percent, or 20% (v/v) FBS (Sigma Aldrich– Merck).

Sterilize using a 0.22 μm cellulose-acetate disposable vacuum-filtration system (Merck – Millipore). Store up to 1 month at 4 °C.

Do not add FBS before shaking to avoid excessive foam formation.

The stock solution of penicillin/streptomycin contains 10,000 U/mL penicillin and 10 mg/mL streptomycin (100 U and 100 μg/mL final, respectively). Store this stock solution up to indicated expiration date in 5 mL aliquots at −20 °C.

Poly-L-Ornithine Solution, 1×

Prepare a 100× stock solution by dissolving 10 mg poly-L-ornithine (Sigma-Aldrich – Merck) in 6.67 mL water. Sterilize with disposable syringe filter (0.22 mm pore size). Store up to 6 months in 1 mL aliquots at −20 °C. Dilute to 1× with water.

Neurobasal Complete Medium

To 500 mL neurobasal (Gibco – Thermo Fisher Scientific), add 5 mL of 10,000 U/mL penicillin and 10 mg/mL streptomycin, 1.25 mL of 200 mM l-glutamine, and 1% B-27 (Invitrogen – Thermo Fisher Scientific). Sterilize in a vacuum-driven disposable filtration system (0.22 mm pore size, cellulose acetate). Store up to 1 month at 4 °C.

Acknowledgments

This research is supported by JPI-HDHL – Selenius – Selenium in early life to enhance neurodevelopment in unfavorable settings.

References

1. Biber K, Neumann H, Inoue K, Boddeke HWGM (2007) Neuronal 'On' and 'Off' signals control microglia. Trends Neurosci 30(11):596–602
2. Fields RD, Stevens-Graham B (2002) New insights into neuron-glia communication. Science 298(5593):556–562
3. Bezzi P et al (1998) Prostaglandins stimulate calcium-dependent glutamate release in astrocytes. Nature 391(6664):281–285
4. Jäkel S, Dimou L (2017) Glial cells and their function in the adult brain: a journey through the history of their ablation. Front Cell Neurosci 11:24
5. Perea G, Sur M, Araque A (2014) Neuron-glia networks: integral gear of brain function. Front Cell Neurosci 8:378
6. Vezzani A, Viviani B (2015) Neuromodulatory properties of inflammatory cytokines and their impact on neuronal excitability. Neuropharmacology 96:70–82
7. Wang F, Yuan T, Pereira A, Verkhratsky A, Huang JH, Huang JH (2016) Glial cells and synaptic plasticity. Neural Plast 2016:5042902
8. Volterra A, Meldolesi J (2005) Astrocytes, from brain glue to communication elements: the revolution continues. Nat Rev Neurosci 6(8):626–640
9. Stogsdill JA, Eroglu C (2017) The interplay between neurons and glia in synapse development and plasticity. Curr Opin Neurobiol 42:1–8
10. Ullian EM, Sapperstein SK, Christopherson KS, Barres BA (2001) Control of synapse number by glia. Science (80-) 291(5504):657–661
11. Schreiner B et al (2015) Astrocyte depletion impairs redox homeostasis and triggers neuronal loss in the adult CNS. Cell Rep 12(9):1377–1384
12. Tsai H-H et al (2012) Regional astrocyte allocation regulates CNS synaptogenesis and repair. Science 337(6092):358–362

13. Allen NJ (2013) Role of glia in developmental synapse formation. Curr Opin Neurobiol 23(6):1027–1033
14. Keenan TM, Folch A (2008) Biomolecular gradients in cell culture systems. Lab Chip 8(1):34–57
15. Wolff A, Antfolk M, Brodin B, Tenje M (2015) In vitro blood-brain barrier models-an overview of established models and new microfluidic approaches. J Pharm Sci 104(9): 2727–2746
16. Marx U et al (2016) Biology-inspired microphysiological system approaches to solve the prediction dilemma of substance testing. ALTEX 33(3):272–321
17. Di Lullo E, Kriegstein AR (2017) The use of brain organoids to investigate neural development and disease. Nat Rev Neurosci 18(10):573–584
18. Viviani B et al (2006) Interleukin-1β released by gp120 drives neural death through tyrosine phosphorylation and trafficking of NMDA receptors. J Biol Chem 281(40): 30212–30222
19. Viviani B, Corsini E, Binaglia M, Galli CL, Marinovich M (2001) Reactive oxygen species generated by glia are responsible for neuron death induced by human immunodeficiency virus-glycoprotein 120 in vitro. Neuroscience 107(1):51–58
20. Viviani B et al (2006) Interleukin-1 beta released by gp120 drives neural death through tyrosine phosphorylation and trafficking of NMDA receptors. J Biol Chem 281(40):30212–30222
21. Banker G, Goslin K (1998) Culturing nerve cells, 2nd edn. MIT, Cambridge, MA
22. McCarthy KD, de Vellis J (1980) Preparation of separate astroglial and oligodendroglial cell cultures from rat cerebral tissue. J Cell Biol 85(3):890–902
23. Giulian D, Baker TJ (1986) Characterization of ameboid microglia isolated from developing mammalian brain. J Neurosci 6(8):2163–2178

Chapter 6

Assessment of Mitochondrial Stress in Neurons: Proximity Ligation Assays to Detect Recruitment of Stress-Responsive Proteins to Mitochondria

Monica Rodriguez-Silva, Kristen T. Ashourian, Anthony D. Smith, and Jeremy W. Chambers

Abstract

Mitochondria are highly integrated organelles that must readily alter organelle physiology to adapt to the changing environment of neurons. Failure in the mechanisms regulating organelle adaptation and homeostasis manifests as perturbations in bioenergetics, Ca^{2+} buffering, and mitochondrial dynamics, which ultimately affect the integrity of organelle membranes, DNA, and proteins. Collectively, these anomalies in organelle function are referred to as mitochondrial stress. While elegant methods have been developed to measure fundamental mitochondrial physiology, only recently have new strategies emerged to investigate the regulatory mechanisms responsible for mitochondrial stress responses. The emergence of cytosolic, stress-responsive protein kinases and phosphatases demonstrates the importance of neuron-mitochondrial cross talk for regulating organelle health and quality. The magnitude of signaling cascades on the outer mitochondrial membrane (OMM) can greatly influence organelle form and function. Thus, interpreting OMM signaling events in the context of mitochondrial function is critical to understanding the role of stress-responsive protein kinases and phosphatases in health and disease.

In this chapter, we will provide a brief review of standard approaches to assess mitochondrial physiology and stress in neurons. The sources of neuronal mitochondria and the techniques used to measure bioenergetics, Ca^{2+} flux, organelle dynamics, radical production, and the integrity of fundamental organelle processes are discussed. The emphasis of the chapter pertains to methods that identify and validate the presence of stress-responsive signaling proteins (i.e., kinases and phosphatases) on the OMM in cultured neurons and fixed CNS tissues. We will describe our approach to proximity ligation assays for evaluating mitochondrial stress responses, specifically c-Jun N-terminal kinase (JNK) OMM signaling, in cells and brain sections.

Key words Calcium, Mitochondria, Mitochondrial dynamics, Mitophagy, Microscopy, Neurons, Protein-protein interactions, Protein import, Proteostasis, Proximity ligation assay, Reactive oxygen species, Respirometry, Stress response

1 Methods to Assess Mitochondrial Stress in Neurons

1.1 Introduction

Maintaining the electrochemical gradients and sustaining the synaptic transmission necessary for optimal neuronal communication and function creates a high energetic demand within the cell [1].

Michael Aschner and Lucio Costa (eds.), *Cell Culture Techniques*, Neuromethods, vol. 145,
https://doi.org/10.1007/978-1-4939-9228-7_6,

Consequently, mitochondria are crucial to neuronal function because these organelles are responsible for buffering intracellular calcium (Ca^{2+}) and generating cellular ATP [2, 3]. The energetic needs and electrochemical gradients are not uniformly distributed, spatially or temporally, within neurons [4]. The variability in bioenergetic demand and Ca^{2+} buffering within neurons requires an adaptable and dynamic mitochondrial network to support discrete subcellular activities. In fact, mitochondria regularly alter their spatial configuration and morphology in response to changes in the neuronal environment [5]. Because the restitution of adult neurons is difficult, mechanisms regulating mitochondrial quality control are of distinct importance to neurons, and perturbations in basal mitochondrial physiology (i.e., stress) can be detrimental to neurological processes.

Mitochondrial stress, or perturbations in nominal organelle physiology, can significantly affect the functions of neurons. The impact of mitochondrial stress varies depending upon the cellular context. Mitochondrial stress responses can be necessary for neuronal development and pathfinding [6, 7]. Alternatively, failure to curb mitochondrial stress or adverse environmental exposures can result in mitochondrial dysfunction [8–10]. While defects in mitochondrial physiology are well-documented in peripheral neuropathies and mitochondrial encephalomyopathies [11], mitochondrial dysfunction has recently emerged as a common component of neurological diseases, including autism [12], Alzheimer's disease [13], depression [14], Parkinson's disease [15], and schizophrenia [16]. The abnormalities in mitochondrial function are distinct among disorders; defects in bioenergetics, Ca^{2+} transport, organelle dynamics (fission/fusion), membrane organization, proteostasis, radical generation, and mitophagy are all associated with neurodegenerative disease pathophysiology [17, 18]. However, if the etiology of these disorders was simply diminished ATP production, one might expect a more homogeneous symptomology among the aforementioned neurologic conditions. The variability in neurological disease onset and manifestation may imply that the distinct pathogeneses arise from discrete capacities of different neuron types to handle specific mitochondrial stressors in the CNS. Wherein, specific classes of neurons, such as dopaminergic neurons, may be more susceptible to the effects induced by mitochondrial toxins, such as rotenone, than other types of neurons in the brain [19, 20].

The term mitochondrial stress is used collectively to describe distinct changes in mitochondrial form and function. Therein lies the issue, as it is difficult to determine the events responsible for pathogenesis and those attempting to maintain homeostasis. Therefore, a complete assessment of mitochondrial function may be required to ascertain the root of organelle stress and the dysfunction associated with disease progression. While it is important to assess mitochondrial function, it is also imperative to remember that maintaining mitochondrial health is a concerted process between organelle and neuron [21, 22]. Consequently, mitochondrial physiology should be examined in the context of the capacities for cellular processes to address distinct defects in organelle function. For example, Protein kinase A (PKA) signaling on mitochondria promotes organelle efficiency and neuronal survival [23–26], while excessive mitochondrial c-Jun N-terminal kinase (JNK) activity facilitates mitochondrial dysfunction and neurodegeneration [27–30]. The biological outcomes of mitochondrial PKA and JNK signaling events are influenced by the relative abundance of outer mitochondrial membrane scaffold proteins A-kinase anchoring protein-1 (AKAP-1) and Sab (or SH3-binding protein 5, SH3BP5), respectively [31–33]. Recently, AMP-dependent protein kinase (AMPK) has emerged as a signaling kinase with discrete localization on mitochondria, which ultimately impacts organelle dynamics and functions [34, 35]. These examples illustrate the importance of cellular pathways in the regulation and adaptation of mitochondrial function in neurons. Furthermore, we contend that analysis of mitochondrial form and function in concert with the assessment of active signaling pathways at the organelle level is necessary to determine if individual mitochondria are responding, or succumbing, to stress.

In this chapter, we will briefly review the methods used to assess mitochondrial form and function in neurons. Moreover, we will specifically address how the recruitment and abundance of stress-responsive proteins can be observed on the mitochondria of primary neurons and brain sections using proximity ligation assays (PLAs).

1.2 Sources of Neuronal Mitochondria

The analysis of mitochondria in neurons, especially changes in organelle physiology, has become commonplace in neurological studies, but reconciling the findings can be challenging due to the diversity of studies and the distinct sources of mitochondria. Each source of neuronal mitochondria identified in Table 1 has benefits and drawbacks that will be discussed below. Homogenates of the whole brain or of dissected areas of the brain can yield large amounts of mitochondria suitable for biochemical, enzymatic, and protein analyses; unfortunately, the regional and cellular context of the mitochondria is lost, and the findings can mask cell-type specific differences in mitochondria that may exist within a region of the brain [36, 37]. The drawbacks from the homogenates can be overcome in part by the use of refined isolation techniques, such as microdissections, synaptosome isolations [38], or preparation of brain sections [36, 37]. The isolation of synaptosomes can provide insights into mitochondria at discrete synapses; however, the synaptosomes are a pooled sample of synapses from many types of neurons in a particular area of the brain. Additionally, synaptosome isolates do not account for the nonautonomous cellular effects of astrocytes and glia and can be considered an

Table 1
Sources of neuronal mitochondria

Source	Acquisition/analysis	Benefits	Drawbacks
Brain homogenates	Ultracentrifugation Post-acquisition analyses	Assess relative mitochondrial activity	Cannot discern cells of origin or segregate phenotypes
Brain sections	Slicing fresh, fresh frozen, and fixed tissues	Examine region-specific differences in situ	Limited applications are compatible with sections
iPSC-derived neurons	Cellular analyses Ultracentrifugation	Assess mitochondria in specific neurons	Cell-autonomous effects only Difficult to generate sufficient for organelle isolation
Primary neuron cultures	Cellular analyses Ultracentrifugation	Assess mitochondria in specific neurons	Cell autonomous effects only Difficult to generate sufficient quantities for organelle isolation
Synaptosomes	Ultracentrifugation Post-acquisition analyses	Examine mitochondria physiology from synapses	Only accounts for local mitochondrial phenotypes

endpoint analysis for mitochondrial physiology. The use of brain sections will also allow the examination of mitochondria within neurons in the context of a select brain region when utilized with microscopy or other imaging approaches. Moreover, analysis of mitochondria can be performed within subcellular regions of individual neurons and, when combined with complementary analyses, can account for changes in astrocytes and glia that may affect mitochondrial functions in neurons. The use of primary neuronal cultures or neurons obtained from patients (or healthy controls) and derived pluripotent stem cells has also been used to assess mitochondrial form and functions [39]. While these cell-based approaches can provide specific answers regarding mitochondrial function in neurons, primary neurons and iPSC-derived neurons often have small amounts of mitochondria for biochemical analyses and isolations [40–42]; furthermore, the growth of "pure" neuronal cultures does not account for nonautonomous cellular and mitochondrial influences. Therefore, considering studies across multiple mitochondrial formats may be useful to understanding the growing body of literature surrounding neuronal mitochondrial physiology and pathophysiology. Additionally, using a combination of the mitochondrial sources mentioned in Table 1 can improve the quality of an investigation.

1.3 Approaches to Measure Mitochondrial Health in Neurons

1.3.1 Bioenergetics

Mitochondria are responsible for the generation of ~90% of the ATP produced in neurons and are consequently the prevailing sources of energy for neuronal activity [43]. Synaptic activities such as vesicle recycling, Ca^{2+} clearance, and transmitter import require significant amounts of energy; therefore, maintaining energy levels is critical to sustaining neurological functions [44–48]. For neuronal stress-related ATP decline, any decrease in cellular ATP concentration beyond 20% will likely induce apoptosis [49]. The assessment of bioenergetics of neurons justifiably focuses on the monitoring of cellular ATP levels and the functions of respiratory chain enzymes.

The measurement of ATP concentrations in neurons is performed primarily by two approaches as seen in Table 2: direct

Table 2
Methods used to assay discrete types of mitochondrial stress phenotypes

	Method(s)	Benefits	Drawbacks
Bioenergetics	ATP production		
	Direct quantitation	Allows for accurate determination of [ATP]	Do not identify mechanisms responsible for changes
	FRET	Evaluate ATP production in discrete locales	
	Complex analysis		
	Blue native gels	Identify changes in complex concentrations	
	Enzyme reactions	Determine relative enzymatic activity	
	Respirometry		
	Clark electrode	Determine total and complex-specific activities	Seahorse analyses are not compatible with fluorescent cells
	Seahorse analyzer		
Calcium homeostasis	Fluorescent dyes		Most mitochondrial dyes for Ca^{2+} have significant cytosolic levels
	Calcium electrode	Identify relative organellar changes in Ca^{2+} levels	
Membrane potential	Fluorescent dyes		Fluorescent dyes are qualitative and require significant optimization
	Microscopy	Visually presents relative membrane potential changes	
	TPP^+ electrode	Quantitatively determines membrane potential	
Mitophagy	Mito-QC mouse	Examine mitophagy in vivo	Methods require secondary analysis of mitochondrial components to confirm loss
	Microscopy	Detect abundance, localization, or ubiquitination	

(continued)

Table 2
(continued)

	Method(s)	Benefits	Drawbacks
Morphology/ architecture	Microscopy Electron Fluorescent	Visualize organelle form Evaluate specific protein functions	Mitochondrial dyes require optimization and can be leaky; genetically encoded probes remain the best option to assess form
	Mito-QC mouse	Assess architecture in vivo	Mitochondrial morphology in neurons is highly variable at discrete cellular locations, so architecture and form should be assessed carefully
Movement	Microscopy Live Post-fixed	Determine movement rates Reveal organelle location	Speed is difficult to assess and can vary depending on neuronal needs
mtDNA abundance/ mutation	RT-PCR Sequencing	Assess the integrity of the organelle genome	mtDNA is variable; experiments must be tightly controlled
Permeability transition pore	Microscopy/ depolarization Protein localization	Reveal mitochondrial membrane integrity	The induction of the mPTP differs among cell types and culture methods
Protein-protein interactions	Co-IP B/FRET	Identify specific interactions with mitochondria that may alter function	PPIs are highly dependent upon cellular context; therefore, replication and validation by multiple methods is important
	APEX/ biotinylation PLA	PLA works in fixed tissues and cells	
Proteostasis	Aggregation	Protein import and maintenance are necessary for proper organelle function	Defects in proteostasis can stem from the use of overexpression models
	Protein import UPR^{mt}	Determine the extent of cellular recognition for mitochondrial stress	
Reactive species generation	Fluorescent dyes	Determine the relative abundance of ROS/ RNS	Most ROS/RNS probes are not specific for a particular species
	Molecular probes	MS-based approaches can be quantitative and specific	MS-based approaches require significant expertise

quantitation [50, 51] and fluorescence resonance energy transfer (FRET) systems [52]. For direct quantitation, enzymatic reactions, such as luciferase-based reactions, can be used to determine ATP concentrations through the propagation of energy-dependent signals (luminescence and fluorescence). The accuracy of these assays is predicated on the use of robust standard curves and controls in order to demonstrate the contribution of mitochondrial ATP production, namely, oligomycin, which inhibits ATP synthase. A pitfall of direct ATP quantitation assays is that they cannot account for subcellular distributions of energy production in neurons. This drawback was addressed by the development of FRET-based systems that can localize ATP production at discrete subcellular sites. A FRET-based system was generated to detect ATP by linking a cyan fluorescent protein (CFP) to a yellow fluorescent protein variant, Venus, by the ATP-binding domain of bacterial ATP-synthase [52]. In the presence of ATP, a conformational change in the ATP-binding domain brings CFP and Venus close enough for photon transfer. Thus, exciting CFP (435 nm) will transfer a photon (475 nm) to excite Venus, which will emit at ~525 nm [52]. When this technology is combined with imaging techniques, such as confocal microscopy, one can assess the subcellular production of ATP, specifically in axons and synapses, by monitoring FRET-evoked fluorescence. The drawback to this system is that it requires genetically tractable cells.

In addition to monitoring ATP levels, assessing the abundance and activities of complexes comprising the electron transport chain (ETC) can reveal specific insights into bioenergetic stress. Defects in ETC components and oxidative phosphorylation (OXPHOS) are associated with neurological disease [53, 54]. While perturbations in OXPHOS can manifest as decreased ATP output or increased reactive oxygen species (ROS) [29, 55], the discrete alterations of ETC complexes can lead to distinct pathological effects [56]. Monitoring the relative abundance of ETC constituents and super-complexes is commonly performed using blue native gel electrophoresis [57]. However, blue native gels cannot always discern the specific proteins that are lacking without a complementary Western blot analysis. Native electrophoresis of ETC components can be coupled with in-gel enzymatic assays to determine the relative activities of specific enzymatic assemblies [58]. Traditionally, these assays have been conducted using cellular extracts in a spectrophotometric format [59]. A drawback to using blue native gels and enzymatic analyses is that they require a robust amount of material, which can be problematic when dealing with specific neuronal species and small amounts of sample.

Respirometry has been revolutionized in recent years with the development of metabolic analyzers that can assess respiration in small sample sizes [60, 61]. Modern Clark electrodes and instruments like the Seahorse extracellular flux analyzer permit researchers to assess mitochondrial function with small samples and in real time [61]. When combined with established respirometry approaches, investigators can identify specific metabolic lesions in cultured neurons or in mitochondria extracted from minute tissue samples [62]. The earnest is on researchers to establish standardization approaches to compare samples. To date, investigators have used cell number and protein concentration as accepted measures for normalization [63]. For cell number normalization with Seahorse analyzers, it is considered the standard procedure to DAPI stain cells following an assay and to normalize respiratory and glycolytic signals to the number of nuclei per well [63]. Despite the advantages of modern respirometers, a drawback to the Seahorse extracellular flux analyzer is that fluorescent cells cannot be used as the emissions from these cells can saturate the detectors for the oxygen and hydrogen fluorophore probes. Thus, experimental designs should eliminate fluorescence in the GFP and RFP ranges from Seahorse assays; we recommend using near-infrared fluorophores (i.e., RFP670) for studies that require metabolic analysis and gene manipulation [64]. A rigorous approach to bioenergetic analyses, incorporating ATP assessments and the examination of ETC components and activities, can help to identify specific metabolic perturbations in neuronal mitochondria under varying conditions.

1.3.2 Calcium Buffering

Mitochondria are essential to buffering Ca^{2+} during excitation, and the relative capacity of mitochondria to import and release the ions affects the amplitude and duration of signals [65, 66]. Also, the maintenance of Ca^{2+} flux in neurons is critical to bioenergetics, survival, and synaptic plasticity [67]. Mitochondrial Ca^{2+} cycling is commonly measured with fluorescent dyes; however, the most common dye, Rhod-2, and its derivatives (Rhod-5 N and Rhod-FF), while useful, are not exclusively mitochondrial, and the cytosolic signals associated with these dyes should be accounted for during use [68–70]. The recent development of genetically encoded Ca^{2+} probes targeted to mitochondria (such as mito-GCaMP derivatives and mito-R-GECO analogs) can overcome the "leakiness" of the Rhod-2-derived dyes [71–73]. The use of rigorous controls and appropriate Ca^{2+} ionophores should be used to account for specific changes in Ca^{2+} concentrations in discrete neuronal locales [74]. Additionally, bioluminescent approaches and ion-specific electrodes can be used to quantify changes in Ca^{2+} flux in cells [75–77] as mitochondrial stress associated with disturbances in Ca^{2+} cycling leads to changes in organelle uptake and release of Ca^{2+}. Prolonged high or low levels of mitochondrial Ca^{2+} can adversely impact bioenergetics [3]. The use of highly selective methods for monitoring mitochondrial calcium is crucial to determining the intra-organelle calcium flux.

1.3.3 Mitochondrial Membrane Potential

Mitochondrial membrane potential ($\Delta\Psi$) is the proton motive force generated across the inner membrane by electron flow during respiration [78]. Changes in $\Delta\Psi$ can affect electron transport, ATP production, and Ca^{2+} buffering; thus the examination of mitochondrial function in neurons should not be interpreted in the context of $\Delta\Psi$ alone [79, 80]. To assess the physiological impact of $\Delta\Psi$ on the mitochondrial or neuronal function, it is imperative to accurately assess the potential. The evaluation of $\Delta\Psi$ can differ depending on the needs of the investigator; for example, chemical probes may be suitable for isolated organelles or intact cells [81]. To measure $\Delta\Psi$ in mitochondria isolates from synaptosomes, a tetraphenylphophonium (TPP^+)-selective electrode is the most suitable approach to estimate $\Delta\Psi$ by monitoring cationic uptake by organelles [82, 83]. There are drawbacks to using the TPP^+ electrode, including the necessity to avoid high concentrations of hydrophobic compounds; additionally, the reliance upon diffusion across membranes limits the effectiveness of the TPP^+ electrode to gauge $\Delta\Psi$ in intact neurons [82]. The use of fluorescent probes such as JC-1 and tetramethylrhodamine methyl ester (TMRM) has become standard to assess $\Delta\Psi$ in cells. In neurons, the use of TMRM with fluorescent microscopy is the standard approach because of the reliability of TMRM and because microscopy allows for the visualization of the functional diversity of mitochondria in neurons [80, 84]. Despite its widespread use, TMRM requires significant optimization for use in neurons to avoid quenching due to aggregation, and investigators should be mindful of potential nonspecific interactions of TMRM [84]. Older probes, like Rhodamine 123, should be used with extreme caution (if at all) due to extensive non-specific staining [84]. Any change in $\Delta\Psi$, either loss of potential or hyperpolarization, should be considered with respect to other changes in mitochondrial function, like diminished bioenergetics, altered Ca^{2+} buffering, or morphological changes.

1.3.4 Mitochondrial DNA

Although the mitochondrial genome is small relative to genomic DNA, mutations and deletions in mitochondrial DNA (mtDNA) can have a profound impact on organelle function and overall health [85]. For example, the mutator mouse ($POLG^{D257A}$) accumulates mtDNA mutations rapidly due to diminished proofreading capabilities by the mutated polymerase, resulting in advanced-age-like phenotypes such as greying, hunched posture, and muscular decline by 9 months of age [86]. Furthermore, genetic anomalies in mitochondrial DNA lead to a class of disorders known as mitochondrial encephalomyopathies, which manifest as neuromuscular diseases with lactic acidosis and diminished OXPHOS [11]. Therefore, maintaining sufficient levels of healthy mtDNAs is critical. The amount of mtDNA present in neurons can be determined by using RT-PCR-based approaches to normalize mtDNA content to that of genomic DNA [87, 88]. However, mtDNA is typified by heterogeneity, which arises from scantly regulated replication and limited proofreading mechanisms [89–91]. Emerging techniques dedicated to assessing mtDNA at the single cell and subcellular levels represent promising approaches to tackle mtDNA complexity in neurons [92–94]. When assessing the diversity of mtDNA levels in neurons, deep sequencing-based methods can allow for the quantification of low-level variants in mtDNA [93]. As with the assessment of $\Delta\Psi$, mtDNA perturbations should be examined in the context of organelle function.

1.3.5 Mitochondrial Morphology

Mitochondria exist in a highly dynamic network that is in a constant flux of fusion and fission [95]. This dynamic nature is exemplified in neurons, as the size and distribution of mitochondria are highly variable and differ among subcellular regions [96, 97]. Perturbations in mitochondrial dynamics are linked to neurological diseases. For example, mutation, deletion, or duplication of the mitofusin-2 gene (*Mfn2*) leads to the increase in shortened, dysfunctional mitochondria in Charcot-Marie-Tooth Neuropathy Type 2A [98, 99]. Because of the complexity and diversity of mitochondrial shapes and sizes in neurons, a rigorous approach to assessing mitochondrial shape, involving molecular probes and electron microscopy (EM), should be employed with great respect to the cellular localization of mitochondria in neurons (axon, dendrite, synapse, or soma) [100]. Commercially available dyes, such as MitoTracker (Life Technologies), are commonly used; however, these molecules can be non-specific and are often used at concentrations detrimental to mitochondrial function. In neurons, genetically encoded probes (mito-GFP, mito-BFP, etc.) targeted to mitochondria should be used to assess mitochondrial morphology under fluorescent conditions. The use of these probes should be accompanied by quantitative measurements using algorithms in software, such as ImageJ [101]. It is important to consider the subcellular localization of mitochondria when assessing form; therefore, all pathological or treatment-induced morphological phenotypes should be compared with healthy controls at the same location within neurons [102]. Any morphological differences among mitochondria in neurons should be examined with EM, which will provide insight into organelle superstructure as well as shape and organization [103]. Collectively, this information will provide a holistic explanation of mitochondrial form when combined with cellular assays to assess fission and fusion [104–107]. Again, visualizing fusion and fission events is more reliable with genetically encoded mitochondrial probes using real-time measurements than is post hoc staining.

1.3.6 Mitophagy

The turnover of aged, damaged, or excessive mitochondria is critical to maintaining neuronal health, and defects in mitophagy can lead to neurological dysfunction and neurodegeneration

[108]. In fact, loss-of-function mutations in the mitophagy components PTEN-inducible kinase 1 (PINK1) and the E3-Ubiquitin ligase, Parkin, are associated with familial Parkinson's disease [15]. Thus, maintaining regulation of organelle turnover and replacement is crucial to neuronal health. The development of MitoTimer [109], mt-Keima [110], and the Mito-QC mouse [111] now permit investigators to observe and quantify mitophagic events in sensitive cells and tissues, including neurons [112]. MitoTimer is a COX8-fused variant of DsRed1-E5 that undergoes an irreversible conversion from a green-emitting fluorophore to a red-emitting fluorophore as the protein matures. In most cases, the change in emission occurs within 48 hours of expression [109]. Consequently, MitoTimer can be used to follow older organelles through their lifecycle and can be utilized as a means to assay biogenesis by monitoring the appearance of green mitochondria [109]. These strategies make MitoTimer a useful tool to assess mitochondrial turnover and replacement. Mt-Keima is another COX8 fusion protein used to quantify mitophagy. The mt-Keima protein changes its emission from green in a neutral pH environment to red in an acidic environment, such as the autophagolysosome [110]. The red fluorescence emitting mt-Keima-containing mitochondria are likely undergoing mitophagy [110]. The mito-QC mouse, like mt-Keima, is a pH-sensitive fluorescent sensor localized to the mitochondrial surface. The probe in the mito-QC mouse is a mCherry-GFP fusion protein with a FIS1 targeting sequence to the outer membrane. Under normal conditions, both red emission (from mCherry) and green emission (GFP) can be observed; however, upon autophagosome-lysosome fusion, the acidic environment quenches the GFP leaving only the red emission of mCherry [111]. Thus, the mito-QC mouse is a useful tool for assaying mitophagy in vivo and, when under the control of a neuronal-specific promoter, can be used to monitor mitophagy within neurons [113]. Because the changes in fluorescence of the above probes may not indicate specific pathological changes in mitochondria or represent particular damage or stress, these probes should be supported by other studies and rigorous controls such as inhibitors of organelle biogenesis and mitophagy.

1.3.7 Mitochondrial Movement

In neurons, the transport of mitochondria from the soma to dendrites and synapses is critical to neurotransmission, and the proper distribution of mitochondria in neurons assures that the discrete energetic demands of subcellular regions are met [114]. Failure to translocate mitochondria to their destinations along the microtubule network has various consequences to neuronal activity and neurological disease [115]. The common approach to measure mitochondrial motility in neurons is to utilize a genetically encoded mitochondrial fluorophore controlled by a neural-specific promoter to label mitochondria selectively in neurons [116, 117].

Using time-lapse fluorescent confocal microscopy of primary neurons or explanted neurons, one can monitor the movement of mitochondria in real time. Software is available to track individual organelles (Volocity – Perkin-Elmer) or kymographs can be generated and quantified using the Straighten and Manual Tracking plugins for ImageJ, respectively [117]. Inhibitors of molecular motor proteins, accessory proteins, and microtubule dynamics should be considered when designing experiments [118].

1.3.8 Permeability Transition Pore

Oxidative stress and elevated intraorganellar Ca^{2+} concentrations can induce the formation of a large protein-based pore in the inner mitochondrial membrane, known as the mitochondrial permeability transition pore (PTP) [119, 120]. Ultimately, the PTP results in mitochondrial depolarization and swelling, which inhibits Ca^{2+} uptake and induces apoptosis by rupturing the outer membrane, respectively [121, 122]. The majority of the approaches to assess PTP opening have centered on Ca^{2+} probes (discussed earlier), but because many factors can influence Ca^{2+} movement, it is necessary to also consider the mitochondrial changes, specifically the acidification of the matrix during PTP opening [123, 124]. Matrix pH has been monitored using specialized dyes, namely, SNARF-1 derivatives, or pH-sensitive fluorescent proteins (mito-eYFP) [125]. By simultaneously recording Ca^{2+} release and matrix pH changes, it is possible to observe PTP openings and, by extrapolation, determine if there are perturbations [120, 123]. The use of established inhibitors of PTP opening, namely, cyclosporin A, is a useful control for these studies [126]. A major obstacle to assessing the PTP is the optimization of the pH-sensitive probes, and great care should be taken both in selecting a probe that fits the pH range and in validating the probe function in neurons under distinct pH conditions.

1.3.9 Proteostasis

Because the mitochondrial genome only encodes 13 proteins, the remaining proteins must be produced outside of the mitochondria and then incorporated into the organelles using sophisticated protein transport machinery and chaperones [127]. Problems in the transport and folding of proteins can result in bioenergetic and physiological problems for mitochondria [128]. A standard assay for measuring mitochondrial protein import is quantifying radiolabeled ornithine transcarbamylase (OTC) within neurons. Incubation of primary neurons with L-[^{35}S]-methionine will label newly made proteins [129]; furthermore, ectopic expression of OTC ensures that a significant amount of the radiolabel is incorporated into the cellular OTC pool. As OTC is transported into mitochondria, the polypeptide undergoes a cleavage event that decreases its molecular weight [129]. This change in mass can be observed following resolution by denaturing protein gel electrophoresis. Fluorescent strategies have been proposed as well to monitor protein import into mitochondria [130]. Regardless of the approach used, uncouplers, such as 2,4-dinitrophenol [131], can be used as inhibitors of protein import and should be considered in experimental designs. The recent descriptions of the mitochondrial unfolded protein response (UPR^{mt}) demonstrate that the regulation of mitochondrial proteostasis is a cellular endeavor [132, 133]. Indeed, the modulation of gene expression programs has been reported and includes increasing mitochondrial chaperones and proteases to handle potential problems in protein homeostasis. Therefore, analysis of protein levels and mRNA levels can be used to determine the relative engagement of UPR^{mt} [134]. The examination of deficits in mitochondrial proteostasis should be considered in light of the neuronal context.

1.3.10 Reactive Species Generation

The generation of reactive oxygen and nitrogen species (ROS/RNS) above the physiological norms can indicate problems with OXPHOS and ETC components [55]. Again, fluorescent dyes have emerged as the leading tool in the field largely due to the relative ease of use. The most commonly used mitochondrial ROS detecting dye is MitoSOX (Life Technologies), which detects mitochondrial ROS species from superoxide to hydrogen peroxide [29, 135]. The dye should be used with caution as it can be subject to quenching at high concentrations and improper storage can lead to false positives. Appropriate controls and optimization should be considered when using redox-sensitive dyes to assure that measures are taken within a reliable portion of the dynamic range [135]. Thus, fluorescent changes would be best supported by complementary assessments related to oxidant production. There are redox -sensitive probes that can be ectopically expressed in neurons to detect the relative levels of reactive species [136]. However, the emergence of ratiometric redox-sensitive probes targeted to mitochondria represents a quantitative approach to assess oxidant levels [137, 138]. It is recommended that ROS/RNS levels be examined in the context of oxidation-induced changes to proteins (carbonylation or nitrosylation), lipids (peroxidation), and mtDNA damage to determine whether the oxidant production is truly pathological [139]. It is important to analyze the production of radicals in concert with bioenergetics, morphometry, and other analyses that can provide insights into the physiology driving oxidant generation.

1.4 Protein-Protein Interactions on Mitochondria

The outer mitochondrial membrane (OMM) is the interface between a mitochondrion and other cellular compartments; perhaps, more importantly, the OMM is a critical site for the integration of cytosolic and mitochondrial signaling events [140, 141]. The highly integrative nature of mitochondria in neurons can in part be explained by the coordination of signaling cascades on the OMM. An increasing number of reports continue to identify cytosolic protein kinases and phosphatases with pronounced mitochondrial localization, which demonstrates the level of synchronization between mitochondria and neurons necessary to achieve optimal neuronal function [142–144]. Protein kinase A (PKA) was one of the first protein kinases reported to have mitochondrial localization, and subsequent studies reveal that mitochondrial PKA signaling events enhance organelle efficiency, influence mitochondrial dynamics, and promote cellular survival [145]. The recent discovery of AMP-dependent protein kinase (AMPK) on the OMM demonstrates a clear link between local energetic status and cellular adaptations [34, 35]. From a pathological perspective, the mitochondrial activities of discrete c-Jun N-terminal Kinase (JNK) isoforms have been shown to impair bioenergetics, induce mitophagy, and trigger apoptosis [27–29]. Each one of the aforementioned kinases responds to a particular state of cellular or organelle physiology and induces relevant responses in accordance with the status of mitochondria, neurons, or both. With the recent boom in the identification of new mitochondrial protein kinases and phosphatases, the methods associated with detecting distinct subcellular interactions have evolved to characterize these important signaling nexuses.

1.4.1 Subcellular Fractionation/ Co-immunoprecipitation

The classic approach to demonstrating the presence of a protein on or within mitochondria has been to isolate the organelles and perform Western blots or proteomics to determine the presence of a protein [146–150]. While mitochondrial preparations can contain other cellular components, recent developments have improved the quality of organelle separations and increased the purity of mitochondrial isolations [151, 152]. Subcellular fractionations are often coupled with co-immunoprecipitations (coIPs) to identify potential interacting partners. Indeed, coIPs work well when coupled with proteomic data, which can be used to narrow the number of potential interacting partners. Despite the widespread use of these approaches, coIPs of mitochondrial preparations only provide a snapshot of the interactions and do not necessarily account for the kinetics and stability of the interactions. In fact, due to the rigorous isolation protocols required to obtain highly pure mitochondrial preparations, many transient or weak protein-protein interactions (PPIs) may be lost during the acquisition. This may partly explain why many of the cytosolic protein kinases and phosphatases identified on mitochondria do not appear in the MitoCarta or other protein databases. Nonetheless, subcellular fractionation and coIPs remain a useful means to validate protein presence and PPIs on the mitochondrial surface.

1.4.2 Proximity-Dependent Labeling Techniques

In part, due to the concerns regarding contamination of mitochondrial isolates from subcellular fractionations, refinements in mass spectrometry (MS)-based techniques have been made to reduce the occurrences of false positives in proteomic analyses of mitochondria. Proximity-labeling approaches utilize enzymes to convert nearby proteins into radicals, which permits the covalent attachment of biotin [153, 154]. The biotinylated proteins can then be identified using MS-based approaches, and the levels of the proteins can be quantified through the addition of protein labeling strategies, such as stable isotope labeling by amino acids in cell culture (SILAC). This technique is useful in live or fixed cells and in tissue samples [154]. The three prevailing approaches for proximity labeling are distinguished by the enzyme system used to modify the proteins and link the biotin; these are bacterial biotin ligase (BirA) (BioID) [155], engineered ascorbate peroxidase (APEX) [156], and horseradish peroxidase (HRP) [157]. The spatial distribution of biotin addition is determined by fusing the enzyme to a protein of interest or to a targeting peptide. In contrast to subcellular fractionation-based proteomic endeavors, proximity-dependent ligation approaches are performed with intact cells and tissues; this feature allows interactions to be assessed within the cellular context. Of additional benefit, proximity-dependent labeling can identify transient or weak interactions often lost with mitochondrial isolation. The ability of these techniques to circumvent the tedious and problematic aspects of subcellular fractionation offers a means to assess mitochondrial PPIs in physiologically relevant situations.

1.4.3 Bioluminescent/Fluorescence Resonance Energy Transfer (B/FRET)

While proximity-dependent labeling methods can determine if proteins are close to one another, the spatial coverage of the labeling can tag proteins that are not directly interacting partners with the protein or are not present on mitochondria. Therefore, rigorous validation strategies should be employed to determine if proteins directly associate with one another. As mentioned above, coIPs are one option for assessing direct PPIs, but coIPs are done outside of the cellular context and may miss transient and weak interactions. For this reason, we recommend supporting coIPs and proximity-labeling strategies with proximity-based fluorescent assays, such as BRET and FRET, using ectopically expressed or genetically modified proteins fused to either a luciferase enzyme or a fluorophore [158]. If the fused proteins are interacting (or very close <10 nm), the activation of bioluminescence or excitation of the fluorophore will cause a photon from the donor to excite the fluorophore of the accepting fusion protein. The use of BRET or FRET can allow for the detection of PPIs in intact cells and tissues in real time [158]. These approaches are useful because they can detect weak and transient interactions that are lost in isolations or other labeling techniques. In addition to BRET and FRET, similar approaches are available to assess PPIs; these can include functional ligand-binding identification by Tat-based recognition of associating proteins (FLI-TRAP) or similar technology designed to fuse complementary portions of fluorophores [159]. A prevailing limitation to these validation approaches is that BRET/FRET will not work in fixed tissues, such as preserved patient samples, limiting their use in clinical-based studies.

1.4.4 Proximity Ligation Assays

PPIs can be detected in fixed and permeabilized cells and tissue samples using proximity ligation assays (PLAs) [160]. PLAs are antibody-based approaches useful for visualizing proteins near one another in fixed cells and tissues [161, 162]. PLAs are dependent upon the binding, circularization, and amplification of oligonucleotides on probes that are localized on adjacent proteins [163]. The amplified product can then be detected and visualized following the addition of fluorescently-labeled complementary oligonucleotides (Fig. 1). Briefly, oligonucleotides are conjugated to antibodies (either primary or secondary), and if the oligonucleotides are close (less than 40 nm apart), they can be joined in a circular fashion using a linker oligonucleotide (often called a PLA probe) [164]. The circular substrate, once ligated, serves as the template for rolling circle amplification. The addition of specific DNA polymerases copies the circular substrate into approximately 1000 copies connected to the PLA probe. Finally, fluorescently conjugated oligonucleotides complementary to the PLA probe are introduced and bind to the amplified product. Excitation of the fluorophore will reveal where PPIs have occurred [164]. PLAs are contingent

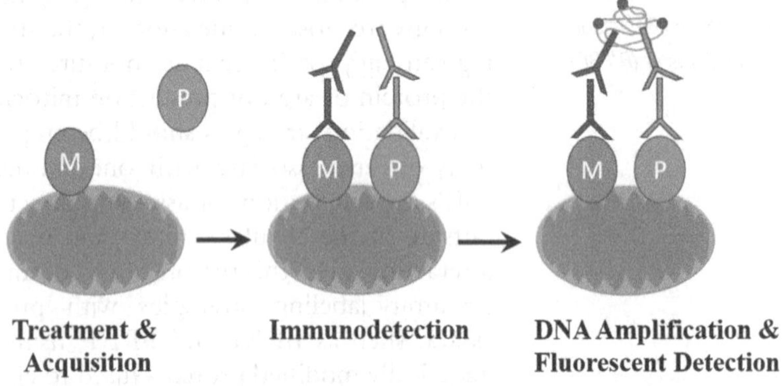

Fig. 1 A schematic representation of a proximity ligation assay (PLA). If a cytosolic protein (P) translocates to mitochondria, it may be near a mitochondrial resident protein (M), such as TOM20 or VDAC. Following cell and tissue acquisition (and fixation), the two proteins, M and P, can be recognized by specific primary antibodies. Secondary antibodies can be used to recognize the primary antibodies, and oligonucleotides conjugated to the secondary antibodies are linked by a PLA probe. The PLA probe is then amplified by a polymerase, and the amplified product is incubated with fluorescent oligonucleotides that recognize the amplified PLA product. The fluorescence can then be detected and quantified by a fluorimeter, and localization can be determined with fluorescent or confocal microscopy

upon well-validated antibodies with minimal non-specific interactions for PPIs to be reliably detected. However, if suitable antibodies are available, PLAs represent a quick and sensitive assay to detect PPIs. For the purposes of detecting PPIs on mitochondria, within cells or in tissues expressing a fluorescent mitochondrial protein or immunofluorescent colocalization with established mitochondrial proteins (Tom20, VDAC, NDUFB8, etc.) can be used with PLAs to determine the relative overlap of PPIs with mitochondria. This approach will be discussed below using cells and tissue samples. In addition to microscopy-based methods, recent PLA applications have combined the PLA with real-time PCR techniques to enhance the quantitative potential of PLAs [165]; however, these RT-PCR-based PLAs do not allow for the subcellular resolution of PPIs, which we will describe in our methods below. Nonetheless, PLAs represent a convenient and versatile means to assess PPIs in fixed cells and tissues. We will describe our protocols for PLAs in cells and brain sections in the protocol below.

2 Materials

2.1 Cell Culture

1. We have used the approach with established neuron-like cell lines (i.e., SH-SY5Y) and primary neuronal cultures.
   (a) SH-SY5Y cells were maintained at sub-confluency between passages 3 and 20 in Dulbecco's Minimal Essential Medium with F12 supplement mix (DMEM:F12) with 5% fetal bovine serum, penicillin, streptomycin, and plasmocin. A stable cell line expressing a humanized green fluorescent protein localized to mitochondria with a TOM20 localization sequence (mito-GFP) was used for our method below.
   (b) Alternatively, primary cortical neurons from 16-day embryos of C57/BL6 mice (Jackson Laboratories) can be used. However, primary neurons require delivery of mitochondrially localized fusion proteins by lentiviral or adeno-associated viral transduction (*see* **Note 1**).
2. Poly-d-lysine-coated chamber slides.

2.2 Tissue Preparation

1. For acquisition of rodent tissues, animals are overdosed with ketamine and xylazine and then fixed using cardiac perfusion with 0.9% saline followed by 4% paraformaldehyde (PFA) in 0.1 M sodium phosphate buffer, pH 7.4. The brains are removed and post-fixed an additional 24 hours in 4% PFA at 4 °C, and are placed in cryoprotectant (30% sucrose in phosphate-buffered saline (PBS)) for 4 days. Brains are stored at −80 °C until sectioning. If desired, brains may also be embedded prior to storing at −80 °C.

2. The brains are sliced symmetrically into 20-μm thick sections using a cryostat and are rinsed in PBS prior to beginning the PLA.

2.3 PLA Supplies

1. Primary antibodies (*see* **Note 2**).
2. Reliable PLA kit (such as DuoLink [Sigma-Aldrich]; described below) containing optimized secondary antibodies conjugated to oligonucleotides and a PLA probe (*see* **Note 3**) with DNA ligase, polymerase, and fluorescent oligonucleotides complementary to PLA probe. Some kits may contain a blocking buffer; however, one may use a blocking buffer previously optimized with the primary antibody (*see* **Note 4**).
3. Ice bucket or freeze block.
4. Multi-well plates for staining (for brain sections, we recommend 12- or 24-well formats).
5. Shakers.
6. Coverslips and slides.
7. Light-protected boxes for staining (aluminum foil-wrapped lids from tip boxes).
8. Humidity chamber.
9. Temperature controlled incubator.
10. Inverted fluorescent or confocal microscope 60 and 100× oil immersion lens.
11. Computer with an operating system compatible with an image analysis software.
12. Image analysis software (ImageJ, Photoshop/BlobFinder, or DuoLink ImageTool).

3 Methods

3.1 Cell-Based PLA

1. After the cells have been grown in sterile chamber slides (generally 4-well) and treated according to the experimental design, the cells should be washed three times in Hank's balanced salt solution (without calcium and magnesium). SH-SY5Y cells and primary cortical neurons will require chamber slides that have been coated with poly-D-lysine to adhere.
2. The cells can then be fixed for 30 minutes in 4% paraformaldehyde (PFA) in phosphate-buffered saline (PBS) at room temperature.
3. Block the slides in blocking solution for 30 minutes at 37 ° C in a humid chamber, and gently tap the blocking solution from the slides. A permeabilizing agent, such as 0.1% Triton X-100, is often included to improve the penetrance of antibodies (*see* **Note 4**).

4. Perform a glycine quench to reduce background fluorescence with 0.2 M glycine for 10 minutes at room temperature (*see* **Note 5**).
5. Wash in blocking solution once for 5 minutes to remove residual quenching agent.
6. Incubate the slides immediately with primary antibodies diluted in blocking buffer overnight at 4 °C in a humid chamber. For the experiment shown in Fig. 2, a monoclonal antibody for Sab (Novus Biologicals; H00009467-M01) was diluted 1:200 in blocking buffer.

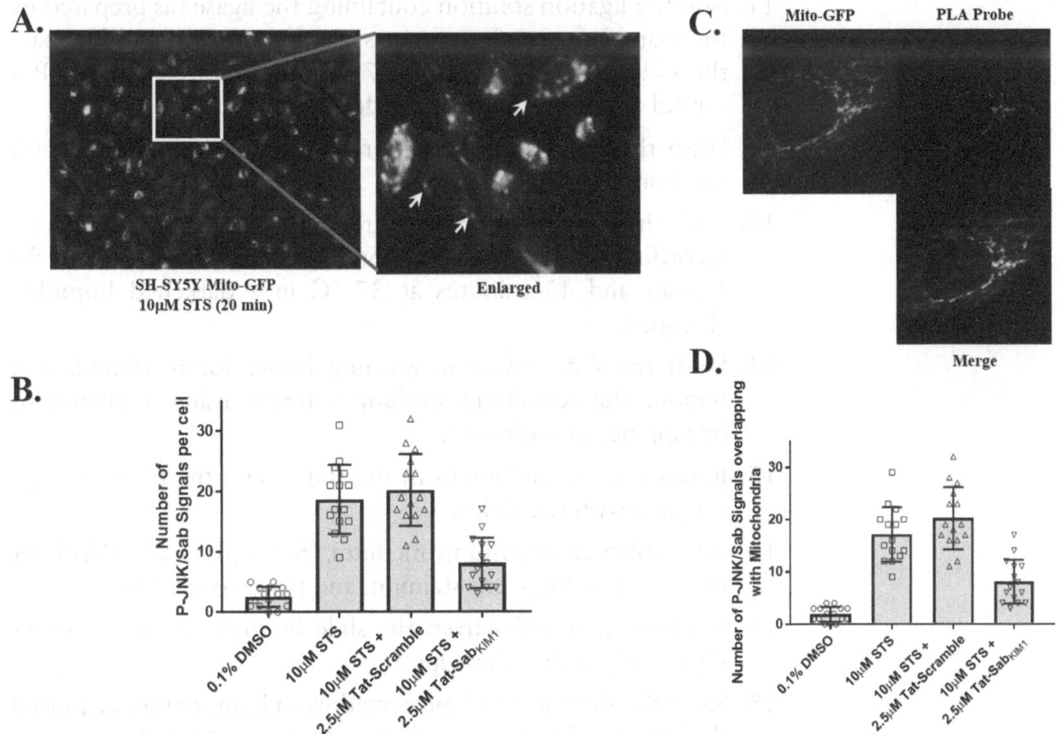

Fig. 2 Cell-based PLA for mitochondrial-localized cytosolic kinases. We stably expressed a mitochondria-localized GFP (mitoGFP) in human SH-SY5Y cells and treated the cells with 10 μM staurosporine (STS) for 20 minutes to induce JNK migration to mitochondria (a). The presence of JNK on mitochondria was determined by PLA of JNK and its mitochondrial scaffold protein Sab (red) (a). The red signals (a, left) were counted per cell for quantification (b). The interaction of JNK and Sab could be blocked using a competitive Tat-Sab$_{KIM1}$ with Tat-Scramble serving as a peptide control (b). The experiments can also be performed at the cellular level to determine the overlap of PLA signals with mito-GFP (c). The red and green overlap can be used to determine the number of events that occur on mitochondria (d) in the presence and absence of the Tat-Sab$_{KIM1}$ peptide. If desired, a ratio can be calculated for the distinct distributions of the interactions between mitochondria and other subcellular compartments. A minimum of ten cells per focal plane were counted, and statistical differences were assessed using a one-way ANOVA. An asterisk (*) demonstrates a significant deviation from untreated cells ($p < 0.01$), and a double asterisk (**) represents a difference from the cells exposed to STS ($p < 0.01$)

7. Gently remove the primary antibody solution from the coverslips by gently tapping on a Kimwipe.
8. Wash the slides twice carefully for 5 minutes with the wash solution, which is often 1× Tris-buffered saline (TBS) with 1% goat serum (or 2.5% BSA). Using Tris-based buffers over phosphate buffers can aid in quenching.
9. Incubate the slides with the PLA probes diluted by blocking buffer without 0.1% Triton X-100 (or permeabilization agent) for 1 hour at 37 ° C in humidity chamber.
10. Carefully, wash the slides three times for 5 minutes with the washing solution.
11. Add the ligation solution containing the ligase (as prepared by the manufacturer's instructions) to the samples, and incubate the slides for 30 minutes at 37 °C in humidity chamber. Pay careful attention that samples do not dry.
12. Wash the slides twice, gently rocking for 5 minutes in wash solution.
13. Add the amplification mixture containing the polymerase according to the kit instructions, and incubate the slides for 1 hour and 45 minutes at 37 °C in a darkened humidity chamber.
14. Wash the slides twice in washing buffer for 5 minutes and remove the remaining washing solution using a pipette or by tapping on a Kimwipe.
15. Remove the chambers from the slide using the plastic wedge supplied with the slides.
16. Add sufficient mounting medium (5–10 μL) with DAPI for mounting and nucleus staining, and place cover slips.
17. Remove air bubbles from the slide by pushing on the cover slips gently with a Kimwipe.
18. Store the slide at 4 ° C, protected from light, before acquiring the images using a confocal microscope (*see* **Note 6**).

3.2 In Situ PLA

1. Place individual brain slices in separate wells of a 12-well plate containing TBS.
2. Remove the TBS and replace it with 0.3% hydrogen peroxide (H_2O_2) in TBS. Incubate each slice in the solution for 15 minutes.
3. Remove the H_2O_2 solution, and place the slices in blocking buffer (10% goat serum in TBS with 0.15% Triton X-100), and then incubate at room temperature while gently rocking.
4. Wash the sections in blocking buffer for 5 minutes at room temperature.

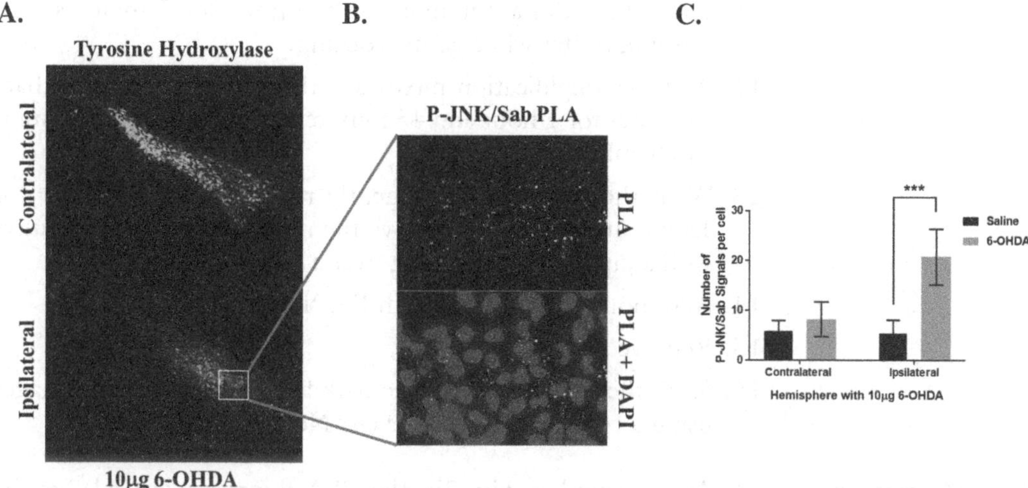

Fig. 3 Tissue-based PLA for phospho-JNK translocation to mitochondria in a PD model. We previously demonstrated that mitochondrial JNK signaling was necessary for dopaminergic (DA) neuron loss in the rat 6-hydroxydopamine (6-OHDA) model of PD. Following a unilateral injection of 6-OHDA, animals were perfused, and brain sections were acquired. Sections from the ventral midbrain containing the substantia nigra pars compacta (SNpc) were immunostained for tyrosine hydroxylase (TH) to indicate the presence of DA neurons on the non-lesioned (contralateral) and lesioned (ipsilateral) hemispheres (**a**). A PLA was performed simultaneously to detect phospho-JNK and Sab on the ipsilateral side (**b**). Nuclei were stained with DAPI. The interactions were then quantified for both sides (**c**). Statistical comparisons were performed using a one-way ANOVA, and a triple asterisk (***) illustrates a significant deviation from control and lesioned hemispheres

5. Incubate the slides with primary antibodies diluted in blocking buffer overnight at 4 °C in a humidity chamber while gently rocking. For the experiment shown in Fig. 3, a monoclonal antibody for Sab (Novus Biologicals; H00009467-M01) was diluted 1:200 in blocking buffer, while Phospho-SAPK/JNK (Thr183/Tyr185; Cell Signaling Technology; 4668) was diluted 1:100 in blocking buffer.
6. Gently remove the primary antibody solution from the coverslips by gently tapping on a Kimwipe.
7. Wash the slides twice carefully for 5 minutes with washing solution or blocking solution without permeabilization agents (or use 2.5% BSA or 1% goat serum).
8. Incubate the slides with the PLA probes diluted in blocking buffer without 0.1% Triton X-100 (or permeabilization agent) for 1 hour at 37 ° C in a humidity chamber.
9. Wash the slides at least twice for 5 minutes with washing solution while gently rocking.
10. Add the ligation solution with ligase and incubate the slides for 30 minutes at 37 ° C in a humidity chamber.

11. Wash the slides a minimum of two times for 5 minutes with washing buffer while gently rocking.
12. Add the amplification mixture with polymerase and incubate the slides for 1 hour and 45 minutes at 37 °C in a dark humidity chamber.
13. Wash the slides twice by gently rocking in wash buffer for 15 minutes each, and remove the remaining washing solution with a pipette or by tapping over a Kimwipe.
14. Add mounting medium with DAPI and place coverslips.
15. Remove air bubbles.
16. Store the slide at 4 °C in the dark before acquiring the images using a confocal microscope (*see* **Note 6**).

3.3 Determination of PPI Abundance on Mitochondria

1. As presented in Fig. 2b, the PLA fluorescence (red) can be overlaid onto the green emission of the mitoGFP resulting in a yellow-orange hue.
2. Specialized software (we use BlobFinder) can be used to detect the number of red PLA signatures per cell and then to detect the number of yellow-orange signals to indicate the number of mitochondria-localized interactions.
3. The mitochondrial PPIs per condition are compared across samples and experiments (*see* **Note 7**).

4 Notes

1. We advise that mitochondrial labeling in cells and tissues should be performed using genetically encoded probes. Using genetic approaches will eliminate a staining step with chemical mitochondrial probes, as chemical probes can affect organelle function or perhaps quench other fluorophores. We recommend that investigators consider using a near-infrared emitting fluorophore (i.e., mito-RFP^{670}) to eliminate spillover between filters. Furthermore, the use of fluorophores like RFP^{670} escapes the robust autofluorescence in brain tissue that can obscure modest outputs from molecules that emit at ~488 nm. We often combine mito-RFP^{670} with a red-emitting oligonucleotide-conjugated fluorophore (584 nm) for optimal signal acquisition. If used properly, the mitochondrial probe can provide information regarding the morphology of the mitochondria on which the PPI occurs.
2. Antibodies are the most critical component of the PLA. It is strongly recommended that affinity-purified antibodies are rigorously validated by Western blotting and immunofluorescence (IF) or immunohistochemical (IHC) analyses for

specificity. Moreover, optimization by IF or IHC prior to use in PLA will help optimize antibody conditions and reveal patterns of localization for individual target proteins. Most PLA kits, including DuoLink®, recommend that IgG-type antibodies (both monoclonal and polyclonal) be used. As such, researchers should include individual antibody controls to detect non-specific binding of each primary antibody to itself, and to determine the optimal concentration of each primary antibody to use in the PLA.

3. Non-specific oligomer interactions can represent an obstacle for PLAs if not properly controlled for during the optimization of the assay. These non-specific interactions can arise when the fluorescent oligonucleotides are in high concentrations or react with other molecules in the cells or tissues. Alternatively, non-specific binding of the PLA probes (often when at higher concentrations) can also create dubious signals. We recommend that a non-primary antibody control be included in each analysis to account for these possibilities, as well as to optimize the concentrations of probes and oligonucleotides needed with each antibody pair.
4. For blocking buffers, most commercial kits come with their own mixtures; however, we have found that the buffers used during our primary antibody validation studies are sufficient for PLA assays, and goat serum or low percentage BSA (2–4%) buffers are generally compatible. For optimal detection of mitochondrial PPIs, we have found it necessary to include permeabilization agents (i.e., Triton X-100) during the blocking steps to improve the penetration of antibodies into the cells. Beginning with the primary antibody incubation, we often omit the permeabilization reagent from future incubation and washes. As with all steps of the PLA, blocking approaches should be optimized for each antibody pair.
5. We usually employ a glycine quench for 10 minutes or longer to reduce background fluorescence (often from formaldehyde) in samples. However, in our brain sections, we often can incur higher background fluorescence. In these instances, we employ ammonium chloride (50 mM) to quench.
6. For our image acquisitions and analyses, we recommend that no less than six images per plane be used for counting in a single culture. For our tissue-based studies, we recommend eight images per plane. The images should be taken at random using a motor-controlled objective. Also, we blind our experimentalists to the sample to prevent bias.
7. While each experiment requires its own statistical approach, we often employ a Mann-Whitney test for analysis of two

condition experiments. For multiple samples (>3), we use the Kruskal-Wallis test and then incorporate a Dunn's correction post hoc to analyze specific sample pairs in an experiment.

5 Summary

In this chapter, we reviewed standard approaches used to assess fundamental aspects of mitochondrial physiology, and we acknowledge that, due to the scope of our article, some approaches and processes were not discussed. These unintentional omissions demonstrate the tremendous growth that has occurred in our field during the last decade. It is crucial that these methods are employed with great appreciation for the neuronal context and localization. An important cellular context to consider is the nature of stress signaling present on the OMM. Because stress-responsive protein kinases and phosphatases can influence mitochondrial form and function, characterizing PPIs on the OMM is important to understanding the regulation of mitochondrial homeostasis and stress responses. While techniques such as co-immunoprecipitation and fluorescent-based emission assays can indicate PPIs, we propose that a rigorous methodology, employing proximity-labeling approaches validated by classical techniques, can limit the number of false-positive associations. Moreover, PLAs in particular can be used to interrogate PPIs in fixed tissues or cell lines from which limited samples may be available. Overall, the analysis of mitochondrial stress in neurons requires a holistic approach that examines organelle form and function in the context of cellular physiology and stress. An important component of the cellular contexts of mitochondrial stress includes the ability to detect changes in signal transduction events on the OMM that influence mitochondrial physiology and stress responses.

Acknowledgments

The authors would like to thank Florida International University and the Robert Stempel College of Public Health & Social Work for the start-up funds that supported the studies discussed in the chapter. MRS was supported by a RISE grant NIH/NIGMS R25 GM06134. The authors would also like to express their gratitude to Philip V. LoGrasso (formerly of the Scripps Research Institute) for the tissue samples and imaging data using the 6-OHDA model. The authors extend their appreciation to the members of the Chambers' Lab who provided helpful comments during the preparation of the chapter.

References

1. Harris JJ, Jolivet R, Attwell D (2012) Synaptic energy use and supply. Neuron 75:762–777
2. Duchen MR (1992) Ca(2+)-dependent changes in the mitochondrial energetics in single dissociated mouse sensory neurons. Biochem J 283(Pt 1):41–50
3. Llorente-Folch I, Rueda CB, Pardo B, Szabadkai G, Duchen MR, Satrustegui J (2015) The regulation of neuronal mitochondrial metabolism by calcium. J Physiol 593:3447–3462
4. MacAskill AF, Kittler JT (2010) Control of mitochondrial transport and localization in neurons. Trends Cell Biol 20:102–112
5. Sheng ZH (2017) The interplay of axonal energy homeostasis and mitochondrial trafficking and anchoring. Trends Cell Biol 27:403–416
6. Smith GM, Gallo G (2018) The role of mitochondria in axon development and regeneration. Dev Neurobiol 78:221–237
7. Coffey ET, Hongisto V, Dickens M, Davis RJ, Courtney MJ (2000) Dual roles for c-Jun N-terminal kinase in developmental and stress responses in cerebellar granule neurons. J Neurosci 20:7602–7613
8. Petruzzella V, Sardanelli AM, Scacco S, Panelli D, Papa F, Trentadue R, Papa S (2012) Dysfunction of mitochondrial respiratory chain complex I in neurological disorders: genetics and pathogenetic mechanisms. Adv Exp Med Biol 942:371–384
9. Chambers JW, Pachori A, Howard S, Ganno M, Hansen D Jr, Kamenecka T, Song X, Duckett D, Chen W, Ling YY, Cherry L, Cameron MD, Lin L, Ruiz CH, Lograsso P (2011) Small molecule c-jun-N-terminal kinase (JNK) inhibitors protect dopaminergic neurons in a model of Parkinson's disease. ACS Chem Neurosci 2:198–206
10. Chambers JW, Pachori A, Howard S, Iqbal S, LoGrasso PV (2013) Inhibition of JNK mitochondrial localization and signaling is protective against ischemia/reperfusion injury in rats. J Biol Chem 288:4000–4011
11. Suomalainen A, Battersby BJ (2017) Mitochondrial diseases: the contribution of organelle stress responses to pathology. Nat Rev Mol Cell Biol 19:77
12. Oliveira G, Diogo L, Grazina M, Garcia P, Ataide A, Marques C, Miguel T, Borges L, Vicente AM, Oliveira CR (2005) Mitochondrial dysfunction in autism spectrum disorders: a population-based study. Dev Med Child Neurol 47:185–189
13. Krstic D, Knuesel I (2012) Deciphering the mechanism underlying late-onset Alzheimer disease. Nat Rev Neurol 9:25
14. Czarny P, Wigner P, Galecki P, Sliwinski T (2018) The interplay between inflammation, oxidative stress, DNA damage, DNA repair and mitochondrial dysfunction in depression. Prog Neuro-Psychopharmacol Biol Psychiatry 80:309–321
15. Ryan BJ, Hoek S, Fon EA, Wade-Martins R (2015) Mitochondrial dysfunction and mitophagy in Parkinson's: from familial to sporadic disease. Trends Biochem Sci 40:200–210
16. Ben-Shachar D (2017) Mitochondrial multifaceted dysfunction in schizophrenia; complex I as a possible pathological target. Schizophr Res 187:3–10
17. Andreux PA, Houtkooper RH, Auwerx J (2013) Pharmacological approaches to restore mitochondrial function. Nat Rev Drug Discov 12:465
18. Cooper-Knock J, Kirby J, Ferraiuolo L, Heath PR, Rattray M, Shaw PJ (2012) Gene expression profiling in human neurodegenerative disease. Nat Rev Neurol 8:518
19. Sherer TB, Betarbet R, Testa CM, Seo BB, Richardson JR, Kim JH, Miller GW, Yagi T, Matsuno-Yagi A, Greenamyre JT (2003) Mechanism of toxicity in rotenone models of Parkinson's disease. J Neurosci 23:10756–10764
20. Sherer TB, Kim JH, Betarbet R, Greenamyre JT (2003) Subcutaneous rotenone exposure causes highly selective dopaminergic degeneration and alpha-synuclein aggregation. Exp Neurol 179:9–16
21. Couvillion MT, Soto IC, Shipkovenska G, Churchman LS (2016) Synchronized mitochondrial and cytosolic translation programs. Nature 533:499
22. Shpilka T, Haynes CM (2017) The mitochondrial UPR: mechanisms, physiological functions and implications in ageing. Nat Rev Mol Cell Biol 19:109
23. Akabane S, Uno M, Tani N, Shimazaki S, Ebara N, Kato H, Kosako H, Oka T (2016) PKA regulates PINK1 stability and Parkin recruitment to damaged mitochondria through phosphorylation of MIC60. Mol Cell 62:371–384
24. Dagda RK, Gusdon AM, Pien I, Strack S, Green S, Li C, Van Houten B, Cherra SJ 3rd, Chu CT (2011) Mitochondrially localized PKA reverses mitochondrial pathology and dysfunction in a cellular model of Parkinson's disease. Cell Death Differ 18:1914–1923

25. Das Banerjee T, Dagda RY, Dagda M, Chu CT, Rice M, Vazquez-Mayorga E, Dagda RK (2017) PINK1 regulates mitochondrial trafficking in dendrites of cortical neurons through mitochondrial PKA. J Neurochem 142:545–559
26. Dickey AS, Strack S (2011) PKA/AKAP1 and PP2A/Bbeta2 regulate neuronal morphogenesis via Drp1 phosphorylation and mitochondrial bioenergetics. J Neurosci 31:15716–15726
27. Chambers JW, Cherry L, Laughlin JD, Figuera-Losada M, Lograsso PV (2011) Selective inhibition of mitochondrial JNK signaling achieved using peptide mimicry of the Sab kinase interacting motif-1 (KIM1). ACS Chem Biol 6:808–818
28. Chambers JW, Howard S, LoGrasso PV (2013) Blocking c-Jun N-terminal kinase (JNK) translocation to the mitochondria prevents 6-hydroxydopamine-induced toxicity in vitro and in vivo. J Biol Chem 288:1079–1087
29. Chambers JW, LoGrasso PV (2011) Mitochondrial c-Jun N-terminal kinase (JNK) signaling initiates physiological changes resulting in amplification of reactive oxygen species generation. J Biol Chem 286:16052–16062
30. Nijboer CH, Bonestroo HJ, Zijlstra J, Kavelaars A, Heijnen CJ (2013) Mitochondrial JNK phosphorylation as a novel therapeutic target to inhibit neuroinflammation and apoptosis after neonatal ischemic brain damage. Neurobiol Dis 54:432–444
31. Wong W, Scott JD (2004) AKAP signalling complexes: focal points in space and time. Nat Rev Mol Cell Biol 5:959–970
32. Wiltshire C, Gillespie DA, May GH (2004) Sab (SH3BP5), a novel mitochondria-localized JNK-interacting protein. Biochem Soc Trans 32:1075–1077
33. Wiltshire C, Matsushita M, Tsukada S, Gillespie DA, May GH (2002) A new c-Jun N-terminal kinase (JNK)-interacting protein, Sab (SH3BP5), associates with mitochondria. Biochem J 367:577–585
34. Toyama EQ, Herzig S, Courchet J, Lewis TL, Losón OC, Hellberg K, Young NP, Chen H, Polleux F, Chan DC, Shaw RJ (2016) AMP-activated protein kinase mediates mitochondrial fission in response to energy stress. Science 351:275–281
35. Hoffman NJ, Parker BL, Chaudhuri R, Fisher-Wellman KH, Kleinert M, Humphrey SJ, Yang P, Holliday M, Trefely S, Fazakerley DJ, Stöckli J, Burchfield JG, Jensen TE, Jothi R, Kiens B, Wojtaszewski JFP, Richter EA, James DE (2015) Global phosphoproteomic analysis of human skeletal muscle reveals a network of exercise-regulated kinases and AMPK substrates. Cell Metab 22:922–935
36. Chinopoulos C, Zhang SF, Thomas B, Ten V, Starkov AA (2011) Isolation and functional assessment of mitochondria from small amounts of mouse brain tissue. In: Manfredi G, Kawamata H (eds) Neurodegeneration: methods and protocols. Humana Press, Totowa, pp 311–324
37. Sodero AO, Rodriguez-Silva M, Salio C, Sassoe-Pognetto M, Chambers JW (2017) Sab is differentially expressed in the brain and affects neuronal activity. Brain Res 1670:76–85
38. Dunkley PR, Jarvie PE, Heath JW, Kidd GJ, Rostas JAP (1986) A rapid method for isolation of synaptosomes on Percoll gradients. Brain Res 372:115–129
39. Kim HJ, Magrané J (2011) Isolation and culture of neurons and astrocytes from the mouse brain cortex. In: Manfredi G, Kawamata H (eds) Neurodegeneration: methods and protocols. Humana Press, Totowa, pp 63–75
40. Wang L, Meece K, Williams DJ, Lo KA, Zimmer M, Heinrich G, Martin Carli J, Leduc CA, Sun L, Zeltser LM, Freeby M, Goland R, Tsang SH, Wardlaw SL, Egli D, Leibel RL (2015) Differentiation of hypothalamic-like neurons from human pluripotent stem cells. J Clin Invest 125:796–808
41. Shi Y, Kirwan P, Livesey FJ (2012) Directed differentiation of human pluripotent stem cells to cerebral cortex neurons and neural networks. Nat Protoc 7:1836
42. Giordano G, Costa LG (2011) Primary neurons in culture and neuronal cell lines for in vitro neurotoxicological studies. In: Costa LG, Giordano G, Guizzetti M (eds) In vitro neurotoxicology: methods and protocols. Humana Press, Totowa, pp 13–27
43. Griffiths EJ, Rutter GA (2009) Mitochondrial calcium as a key regulator of mitochondrial ATP production in mammalian cells. Biochim Biophys Acta (BBA) - Bioenerg 1787:1324–1333
44. Alnaes E, Rahamimoff R (1975) On the role of mitochondria in transmitter release from motor nerve terminals. J Physiol 248:285–306
45. Hall CN, Klein-Flugge MC, Howarth C, Attwell D (2012) Oxidative phosphorylation, not glycolysis, powers presynaptic and postsynaptic mechanisms underlying brain information processing. J Neurosci 32:8940–8951
46. Pathak D, Shields LY, Mendelsohn BA, Haddad D, Lin W, Gerencser AA, Kim H, Brand MD, Edwards RH, Nakamura K

(2015) The role of mitochondrially derived ATP in synaptic vesicle recycling. J Biol Chem 290:22325–22336

47. Verstreken P, Ly CV, Venken KJ, Koh TW, Zhou Y, Bellen HJ (2005) Synaptic mitochondria are critical for mobilization of reserve pool vesicles at Drosophila neuromuscular junctions. Neuron 47:365–378
48. McNay EC, Fries TM, Gold PE (2000) Decreases in rat extracellular hippocampal glucose concentration associated with cognitive demand during a spatial task. Proc Natl Acad Sci U S A 97:2881–2885
49. Izyumov DS, Avetisyan AV, Pletjushkina OY, Sakharov DV, Wirtz KW, Chernyak BV, Skulachev VP (2004) "Wages of fear": transient threefold decrease in intracellular ATP level imposes apoptosis. Biochim Biophys Acta 1658:141–147
50. Yang N-C, Ho W-M, Chen Y-H, Hu M-L (2002) A convenient one-step extraction of cellular ATP using boiling water for the luciferin-luciferase assay of ATP. Anal Biochem 306:323–327
51. Ford SR, Chenault KH, Bunton LS, Hampton GJ, McCarthy J, Hall MS, Pangburn SJ, Buck LM, Leach FR (1996) Use of firefly luciferase for ATP measurement: other nucleotides enhance turnover. J Biolumin Chemilumin 11:149–167
52. Imamura H, Nhat KP, Togawa H, Saito K, Iino R, Kato-Yamada Y, Nagai T, Noji H (2009) Visualization of ATP levels inside single living cells with fluorescence resonance energy transfer-based genetically encoded indicators. Proc Natl Acad Sci U S A 106:15651–15656
53. Nadee N, Moraes TC (2018) Mitochondrial DNA damage and reactive oxygen species in neurodegenerative disease. FEBS Lett 592:728–742
54. Pei L, Wallace DC (2018) Mitochondrial etiology of neuropsychiatric disorders. Biol Psychiatry 83:722–730
55. Murphy M (2009) How mitochondria produce reactive oxygen species. Biochem J 417:1–13
56. Papa S, De Rasmo D (2013) Complex I deficiencies in neurological disorders. Trends Mol Med 19:61–69
57. Pooja J, Xu W, Johan A (2016) Analysis of mitochondrial respiratory chain supercomplexes using blue native polyacrylamide gel electrophoresis (BN-PAGE). Curr Protoc Mouse Biol 6:1–14
58. Beutner G, Porter GA Jr (2017) Analyzing supercomplexes of the mitochondrial electron transport chain with native electrophoresis, in-gel assays, and electroelution. J Vis Exp 124. https://doi.org/10.3791/55738
59. Kirby DM, Thorburn DR, Turnbull DM, Taylor RW (2007) Biochemical assays of respiratory chain complex activity. Methods Cell Biol 80:93–119
60. Gerencser AA, Neilson A, Choi SW, Edman U, Yadava N, Oh RJ, Ferrick DA, Nicholls DG, Brand MD (2009) Quantitative microplate-based respirometry with correction for oxygen diffusion. Anal Chem 81:6868–6878
61. Wu M, Neilson A, Swift AL, Moran R, Tamagnine J, Parslow D, Armistead S, Lemire K, Orrell J, Teich J, Chomicz S, Ferrick DA (2007) Multiparameter metabolic analysis reveals a close link between attenuated mitochondrial bioenergetic function and enhanced glycolysis dependency in human tumor cells. Am J Phys Cell Physiol 292:C125–C136
62. Zhang L, Trushina E (2017) Respirometry in neurons. In: Strack S, Usachev YM (eds) Techniques to investigate mitochondrial function in neurons. Springer, New York, pp 95–113
63. Silva LP, Lorenzi PL, Purwaha P, Yong V, Hawke DH, Weinstein JN (2013) Measurement of DNA concentration as a normalization strategy for metabolomic data from adherent cell lines. Anal Chem 85: 9536–9542
64. Shcherbakova DM, Verkhusha VV (2013) Near-infrared fluorescent proteins for multicolor in vivo imaging. Nat Methods 10:751
65. David G, Barrett EF (2003) Mitochondrial Ca2+ uptake prevents desynchronization of quantal release and minimizes depletion during repetitive stimulation of mouse motor nerve terminals. J Physiol 548:425–438
66. Zengel JE, Sosa MA, Poage RE, Mosier DR (1994) Role of intracellular Ca2+ in stimulation-induced increases in transmitter release at the frog neuromuscular junction. J Gen Physiol 104:337–355
67. Devine MJ, Kittler JT (2018) Mitochondria at the neuronal presynapse in health and disease. Nat Rev Neurosci 19:63–80
68. David G, Talbot J, Barrett EF (2003) Quantitative estimate of mitochondrial [Ca2+] in stimulated motor nerve terminals. Cell Calcium 33:197–206
69. Fonteriz RI, de la Fuente S, Moreno A, Lobaton CD, Montero M, Alvarez J (2010) Monitoring mitochondrial [Ca(2+)] dynamics with rhod-2, ratiometric pericam and aequorin. Cell Calcium 48:61–69
70. Davidson SM, Duchen MR (2012) Imaging mitochondrial calcium signalling with fluorescent probes and single or two photon confocal microscopy. Methods Mol Biol 810:219–234

71. Miyawaki A, Llopis J, Heim R, McCaffery JM, Adams JA, Ikura M, Tsien RY (1997) Fluorescent indicators for Ca2+ based on green fluorescent proteins and calmodulin. Nature 388:882–887
72. Wu J, Prole DL, Shen Y, Lin Z, Gnanasekaran A, Liu Y, Chen L, Zhou H, Chen SR, Usachev YM, Taylor CW, Campbell RE (2014) Red fluorescent genetically encoded Ca2+ indicators for use in mitochondria and endoplasmic reticulum. Biochem J 464:13–22
73. Rose T, Goltstein PM, Portugues R, Griesbeck O (2014) Putting a finishing touch on GECIs. Front Mol Neurosci 7:88
74. Rysted JE, Lin Z, Usachev YM (2017) Techniques for simultaneous mitochondrial and cytosolic Ca2+ imaging in neurons. In: Strack, S, Usachev, YM (eds) Techniques to investigate mitochondrial function in neurons, pp 151–178. https://doi.org/10.1007/978-1-4939-6890-9_8
75. Villalobos C, Alonso MT, García-Sancho J (2009) Bioluminescence imaging of calcium oscillations inside intracellular organelles. In: Rich PB, Douillet C (eds) Bioluminescence: methods and protocols. Humana Press, Totowa, pp 203–214
76. Yamazaki RK, Mickey DL, Story M (1979) The calibration and use of a calcium ion-specific electrode for kinetic studies of mitochondrial calcium transport. Anal Biochem 93:430–441
77. Moreno AJM, Vicente JA (2012) Use of a calcium-sensitive electrode for studies on mitochondrial calcium transport. In: Palmeira CM, Moreno AJ (eds) Mitochondrial bioenergetics: methods and protocols. Humana Press, Totowa, pp 207–217
78. Jackson JB, Nicholls DG (1986) Methods for the determination of membrane potential in bioenergetic systems. Methods Enzymol 127:557–577
79. Budd SL, Castilho RF, Nicholls DG (1997) Mitochondrial membrane potential and hydroethidine-monitored superoxide generation in cultured cerebellar granule cells. FEBS Lett 415:21–24
80. Gerencser AA, Chinopoulos C, Birket MJ, Jastroch M, Vitelli C, Nicholls DG, Brand MD (2012) Quantitative measurement of mitochondrial membrane potential in cultured cells: calcium-induced de- and hyperpolarization of neuronal mitochondria. J Physiol 590:2845–2871
81. Nicholls DG (2012) Fluorescence measurement of mitochondrial membrane potential changes in cultured cells. Methods Mol Biol 810:119–133
82. Palmeira CM, Rolo AP (2012) Mitochondrial membrane potential (ΔΨ) fluctuations associated with the metabolic states of mitochondria. In: Palmeira CM, Moreno AJ (eds) Mitochondrial bioenergetics: methods and protocols. Humana Press, Totowa, pp 89–101
83. Aiuchi T, Matsunaga M, Nakaya K, Nakamura Y (1985) Effects of probes of membrane potential on metabolism in synaptosomes. Biochim Biophys Acta Gen Subj 843:20–24
84. Nicholls DG (2012) Fluorescence measurement of mitochondrial membrane potential changes in cultured cells. In: Palmeira CM, Moreno AJ (eds) Mitochondrial bioenergetics: methods and protocols. Humana Press, Totowa, pp 119–133
85. Carelli V, Chan DC (2014) Mitochondrial DNA: impacting central and peripheral nervous systems. Neuron 84:1126–1142
86. Trifunovic A, Wredenberg A, Falkenberg M, Spelbrink JN, Rovio AT, Bruder CE, Bohlooly-Y M, Gidlöf S, Oldfors A, Wibom R, Törnell J, Jacobs HT, Larsson N-G (2004) Premature ageing in mice expressing defective mitochondrial DNA polymerase. Nature 429:417
87. Ye F, Samuels DC, Clark T, Guo Y (2014) High-throughput sequencing in mitochondrial DNA research. Mitochondrion 17:157–163
88. Wang W, Esbensen Y, Scheffler K, Eide L (2015) Analysis of mitochondrial DNA and RNA integrity by a real-time qPCR-based method. In: Weissig V, Edeas M (eds) Mitochondrial medicine: volume I, probing mitochondrial function. Springer, New York, pp 97–106
89. Holt IJ, Reyes A (2012) Human mitochondrial DNA replication. Cold Spring Harb Perspect Biol 4:a012971
90. Jemt E, Persson Ö, Shi Y, Mehmedovic M, Uhler JP, Dávila López M, Freyer C, Gustafsson CM, Samuelsson T, Falkenberg M (2015) Regulation of DNA replication at the end of the mitochondrial D-loop involves the helicase TWINKLE and a conserved sequence element. Nucleic Acids Res 43:9262–9275
91. Macao B, Uhler JP, Siibak T, Zhu X, Shi Y, Sheng W, Olsson M, Stewart JB, Gustafsson CM, Falkenberg M (2015) The exonuclease activity of DNA polymerase γ is required for ligation during mitochondrial DNA replication. Nat Commun 6:7303
92. Payne BAI, Cree L, Chinnery PF (2015) Single-cell analysis of mitochondrial DNA. In: Weissig V, Edeas M (eds) Mitochondrial medicine: volume I, probing mitochondrial function. Springer, New York, pp 67–76

93. Payne BAI, Gardner K, Coxhead J, Chinnery PF (2015) Deep resequencing of mitochondrial DNA. In: Weissig V, Edeas M (eds) Mitochondrial medicine: volume I, probing mitochondrial function. Springer, New York, pp 59–66
94. Quispe-Tintaya W, White RR, Popov VN, Vijg J, Maslov AY (2013) Fast mitochondrial DNA isolation from mammalian cells for next-generation sequencing. BioTechniques 55:133–136
95. Labbé K, Murley A, Nunnari J (2014) Determinants and functions of mitochondrial behavior. Annu Rev Cell Dev Biol 30:357–391
96. Plucińska G, Paquet D, Hruscha A, Godinho L, Haass C, Schmid B, Misgeld T (2012) In vivo imaging of disease-related mitochondrial dynamics in a vertebrate model system. J Neurosci 32:16203–16212
97. Kopeikina KJ, Carlson GA, Pitstick R, Ludvigson AE, Peters A, Luebke JI, Koffie RM, Frosch MP, Hyman BT, Spires-Jones TL (2011) Tau accumulation causes mitochondrial distribution deficits in neurons in a mouse model of tauopathy and in human Alzheimer's disease brain. Am J Pathol 179:2071–2082
98. Detmer SA, Chan DC (2007) Complementation between mouse Mfn1 and Mfn2 protects mitochondrial fusion defects caused by CMT2A disease mutations. J Cell Biol 176:405–414
99. Züchner S, Mersiyanova IV, Muglia M, Bissar-Tadmouri N, Rochelle J, Dadali EL, Zappia M, Nelis E, Patitucci A, Senderek J, Parman Y, Evgrafov O, Jonghe PD, Takahashi Y, Tsuji S, Pericak-Vance MA, Quattrone A, Battologlu E, Polyakov AV, Timmerman V, Schröder JM, Vance JM (2004) Mutations in the mitochondrial GTPase mitofusin 2 cause Charcot-Marie-Tooth neuropathy type 2A. Nat Genet 36:449
100. Wang DB, Uo T, Kinoshita C, Sopher BL, Lee RJ, Murphy SP, Kinoshita Y, Garden GA, Wang H-G, Morrison RS (2014) Bax interacting factor-1 promotes survival and mitochondrial elongation in neurons. J Neurosci 34:2674–2683
101. Merrill RA, Flippo KH, Strack S (2017) Measuring mitochondrial shape with ImageJ. In: Strack S, Usachev YM (eds) Techniques to investigate mitochondrial function in neurons. Springer, New York, pp 31–48
102. Trevisan T, Pendin D, Montagna A, Bova S, Ghelli AM, Daga A (2018) Manipulation of mitochondria dynamics reveals separate roles for form and function in mitochondria distribution. Cell Rep 23:1742–1753
103. Sasaki S (2010) Determination of altered mitochondria ultrastructure by electron microscopy. In: Bross P, Gregersen N (eds) Protein misfolding and cellular stress in disease and aging: concepts and protocols. Humana Press, Totowa, pp 279–290
104. Bolea I, Gan W-B, Manfredi G, Magrané J (2014) Chapter six – imaging of mitochondrial dynamics in motor and sensory axons of living mice. In: Murphy AN, Chan DC (eds) Methods in enzymology. Academic, Amsterdam, pp 97–110
105. Otera H, Wang C, Cleland MM, Setoguchi K, Yokota S, Youle RJ, Mihara K (2010) Mff is an essential factor for mitochondrial recruitment of Drp1 during mitochondrial fission in mammalian cells. J Cell Biol 191:1141–1158
106. Lovy A, Molina AJA, Cerqueira FM, Trudeau K, Shirihai OS (2012) A faster, high resolution, mtPA-GFP-based mitochondrial fusion assay acquiring kinetic data of multiple cells in parallel using confocal microscopy. JoVE 65:e3991
107. Muñoz JP, Zorzano A (2015) Analysis of mitochondrial morphology and function under conditions of mitofusin 2 deficiency. In: Weissig V, Edeas M (eds) Mitochondrial medicine: volume II, manipulating mitochondrial function. Springer, New York, pp 307–320
108. Ashrafi G, Schwarz TL (2012) The pathways of mitophagy for quality control and clearance of mitochondria. Cell Death Differ 20:31
109. Ferree AW, Trudeau K, Zik E, Benador IY, Twig G, Gottlieb RA, Shirihai OS (2013) MitoTimer probe reveals the impact of autophagy, fusion, and motility on subcellular distribution of young and old mitochondrial protein and on relative mitochondrial protein age. Autophagy 9:1887–1896
110. Sun N, Malide D, Liu J, Rovira II, Combs CA, Finkel T (2017) A fluorescence-based imaging method to measure in vitro and in vivo mitophagy using mt-Keima. Nat Protoc 12:1576
111. McWilliams TG, Prescott AR, Allen GFG, Tamjar J, Munson MJ, Thomson C, Muqit MMK, Ganley IG (2016) *mito*-QC illuminates mitophagy and mitochondrial architecture in vivo. J Cell Biol 214:333–345
112. Rodger CE, TG MW, Ganley IG (2018) Mammalian mitophagy – from in vitro molecules to in vivo models. FEBS J 285:1185–1202

113. McWilliams TG, Prescott AR, Montava-Garriga L, Ball G, Singh F, Barini E, Muqit MMK, Brooks SP, Ganley IG (2018) Basal mitophagy occurs independently of PINK1 in mouse tissues of high metabolic demand. Cell Metab 27:439–449.e435
114. Sheng Z-H, Cai Q (2012) Mitochondrial transport in neurons: impact on synaptic homeostasis and neurodegeneration. Nat Rev Neurosci 13:77
115. Wang X, Winter D, Ashrafi G, Schlehe J, Wong YL, Selkoe D, Rice S, Steen J, LaVoie MJ, Schwarz TL (2011) PINK1 and Parkin target Miro for phosphorylation and degradation to arrest mitochondrial motility. Cell 147:893–906
116. Zhou B, Lin M-Y, Sun T, Knight AL, Sheng Z-H (2014) Chapter five – characterization of mitochondrial transport in neurons. In: Murphy AN, Chan DC (eds) Methods in enzymology. Academic, New York, pp 75–96
117. Course MM, Hsieh C-H, Tsai P-I, Codding-Bui JA, Shaltouki A, Wang X (2017) Live imaging mitochondrial transport in neurons. In: Strack S, Usachev YM (eds) Techniques to investigate mitochondrial function in neurons. Springer, New York, pp 49–66
118. Edelman DB, Owens GC, Chen S (2011) Neuromodulation and mitochondrial transport: live imaging in hippocampal neurons over long durations. JoVE:e2599 52. https://doi.org/10.3791/2599
119. Barrientos SA, Martinez NW, Yoo S, Jara JS, Zamorano S, Hetz C, Twiss JL, Alvarez J, Court FA (2011) Axonal degeneration is mediated by the mitochondrial permeability transition pore. J Neurosci 31:966–978
120. Brustovetsky N, Brustovetsky T, Jemmerson R, Dubinsky JM (2002) Calcium-induced cytochrome c release from CNS mitochondria is associated with the permeability transition and rupture of the outer membrane. J Neurochem 80:207–218
121. Bernardi P, Krauskopf A, Basso E, Petronilli V, Blachly-Dyson E, Di Lisa F, Forte MA (2006) The mitochondrial permeability transition from in vitro artifact to disease target. FEBS J 273:2077–2099
122. Bernardi P, Rasola A (2007) Calcium and cell death: the mitochondrial connection. Subcell Biochem 45:481–506
123. Brustovetsky T, Brustovetsky N (2017) Monitoring of permeability transition pore openings in mitochondria of cultured neurons. In: Strack S, Usachev YM (eds) Techniques to investigate mitochondrial function in neurons. Springer, New York, pp 239–248
124. Abramov AY, Duchen MR (2011) Measurements of threshold of mitochondrial permeability transition pore opening in intact and permeabilized cells by flash photolysis of caged calcium. In: Manfredi G, Kawamata H (eds) Neurodegeneration: methods and protocols. Humana Press, Totowa, pp 299–309
125. Ramshesh VK, Lemasters JJ (2012) Imaging of mitochondrial pH using SNARF-1. In: Palmeira CM, Moreno AJ (eds) Mitochondrial bioenergetics: methods and protocols. Humana Press, Totowa, pp 243–248
126. Friberg H, Ferrand-Drake M, Bengtsson F, Halestrap AP, Wieloch T (1998) Cyclosporin A, but not FK 506, protects mitochondria and neurons against hypoglycemic damage and implicates the mitochondrial permeability transition in cell death. J Neurosci 18:5151–5159
127. Berendzen KM, Durieux J, Shao L-W, Tian Y, Kim H-e, Wolff S, Liu Y, Dillin A (2016) Neuroendocrine coordination of mitochondrial stress signaling and proteostasis. Cell 166:1553–1563.e1510
128. Jensen MB, Jasper H (2014) Mitochondrial proteostasis in the control of aging and longevity. Cell Metab 20:214–225
129. Yano M, Kanazawa M, Terada K, Takeya M, Hoogenraad N, Mori M (1998) Functional analysis of human mitochondrial receptor Tom20 for protein import into mitochondria. J Biol Chem 273:26844–26851
130. Yano M, Kanazawa M, Terada K, Namchai C, Yamaizumi M, Hanson B, Hoogenraad N, Mori M (1997) Visualization of mitochondrial protein import in cultured mammalian cells with Green fluorescent protein and effects of overexpression of the human import receptor Tom20. J Biol Chem 272:8459–8465
131. Korde AS, Pettigrew LC, Craddock SD, Maragos WF (2005) The mitochondrial uncoupler 2,4-dinitrophenol attenuates tissue damage and improves mitochondrial homeostasis following transient focal cerebral ischemia. J Neurochem 94:1676–1684
132. Nargund AM, Fiorese CJ, Pellegrino MW, Deng P, Haynes CM (2015) Mitochondrial and nuclear accumulation of the transcription factor ATFS-1 promotes OXPHOS recovery during the UPRmt. Mol Cell 58:123–133
133. Fiorese CJ, Haynes CM (2017) Integrating the UPRmt into the mitochondrial maintenance network. Crit Rev Biochem Mol Biol 52:304–313
134. Fiorese CJ, Schulz AM, Lin Y-F, Rosin N, Pellegrino MW, Haynes CM (2016) The transcription factor ATF5 mediates a mam-

malian mitochondrial UPR. Curr Biol 26:2037–2043

135. Mukhopadhyay P, Rajesh M, Haskó G, Hawkins BJ, Madesh M, Pacher P (2007) Simultaneous detection of apoptosis and mitochondrial superoxide production in live cells by flow cytometry and confocal microscopy. Nat Protoc 2:2295

136. Wagener KC, Kolbrink B, Dietrich K, Kizina KM, Terwitte LS, Kempkes B, Bao G, Müller M (2016) Redox indicator mice stably expressing genetically encoded neuronal roGFP: versatile tools to decipher subcellular redox dynamics in neuropathophysiology. Antioxid Redox Signal 25:41–58

137. Cochemé HM, Quin C, McQuaker SJ, Cabreiro F, Logan A, Prime TA, Abakumova I, Patel JV, Fearnley IM, James AM, Porteous CM, Smith RAJ, Saeed S, Carré JE, Singer M, Gems D, Hartley RC, Partridge L, Murphy MP (2011) Measurement of H2O2 within living Drosophila during aging using a ratiometric mass spectrometry probe targeted to the mitochondrial matrix. Cell Metab 13:340–350

138. Shchepinova MM, Cairns AG, Prime TA, Logan A, James AM, Hall AR, Vidoni S, Arndt S, Caldwell ST, Prag HA, Pell VR, Krieg T, Mulvey JF, Yadav P, Cobley JN, Bright TP, Senn HM, Anderson RF, Murphy MP, Hartley RC (2017) MitoNeoD: A mitochondria-targeted superoxide probe. Cell Chem Biol 24:1285–1298.e1212

139. Jomova K, Vondrakova D, Lawson M, Valko M (2010) Metals, oxidative stress and neurodegenerative disorders. Mol Cell Biochem 345:91–104

140. Chandel NS (2014) Mitochondria as signaling organelles. BMC Biol 12:34

141. Wilson TJ, Slupe AM, Strack S (2013) Cell signaling and mitochondrial dynamics: implications for neuronal function and neurodegenerative disease. Neurobiol Dis 51:13–26

142. Pagliarini DJ, Dixon JE (2006) Mitochondrial modulation: reversible phosphorylation takes center stage? Trends Biochem Sci 31:26–34

143. Horbinski C, Chu CT (2005) Kinase signaling cascades in the mitochondrion: a matter of life or death. Free Radic Biol Med 38:2–11

144. Lim S, Smith KR, Lim S-TS, Tian R, Lu J, Tan M (2016) Regulation of mitochondrial functions by protein phosphorylation and dephosphorylation. Cell Biosci 6:25

145. Merrill RA, Strack S (2014) Mitochondria: a kinase anchoring protein 1, a signaling platform for mitochondrial form and function. Int J Biochem Cell Biol 48:92–96

146. Boja ES, Phillips D, French SA, Harris RA, Balaban RS (2009) Quantitative mitochondrial phosphoproteomics using iTRAQ on an LTQ-Orbitrap with high energy collision dissociation. J Proteome Res 8: 4665–4675

147. Calvo SE, Clauser KR, Mootha VK (2016) MitoCarta2.0: an updated inventory of mammalian mitochondrial proteins. Nucleic Acids Res 44:D1251–D1257

148. Deng WJ, Nie S, Dai J, Wu JR, Zeng R (2010) Proteome, phosphoproteome, and hydroxyproteome of liver mitochondria in diabetic rats at early pathogenic stages. Mol Cell Proteomics 9:100–116

149. Foster LJ, de Hoog CL, Zhang Y, Zhang Y, Xie X, Mootha VK, Mann M (2006) A mammalian organelle map by protein correlation profiling. Cell 125:187–199

150. O'Rourke B, Van Eyk JE, Foster DB (2011) Mitochondrial protein phosphorylation as a regulatory modality: implications for mitochondrial dysfunction in heart failure. Congest Heart Fail 17:269–282

151. Chen WW, Freinkman E, Wang T, Birsoy K, Sabatini DM (2016) Absolute quantification of matrix metabolites reveals the dynamics of mitochondrial metabolism. Cell 166:1324–1337.e1311

152. Ahier A, Dai C-Y, Tweedie A, Bezawork-Geleta A, Kirmes I, Zuryn S (2018) Affinity purification of cell-specific mitochondria from whole animals resolves patterns of genetic mosaicism. Nat Cell Biol 20:352–360

153. Chiao-Lin C, Norbert P (2017) Proximity-dependent labeling methods for proteomic profiling in living cells. Wiley Interdiscip Rev Dev Biol 6:e272

154. Kim DI, Roux KJ (2016) Filling the void: proximity-based labeling of proteins in living cells. Trends Cell Biol 26:804–817

155. Peipei L, Jingjing L, Li W, Li-Jun D (2017) Proximity labeling of interacting proteins: application of BioID as a discovery tool. Proteomics 17:1700002

156. Lee S-Y, Kang M-G, Park J-S, Lee G, Ting AY, Rhee H-W (2016) APEX fingerprinting reveals the subcellular localization of proteins of interest. Cell Rep 15:1837–1847

157. Susan RJ, Xue-Wen L, Sarah P, Susan LK, Philip JA (2015) Selective proteomic proximity labeling assay using tyramide (SPPLAT): a quantitative method for the proteomic analysis of localized membrane-bound protein clusters. Curr Protoc Protein Sci 80:19.27.11–19.27.18

158. Lohse MJ, Nuber S, Hoffmann C (2012) Fluorescence/bioluminescence resonance energy transfer techniques to study G-protein-coupled receptor activation and signaling. Pharmacol Rev 64:299–336

159. Waraho D, DeLisa MP (2009) Versatile selection technology for intracellular protein–protein interactions mediated by a unique bacterial hitchhiker transport mechanism. Proc Natl Acad Sci 106:3692–3697
160. Chen T-C, Lin K-T, Chen C-H, Lee S-A, Lee P-Y, Liu Y-W, Kuo Y-L, Wang F-S, Lai J-M, Huang C-YF (2014) Using an in situ proximity ligation assay to systematically profile endogenous protein–protein interactions in a pathway network. J Proteome Res 13:5339–5346
161. Paul M, Skalli O (2016) Chapter nineteen – synemin: molecular features and the use of proximity ligation assay to study its interactions. In: Omary MB, Liem RKH (eds) Methods in enzymology. Academic, New York, pp 537–555
162. Jaume T, Víctor FD, Francisco C (2015) Visualizing G protein-coupled receptor-receptor interactions in brain using proximity ligation in situ assay. Curr Protoc Cell Biol 67:17.17.11–17.17.16
163. Bagchi S, Fredriksson R, Wallén-Mackenzie Å (2015) In situ proximity ligation assay (PLA). In: Hnasko R (ed) ELISA: methods and protocols. Springer, New York, pp 149–159
164. Söderberg O, Gullberg M, Jarvius M, Ridderstråle K, Leuchowius K-J, Jarvius J, Wester K, Hydbring P, Bahram F, Larsson L-G, Landegren U (2006) Direct observation of individual endogenous protein complexes in situ by proximity ligation. Nat Methods 3:995
165. Swartzman E, Shannon M, Lieu P, Chen S-M, Mooney C, Wei E, Kuykendall J, Tan R, Settineri T, Egry L, Ruff D (2010) Expanding applications of protein analysis using proximity ligation and qPCR. Methods 50:S23–S26

Chapter 7

Epigenetic Changes in Cultures: Neurons and Astrocytes

David P. Gavin, Xiaolu Zhang, and Marina Guizzetti

Abstract

Over the last decade, epigenetic gene regulation has garnered much attention. One of the important discoveries from these studies has been that neural cells have different epigenetic landscapes from each other as well as compared to peripheral tissue. For example, in neurons the so-called sixth base, 5-hydroxymethylcytosine (5hmC) and DNA methylation outside of a CpG context (CpH methylation), is much more common than in other cells. These unique features of neural cell epigenetic mechanisms underscore the importance for examining how various stimuli and developmental programs affect epigenetic marks in brain cells, outside of whole-body physiology, and in isolation from other neural cells. In this book chapter, we describe epigenetic mechanisms, primarily histone modifications and DNA methylation, and indicate the unique aspects of studying these molecular mechanisms in neural cells. We describe methods for performing primary neuron and primary astrocyte cultures from rodents. We also provide methods for measuring DNA methylation (methylated DNA immunoprecipitation) and histone modifications (chromatin immunoprecipitation). Studies utilizing these methods have resulted in important scientific discoveries and continue to be relevant models moving forward.

Key words Epigenetic, DNA methylation, Histone, Histone acetylation, 5-methylcytosine, 5-hydroxymethylcytosine, Neuron, Astrocyte, Cell culture

1 Introduction

The use of primary cell cultures to study epigenetic mechanisms has contributed much to our understanding of gene regulation in the brain. Isolation and culturing of neural cells from rats and mice have allowed for the testing of novel compounds outside of the context of whole-body physiology. In this chapter, we describe methods for studying epigenetic gene regulation using primary astrocyte and neuronal cell culture.

Neurons have several features which make primary neuron culture especially valuable in the study of epigenetics. Neuronal epigenetic mechanisms have been hypothesized to underlie stable changes in gene expression under both normal conditions (e.g., learning and memory) and in neuropsychiatry illnesses, such as depression, drug addiction, schizophrenia, among others [1–13].

Michael Aschner and Lucio Costa (eds.), *Cell Culture Techniques*, Neuromethods, vol. 145,
https://doi.org/10.1007/978-1-4939-9228-7_7,

Further, because neurons in culture are postmitotic and fairly homogenous, distinctions can be made between epigenetic mechanisms in dividing vs. nondividing cells. This is important as the available transcriptional control armamentarium differs depending on the ability of a cell to replicate. Overall, neuron culture has provided an invaluable tool for studying epigenetic parameters.

Far from being merely an accessory to neurons, astrocytes are just as integral to proper brain function as neurons. Glia outnumber neurons by approximately 1.5:1 in the brain, and astrocytes constitute anywhere from 20% to 60% of glia [14]. Astrocytes are involved in a variety of brain functions, including the regulation of ion homeostasis, reuptake and breakdown of neurotransmitters, maintenance of the blood–brain barrier, synthesizing cholesterol, and controlling synaptogenesis and neuronal plasticity [15–19]. Astrocytes reveal remarkable adaptive plasticity in the functional maintenance, development, and aging of the CNS [20]. Astrocytes increase the number of mature, functional synapses through both secreted factors and direct contact with synapses in the brain [21]. In the mature CNS, adult hippocampal astrocytes regulate synaptic transmission by instructing neuronal stem cells to adopt a neuronal fate [22].

Considering the dramatically different functions and characteristics of astrocytes and neurons, it is essential to consider each cell type in isolation in order to understand CNS function as a whole. For example, in whole brain samples, epigenetic changes may be dramatically different depending on cell type. Neurons are postmitotic, while glia (including astrocytes) proliferate during brain development and retain the ability to reenter the cell cycle even in the adult brain. Therefore, epigenetic changes may be longer lasting in neurons compared to glia. Secondly, the genes affected by neurons and astrocytes by a given stimulus are also different. For example, RE1-silencing transcription factor (REST) suppresses neuronal gene expression in non-neuronal cells, such as astrocytes, through epigenetic mechanisms [23]. Epigenetic mechanisms are therefore intrinsically different in the brain and between different neural cell types. Through the use of primary cell culture, the epigenetic mechanisms and genes affected by these mechanisms in each cell type can be revealed.

1.1 An Introduction into Epigenetics

Several definitions of epigenetics exist, varying from the most conservative definition as mitotically or meiotically heritable nongenetic mechanisms that affect gene expression to the most inclusive definition as molecular mechanisms that alter transcription [24–26]. The former definition is limiting in that only DNA methylation and perhaps some histone methylation marks would be considered "epigenetic." On the other hand, the latter definition could include not only noncoding RNA, which is often included in the definition, but also transcription factors, which are usually not included in the definition. In the current chapter, we have chosen a compromise definition as any molecular mechanism that directly alters chromatin structure, primarily histone modifications and DNA methylation.

The fundamental unit of chromatin structure is the nucleosome. The nucleosome is comprised of a histone octamer core consisting of two each of H2a, H2b, H3, and H4 proteins as well as 146 base pairs of DNA [27]. Promoter regions of actively transcribed genes are often devoid of nucleosomes. The removal of nucleosomes from genes that are being transcribed is accomplished via the nucleosome sliding process, which uses ATP. Specific histone protein posttranslational modifications (PTMs) are more or less associated with gene expression.

There are many documented histone modifications; several of them can be present at a given promoter generating a very high number of histone mark combinations. These modifications can be broadly categorized as those associated with gene regulation and those not associated with gene regulation. Importantly, the cause and effect regarding histone modifications is often difficult to ascertain as in many instances, transcriptional changes precede histone PTMs at a given gene [28].

Histone modifications associated with gene regulation include histone lysine acetylation, histone methylation, phosphorylation, and ADP-ribosylation, among others [29, 30]. Histone lysine acetylation, phosphorylation, and ADP-ribosylation are generally associated with gene expression, while methylation has different effects depending on the lysine methylated and how many methyl groups are bound to a single lysine residue.

Histone lysine acetylation is perhaps the best studied of the histone modifications. In fact, there are already several FDA-approved drugs targeting this pathway. The addition of histone acetylation is mediated by histone acetyltransferases (HAT), and it is removed by histone deacetylases (HDACs). There are currently three FDA-approved HDAC inhibitors: vorinostat and romidepsin for the treatment of cutaneous T-cell lymphoma and panobinostat for multiple myeloma.

Histone lysine methylation has a more complex relationship with gene expression than histone acetylation. On the histone H3 tail alone, lysines 4, 9, 27, 36, and 79 can be methylated to varying degrees and with different relationships with gene expression [31]. For instance, trimethylation of lysine 4 of histone 3 (H3K4me3) is associated with active gene promoters, but H3K4me2 has a broader distribution that does not always correlate with gene activity [32]. Additionally, di- or trimethylation of histone H3 at K9 is associated with gene silencing and provides a binding surface for heterochromatin protein 1 (HP1), which leads to propagation and stabilization of heterochromatin [33]. Lysine 27 of histone 3 methylation is associated with reduced transcriptional elongation and is therefore a mark of gene silencing in a euchromatic environment [34].

DNA cytosine methylation is another epigenetic mechanism utilized by cells. DNA methylation in neurons is perhaps the longest lasting form of gene regulation in the body, with no requirement for turnover over the lifetime of the cell [35]. DNA methylation has important roles in embryonic development, maintenance of pluripotency, X-chromosome inactivation, genomic imprinting, transcriptional regulation, chromatin structure, and chromosome stability [36]. There are several documented DNA base cytosine modifications within cells: these include 5-methylcytosine (5mC), 5-hydroxymethylcytosine (5hmC), 5-formylcytosine (5fC), 5-carboxylcytosine (5CaC), and 5-hydroxymethyluracil (5hmU). DNA methyltransferases (DNMTs) catalyze the reaction that adds methyl groups to DNA forming 5mC, and Tet proteins catalyze the reactions from 5mC to 5hmC, 5fC, and 5CaC. The most abundant 5mC derivative is 5hmC [37]. Additionally, 5hmC does not coincide with lower 5mC levels at gene promoters, indicating that 5hmC is itself a stable modification and not necessarily an intermediate base in DNA demethylation [38, 39]. 5fC and 5CaC are generally considered DNA demethylation intermediates.

The brain is unique in terms of cytosine modifications in several respects. Firstly, 5hmC and 5fC are particularly abundant in the brain [37, 40]. Secondly, 5mC not preceding guanine (CpH methylation) is much more common in the brain than elsewhere in the body [41]. Thirdly, neurons, unlike most other cells including astrocytes, are mostly postmitotic and incapable of replicating. Therefore, while other cells, such as astrocytes, can lose cytosine methyl groups merely by replicating and not remethylating previously methylated sites (passive DNA demethylation), postmitotic neurons can only discard methyl groups through active DNA demethylation. Because primary neuron cultures contain postmitotic cells, neuron culture has provided an essential tool in determining the components of mammalian active DNA demethylation [42–45].

There is also a great deal of cross-talk between cytosine and histone modifications. For example, the 5mC binding protein, MeCP2, can recruit corepressors mSin3A and histone deacetylases (HDACs) [46, 47]. MeCP2 also enhances histone H3 lysine 9 (H3K9) methylation [48].

Fewer studies have been conducted examining epigenetic gene regulation in astrocytes. However, several key findings have begun to emerge clarifying the role of DNA methylation and histone modifications in these cells. Astrocyte differentiation has been shown to be largely reliant upon appropriately timed DNA methylation and histone acetylation changes [49–52]. Further, aberrant

DNA methylation in astrocytes has been documented in many neuropsychiatric disorders. Nagy et al. examined genomic DNA methylation patterns that may contribute to astrocyte dysfunction in depression [8]. Other studies demonstrated reduced astrocytic 5mC, 5hmC, and global DNA methylation in Alzheimer disease [53, 54]. Bailey et al. identified that the astrocyte acetylation levels of histone H2b, H3, and H4 were decreased following blast exposure injury [55]. Dresselhaus et al. found class I HDAC inhibition regulates astrocytic apolipoprotein E expression and secretion, which is related to neurodegenerative diseases such as Alzheimer disease, traumatic brain injury, and multiple sclerosis [56, 57]. Moreover, Wu et al. indicated that histone deacetylase inhibitors increase glial cell line-derived neurotrophic factor and brain-derived neurotrophic factor in astrocytes to protect dopaminergic neurons in neuron-glia cultures [58]. Additionally, there is evidence from animal studies that Rett syndrome which is the result of a MeCP2 mutation in humans is largely the result of a deficit in normal astrocytic MeCP2 [59]. In primary astrocyte culture, alcohol exposure reduced DNMT activity and expression with a corresponding decrease in DNA methylation levels at the tissue plasminogen activator gene promoter [60].

2 Neuron Culture Methods (Fig. 1)

2.1 Source of Neuronal Material

Neural cell cultures are increasingly recognized as powerful tools for neurobiological research. Neuron culture allows examination of gene expression, protein trafficking, and neuronal morphological studies outside of the context of whole brain physiology. Enriched neuronal cultures can be derived from multiple sources: pluripotent stems cells, embryonic stem cells, immortalized neuroblastoma cancer cell lines, and primary neuron cultures from the rodent brain. The advantage of deriving neurons from embryonic stem cells is it is possible to induce differentiation into a fairly homogenous population of neuronal subtypes (e.g., GABAergic, dopaminergic, serotonergic, etc.). The advantage of neuroblastoma cell lines or differentiated pluripotent stem cells is that the neurons produced can be of human origin. The advantage of primary cultures is that they are more likely to recapitulate the properties of neuronal cells in vivo. Also, neuronal primary cultures can be generated from knockout mice.

With few exception (i.e., cerebellar granule neurons), rodent primary neuron cultures are generated from prenatal animals. This is because neurons generated during fetal development and younger are less susceptible to damage during the tissue dissociation process than more mature neurons [61, 62]. Also, the embryonic brain contains fewer glial cells, reducing the risk for glial contamination in neuron cultures. Primary neuron cultures can be

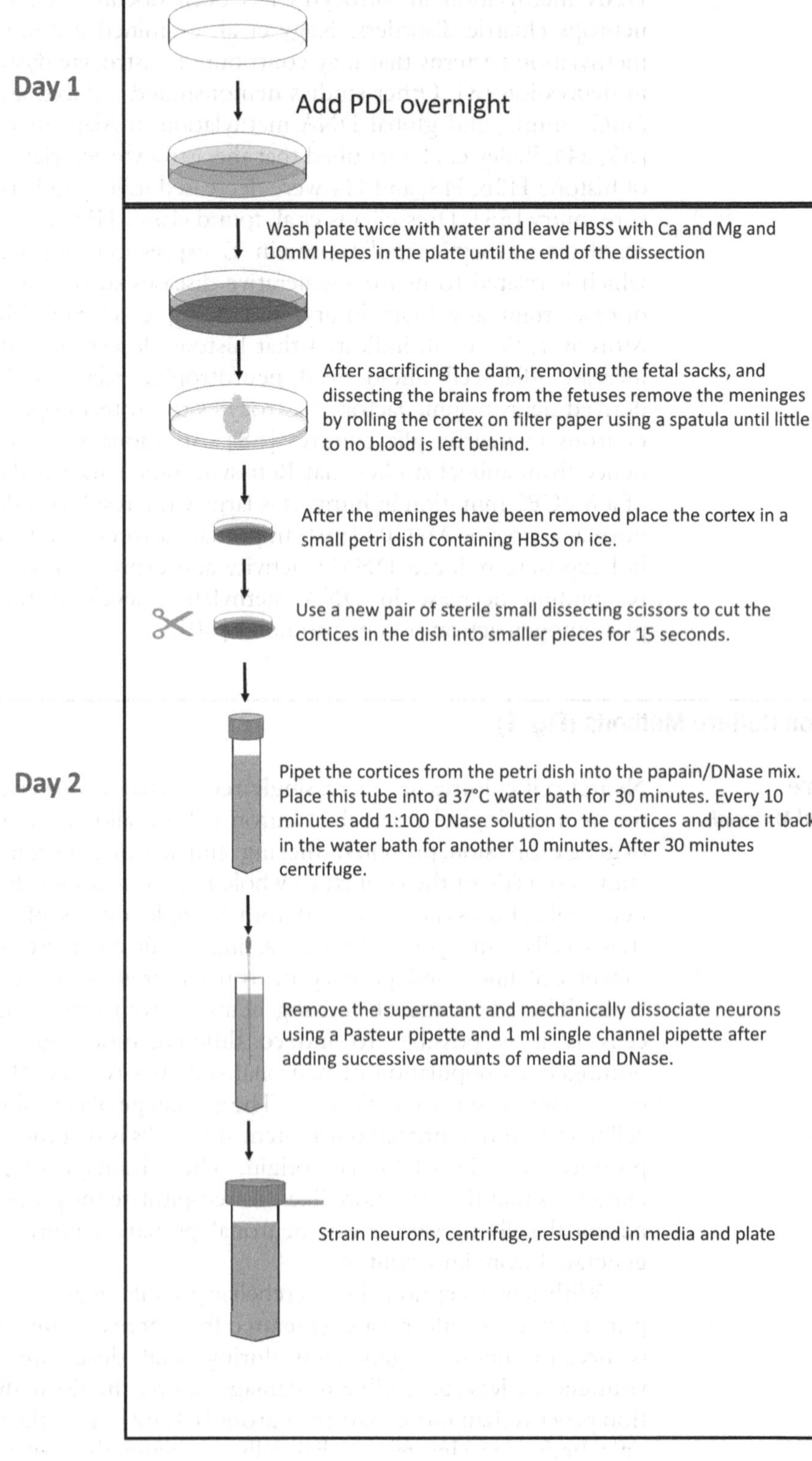
Day 1
Add PDL overnight
Wash plate twice with water and leave HBSS with Ca and Mg and 10mM Hepes in the plate until the end of the dissection
After sacrificing the dam, removing the fetal sacks, and dissecting the brains from the fetuses remove the meninges by rolling the cortex on filter paper using a spatula until little to no blood is left behind.
After the meninges have been removed place the cortex in a small petri dish containing HBSS on ice.
Use a new pair of sterile small dissecting scissors to cut the cortices in the dish into smaller pieces for 15 seconds.
Day 2
Pipet the cortices from the petri dish into the papain/DNase mix. Place this tube into a 37°C water bath for 30 minutes. Every 10 minutes add 1:100 DNase solution to the cortices and place it back in the water bath for another 10 minutes. After 30 minutes centrifuge.
Remove the supernatant and mechanically dissociate neurons using a Pasteur pipette and 1 ml single channel pipette after adding successive amounts of media and DNase.
Strain neurons, centrifuge, resuspend in media and plate

prepared from any region of the brain or spinal cord; here we describe the preparation of cortical primary neuron cultures. Cortical neuron cultures are generally prepared from mice or rats between embryonic day 13 and embryonic day 21 and can be kept in culture for 3–21 days.

2.2 Dish Coating for Neuronal Adhesion

In the absence of astrocyte support, neurons need an exogenous substrate that helps them to adhere to the culture dish, extend neurites, and develop. Poly-d-lysine (PDL) is the most widely used cell adhesion molecule [63, 64]. It attracts neurons through its positive charge electrostatically interacting with the negatively charged cell membrane. The attachment is followed by cell proliferation, histiocytic organization, and neurite formation which are characteristic signs of neural tissue differentiation.

To coat plates for neuronal cultures, make a 10 mg/ml PDL stock solution (Sigma, 1149) using cell culture water. This solution can be frozen at −20 °C. The day before the neuron preparation, PDL should be diluted in the biosafety cabinet to a working concentration of 0.1 mg/ml, filtered using a syringe with attached syringe filter (Sigma, CLS431219) and enough added to the cell culture plates to cover the bottom of the cell culture wells. The plates should then be placed in a cell culture incubator overnight. The following day in the biosafety cabinet, PDL should be removed from the plates, and each well should be washed twice with sterile cell culture water and then filled with Hanks Balanced Salt Solution (HBSS) with Ca^{2+} and Mg^{2+} and 10 mM HEPES until the end of the dissection. These plates can be left in the hood during neuron culture preparation.

2.3 Neuronal Dissection Methods

We use two enzymes to aid in the digestion of tissue to isolate neurons from fetal brains: DNase (Sigma, D4263-5VL) and papain (Sigma, P4762). Trypsin is an alternative to papain, but papain is considered less destructive to neurons. For each eight fetuses (each pregnant dam), we use 8 ml of diluted papain at a concentration of 2 mg/ml diluted in HBSS Ca^{2+}- and Mg^{2}-free. This solution should be syringe filtered for sterility. We also use DNase to prevent the formation of DNA gel that may occur when some of the cells break down during the dissociation process. The stock DNase

Fig. 1 Schematic of primary neuron culture preparation. On day 1, a Poly-d-lysine (PDL) matrix is provided upon which the neurons can grow. On day 2, the basic steps of the neuron preparation are illustrated. First, the cell culture plates which have PDL are washed with water, and HBSS is left in the plate. Next, the dam is sacrificed, fetal sacks are obtained, and the brain is dissected and placed on filter paper. The brain area of interest is then dissected and rolled on the filter paper to remove the meninges. Once this is repeated for all the fetuses, fine scissors are used to cut the brain tissue into smaller pieces. The brain tissue is then placed in a canonical 15 ml tube containing papain and DNase. This tube is incubated for 30 minutes at 37 °C with 1:100 DNase added every 10 minutes. The tissue is then centrifuged. Following centrifugation the neurons are mechanically dissociated using progressively smaller bore pipets and filtered through a 40 μm cell strainer. The cells are then centrifuged, resuspended in media, counted, diluted in additional media, and plated

concentration is 4 mg/ml in HBSS Ca^{2+}- and Mg^{2}-free. This solution should be similarly filtered.

In preparation for the dissection, an ice bucket should be placed in the biosafety cabinet, and a 35 mm cell culture dish containing approximately 3 ml HBSS Ca^{2+}- and Mg^{2}-free should be placed on ice in the cabinet. In addition, several Whatman® qualitative filter papers (Grade 3 (VWR, 28456-065) should be placed in a 100 mm petri dish in the cabinet. The purpose of the filter paper is to remove the meninges (see below). This method of meningeal removal is a tremendous time-saver over alternative methods such as manually separating the meninges using forceps with the aid of an inverted microscope. Another 100 mm petri dish should be brought to the animal facility in which the fetuses will be placed.

Sacrifice the pregnant dam with CO_2 followed by cervical dislocation, and then place the pregnant dam in the supine position, and spray 70% ethanol on its abdomen. Using dissecting scissors and forceps, cut through the skin layer only, and make a vertical incision from perineal region to the top of abdomen. Make another small secondary incision through peritoneal layer, and manually pull the peritoneum apart. Using forceps and dissecting scissors, remove the fetal sacks, and place them in the petri dish. Once complete bring the petri dish with fetuses to the lab. Before placing the petri dish with the fetal sacks into the biosafety cabinet, spray the dish with 70% ethanol. Place blunt forceps, vet scissors, and dissecting scissors in a 50 ml conical tube containing 70% ethanol. Add 80 μl of DNase and 40 μl of 1 M $MgCl_2$ into the papain solution, and place this mix into a 37 °C water bath to incubate for approximately 30 minutes (approximately the time it takes to dissect the cortices from the fetuses).

Using fine and blunt forceps, tear open each fetal sack just prior to cortical dissection. The blunt forceps should be placed in the 70% ethanol after each use. Remove the fetus, and decapitate it using vet scissors. The vet scissors should be placed in the 70% ethanol after each use. Poke the tips of fine forceps through the eyes of the fetal head, and hold the fetal head so that the posterior of the head is visible. Using dissecting scissors, make one vertical incision from the base of the skull anteriorly to the forehead and two lateral incisions from the base of the skull to either side of the skull (make incisions as superficial as possible). Using curved forceps peel back the skull revealing the brain. Using curved forceps pull/scoop out the brain, and place it on the filter paper in the petri dish. Using curved forceps and a spatula, separate the cortex from the rest of the brain. In order to remove the meninges, roll the cortex on the filter paper using the spatula until little to no blood is left behind by the brain tissue. After the meninges have been removed, place the cortex in the small petri dish containing HBSS that had been placed on ice. Repeat this process for all fetuses.

After the dissection is complete, use a new pair of sterile small dissecting scissors to cut the cortices in the dish into smaller pieces for 15 seconds. Remove the papain/DNase mix from the water bath, and use a syringe with attached filter to filter the mix into a new 15 ml tube. Pipet the cortices from the petri dish into the papain/DNase mix. Place this tube into a 37 °C water bath for 10 minutes. After 10 minutes add 1:100 DNase solution to the cortices, and place it back in the water bath for another 10 minutes. Repeat this step, adding 1:100 DNase to the cortices and placing it in the water bath for 10 minutes (total time in the water bath is 30 minutes).

During this period prepare media. Cell media differs considerably based on the derivation of the neuronal culture. Two types of methods have been described for culturing primary neurons from rats and mice [65]. In one method, the neurons are grown sandwiched on an astrocytic feeder layer [66]. This allows for the neurons to not be in direct contact with the astrocytes but to be exposed to factors secreted by astrocytes into the medium. The alternate protocol does not involve the use of a feeder layer, and neurons are maintained in serum-free medium (B27) supplemented with cofactors necessary for neuronal growth and maintenance [67]. The following are the components of the media that are used in our lab: 250 ml Neurobasal media (Thermo, 21103049); 1.25 ml Fungizone (Thermo, 15290018); 0.5 ml gentamicin (Thermo, 15750060); 1 ml D+glucose (Sigma, G8769); 5 ml B27 supplement (Thermo, 17504044); and 2.5 ml glutamax (Thermo, 35050061).

After a total of 30 minutes in the water bath, centrifuge cortices at 300 g for 5 minutes at 4 °C. Remove the supernatant and add 2 ml media and 20 μl of DNase. Use a glass Pasteur pipet with Dropper bulb on the end, and pipet up and down 15 times (avoid making bubbles). Let the cortices rest for 2 minutes. Pipet the translucent supernatant (~ 1 ml), and place on a cell strainer (40 μm) which is placed on a 50 ml conical tube. Add another 2 ml media and 20 μl of DNase to the cortices; cut the tip off a 200 μl pipet tip using sterile scissors such that it is smaller than tip of the Pasteur pipet. Insert a Pasteur pipet into the cut 200 μl pipet tip, and pipet up and down 15 times (avoid making bubbles). Let the cortices rest 2 minutes. Pipet the translucent supernatant (~ 1 ml), and place on the cell strainer. Add another 1 ml of media and 10 μl of DNase to the cortices. Pipet up and down using a 1 ml pipetman (avoid making bubbles). Immediately pipet the remaining cortices onto the cell strainer. Add another 2 ml of media to the cell strainer to push through the remaining cells. Remove the cell strainer, and transfer the cortices into a 15 ml tube. Add media until the total volume is 10 ml. Centrifuge the cortices at 300 g for 5 minutes at 4 °C. Remove the supernatant, and add 10 ml media, and resuspend cortices by pipetting up and down using a serological pipet (avoid making bubbles) (always warm media in water bath before adding to cells). Add 15 μl suspended cells and 15 μl trypan blue into a 0.2 ml tube. Pipet 15 μl onto either end of a hemocytometer. Count the cells. Dilute cells with media to the desired concentration. Generally, we plate cells at a concentration of $1–2 \times 10^6$ cells/ml (always warm media in water bath before adding to cells). Remove the HBSS from the culture plates and pipet the cells into the plates. After 4 hours perform a 100% media change.

2.4 Removal of Glia

At this point the cultured cells are a mix of glia and neurons. There are several methods for ensuring that the cells cultured are primarily neuronal. One method is to use varying amounts of PDL prior to plating to absorb the non-neuronal cells, leaving neurons in the supernatant. An alternative approach used by us is to treat the neurons with cytosine arabinoside 5 μM on the 3rd day in vitro. Twenty-four hours after adding cytosine arabinoside, we replace 100% of the media (always warm media in water bath before adding to cells). Generally we harvest cells 1 week from the dissection using a cell scraper.

3 Astrocyte Culture Methods

3.1 Primary Astrocyte Culture Protocol

Primary astrocyte cultures are a valuable tool to study astrocytes in physiological and pathological states. In this section we described a widely used method to prepare astrocytes developed by McCarthy and de Vellis [68] and modified by us [69, 70].

3.2 Astrocyte Dissection Methods

We prepare our primary astrocyte cultures from gestational day 20–21 fetuses. Tissue culture flasks (Falcon™ Tissue Culture Treated Flasks, 75 cm^2; cat. #353136) are first pre-coated with 10 ml of 40 μg/ml PDL at room temperature for 10 minutes in a biological safety cabinet or a laminar flow tissue-culture hood 1 day before dissection. After 10 minutes PDL can be discarded and the flasks washed once with sterile water. The flasks can then be left to air-dry overnight under the hood.

On the following day, euthanize a pregnant mouse or rat using carbon dioxide followed by decapitation. Spray the abdomen with 70% ethanol, and then using sterile scissors, make a superficial incision through the skin layer and then a second incision through the peritoneal layer. Collect the fetal sacks and place them in a 100 mm petri dish.

In the biosafety cabinet, remove each fetus from its fetal sack just prior to brain dissection. Using large scissors decapitate the fetus, and place these scissors in a 50 ml tube containing 70% ethanol between each use. Carefully cut the skin and skull with fine scissors to expose the brain. Remove the entire brain, and place it on filter paper (Whatman® Qualitative Filter Paper, Grade 3;

#1003-090). Using a scalpel or spatula, separate the cerebellum from the cerebrum, and separate the two hemispheres. Using curved forceps isolate the cortex, and place it on a dry area of the filter paper. Roll the cortices on the filter paper using a spatula in order to remove the meninges. The tissue is then transferred into 60×15 mm petri dish containing 5 ml of Ca^{2+}- and Mg^{2+}-free Hanks Balanced Salt Solution (HBSS, Thermo Fisher #14175-095). Once this has been repeated for all the fetuses, use fine scissors to mechanically cut brain cortices into smaller pieces.

Once all the cortices have been added to the small petri dish, add the cortices to a 50 ml polypropylene tube with 20 ml of 0.25% trypsin (10× dilution of 2.5% trypsin, Gibco, Cat. #15090-046), and incubate for 10 minutes at 37 °C. Trypsin proteolytic activity is stopped by an equal volume of growth medium (20 ml), containing low-glucose Dulbecco's modified Eagle's medium (DMEM; Thermo Fisher #11885-92), 10% fetal bovine serum (FBS; Atlanta Biological #S12450), and 100 units penicillin G plus 100 μg streptomycin per ml (Thermo Fisher #15-140-122). The cell suspension is then centrifuged at 500 g for 10 minutes. Discard the supernatant and resuspend the cells in 20 ml of DMEM/10% FBS/Pen-Strep (pipet up/down 20×), vortex at maximum speed for 1 minute, and centrifuge at 200 g for 10 minutes.

After discarding the supernatant, resuspend the cells in 20 ml of DMEM/10% FBS/Pen-Strep (pipet up/down 20×), and centrifuge at 200 g for 10 minutes. This step is repeated two more times. After three washes, resuspend the cells in 20 ml of the same growth medium as above, and filter to remove aggregates through a nylon mesh of 100 μm pore size (Corning™ Cell Strainers – 100 μm pore size; Yellow; #431752). Cells are then counted and plated in 75 cm^2 flasks pre-coated with PDL at the density of 6×10^6 cells per flask in 10 ml DMEM/10% FBS/Pen-Strep and incubated at 37 °C in a humidified atmosphere of 5% CO_2/95% air.

The next day gently shake the flasks by hand and discard the media. Wash the cells twice with PBS, and add 10 ml of fresh DMEM/10% FBS/Pen-Strep. To remove the layer of non-adherent cells (microglia) growing on top of the flat monolayer, flasks are energetically shaken about 150 times by hand when changing the medium every 2–3 days. Cells are grown until 100% confluent (usually about after 9 days); at this point, astrocytes can be subcultured in vessels appropriate to the experiments.

4 Epigenetic Methods

There are several epigenetic measures in cultured neurons or astrocytes that are commonly performed in laboratories. Among these are changes in DNA methylation and histone modifications.

4.1 DNA Methylation Profiling Methods Introduction

Cytosine methylation levels have been of interest to investigators for the past 40 years. The technology to detect these levels has greatly progressed with tremendous advances on our understanding of the role of DNA methylation in gene expression. There are three basic methods for determining DNA methylation levels at a given locus: enrichment methods, endonuclease digestion, and sodium bisulfite treatment. Enrichment methods include techniques such as MeDIP- and MBD-based assays. Each method has its advantages and disadvantages, and therefore the choice of assay depends on the hypothesis being tested and the scope of the project.

The oldest method used to detect DNA methylation to gain traction used endonucleases, such as HpaII and MspI [71, 72]. The recognition site for both enzymes is CCGG, but the HpaII will not cut if the CpG is methylated. By comparing the amount of intact to digested CCGG sequences, the amount of methylated CpGs could be derived.

In 1970, it was discovered that bisulfite could convert cytosine to uracil [73]. Ten years later, it was found that the conversion from cytosine to uracil was impaired when the cytosine was methylated [74]. [81] described exploiting the differences in the kinetics of this reaction to analyze DNA methylation patterns. The bisulfite treatment of DNA causes the deamination of cytosine forming uracil, which when sequenced is read as thymine. 5mC residues are more resistant to bisulfite deamination causing them to be read as cytosine during sequencing.

Bisulfite sequencing has several limitations and challenges. Alignment of bisulfite-treated DNA can be challenging following next-generation sequencing alignment. This is because bisulfite conversion reduces the number of nucleotides from four nucleotides (A, T, C, G) to three nucleotides (A, T, G) with only the rare methylated cytosine being read as C. Also, the conditions used in the bisulfite treatment is a compromise between ensuring all the non-methylated cytosines are converted to uracil and attempting to avoid the harsh side effects of bisulfite which leads to DNA fragmentation. Previously it was believed that an indicator of incomplete conversion was the persistence of non-CpG cytosines following conversion, but it has now been demonstrated that non-CpG methylation is not as infrequent as once thought particularly in the brain [41]. Another limitation of traditional bisulfite approaches is the fact that both 5mC and 5hmC are read as cytosine. This is not a trivial disadvantage as they may have entirely different distributions across the genome and functions in relation to gene expression.

There are two different methods to enrich for methylated DNA-, MeDIP-, and MBD-based methods. Neither method has single base-pair resolution but rather gives an estimate of methylation in fractionated DNA segments. Despite both enriching for methylated DNA, there are some key differences between the two assays. In MeDIP, an anti-methylcytosine antibody is used to immunoprecipitate methylated DNA fragments. MBD uses recombinant MBD proteins to enrich for methylated DNA. Examples of MBD-based kits are the MethylCap kit (Diagenode), MethylCollector Ultra Kit (Active Motif), MethylMiner (Life Technologies), and CpG MethylQuest DNA Isolation kit (EMD Millipore). MeDIP- and MBD-based methods differ in their abilities to detect non-CpG methylation. This could be particularly important for evaluating the methylome of neurons, which contains abundant non-CpG methylation. Also, MBD methods are better at enriching for CpG islands, while MeDIP is superior in regions with low CpG density [75]. In this chapter we describe in detail the MeDIP procedure.

4.2 MeDIP Procedures

The MeDIP method involves isolating DNA from cells/tissue, fragmenting the DNA, immunoprecipitating 5mC or 5hmC using antibodies, and then eluting and isolating the DNA captured by the antibody. One of the major drawbacks of MeDIP is its inability to pinpoint DNA methylation changes at single base-pair resolution [76]. However, because CpGs in neighboring regions tend to be similarly methylated in healthy cells, some have argued that single base-pair resolution is unnecessary [77]. There are many MeDIP commercial kits available; however below we describe our method for MeDIP that we have found to produce reliable and reproducible results (Fig. 2).

First homogenize cells/tissue in 250 μl SDS lysis buffer (1% SDS, 10 mM EDTA, 50 mM Tris-HCl, pH 8.1). To extract the DNA after sonication, we add 250 μl water to the samples (total volume is then 500 μl). To this we add 10 μl EDTA (0.5 M, pH = 8.0), 20 μl Tris (1 M, pH = 6.5), and 2 μl proteinase K (20 mg/ml). This is incubated at 37 °C overnight. The next day we add 50 μl sodium acetate (3 M, pH = 5.2) to samples and vortex them, and then 500 μl of phenol/chloroform/isoamyl alcohol (25:24:1, v/v) is added. After vortexing, centrifuge the samples at 14,000 RPM for 5 minutes at 4 °C. Remove the aqueous layer, and pipet into a new 1.7 ml tube. Add 1 ml 100% ethanol to samples, vortex the samples, and centrifuge at 14,000 RPM for 30 minutes at 4 °C. Remove the liquid leaving the pellet, and add 500 μl of 70% ethanol and vortex samples. Centrifuge samples at 14,000 RPM for 15 minutes at 4 °C. Remove liquid and leave the pellet. Let the samples air-dry. After pellets are dry, add 30 μl water to the pellets and resuspend. Measure DNA concentration using Qubit or similar method. Resuspend 250 ng of DNA from each sample in 50 μl of nuclease-free water, and place this in a freezer, as this will serve as the input. Resuspend 10 μg of DNA in 250 μl of SDS lysis buffer, and sonicate the DNA into 100–300 base-pair fragments. Resuspend 1–2 μg of this fragmented DNA

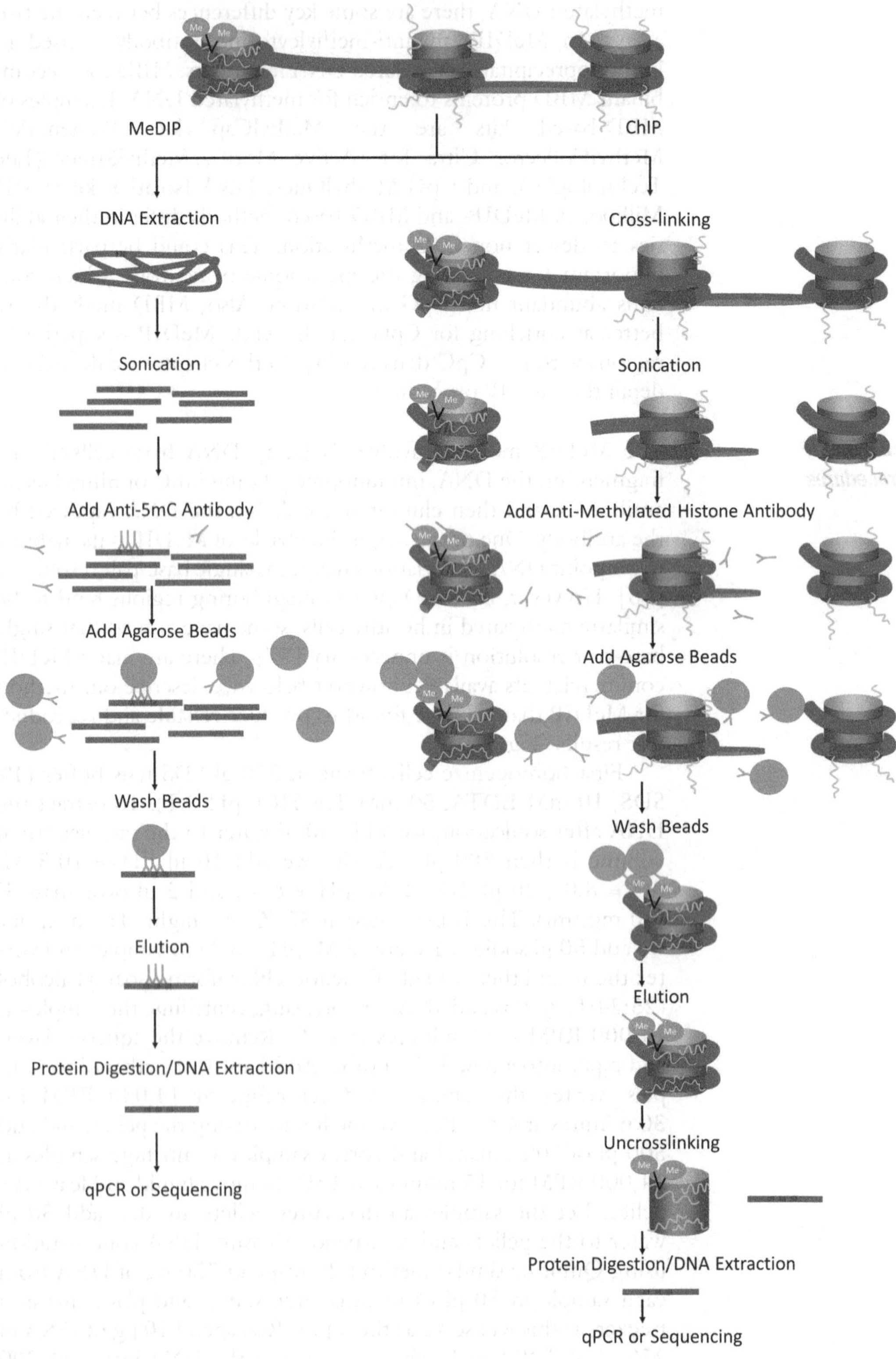
MeDIP
ChIP
DNA Extraction
Cross-linking
Sonication
Sonication
Add Anti-5mC Antibody
Add Anti-Methylated Histone Antibody
Add Agarose Beads
Add Agarose Beads
Wash Beads
Wash Beads
Elution
Elution
Protein Digestion/DNA Extraction
Uncrosslinking
qPCR or Sequencing
Protein Digestion/DNA Extraction
qPCR or Sequencing
Me

in 250 µl of ChIP dilution buffer (0.01% SDS, 1.1% Triton X-100, 1.2 mM EDTA, 16.7 mM Tris-HCl, pH 8.1, 167 mM NaCl). Boil the samples for 10 minutes, and then immediately immerse on ice. After samples cool add 4 µl anti-5mC antibody (diagenode Mab-081-010) or 2 µl anti-5hmC (Active motif, 39791). Incubate at 4 °C on an end-over-end mixer overnight.

The next day, add 40 µl Protein A/G PLUS-Agarose beads (Santa Cruz, sc-2003), and incubate on an end-over-end mixer for 3 hours at 4 °C (do not vortex). After 3 hours centrifuge samples at 1200 RPM for 1 minute at 4 °C. Discard the supernatant and keep the beads. Wash the beads with 750 µl of first low salt wash (0.1% SDS, 1% Triton X-100, 2 mM EDTA, 20 mM Tris-HCl, pH 8.1, 150 mM NaCl), then high salt wash (0.1% SDS, 1% Triton X-100, 2 mM EDTA, 20 mM Tris-HCl, pH 8.1, 500 mM NaCl), then LiCl wash (0.25 M LiCl, 1% IGEPAL-CA630, 1% deoxycholic acid (sodium salt)), and then TE buffer (10 mM Tris-HCl, 1 mM EDTA, pH 8.0) twice. For the wash steps, keep samples on ice between centrifugation steps. After adding each wash, invert the tubes several times, and centrifuge at 1200 RPM for 1 minute at 4 °C. After the washes resuspend the beads in 500 µl elution buffer (100 mM HCO_3, 1% SDS), and place samples in a dry bath set for 67 °C for 2 hours with occasional vortexing. After 2 hours remove the samples, vortex once more, and centrifuge 1200 RPM for 1 minute at room temperature. Remove and keep the supernatant. To the supernatant add EDTA-Tris-proteinase K mix as above, and incubate overnight at 37 °C. The following day perform phenol-chloroform extraction as above, and resuspend samples in 50 µl nuclease-free water. Samples are now ready for PCR quantification of methylation binding as normalized to input.

Fig. 2 Schematic of methylated DNA immunoprecipitation (MeDIP) and chromatin immunoprecipitation (ChIP) procedure. At the top is intact chromatin represented by the histone core as purple cylinders, DNA in red, and histone tails in green. The histone on the far left is more restrictive to gene regulation as a result of methyl groups on the green histone tails and methylated DNA represented by blue stripes on the red DNA. MeDIP begins with protein digestion and DNA extraction. Hence, the histone proteins are no longer present. The now naked DNA is fragmented using ultrasonication. Anti-5-methylcytosine (5mC) antibodies are added which are able to adhere to the methylated DNA. Agarose beads are used to immunoprecipitate the antibodies bound to the DNA, and wash steps are used to remove non-methylated DNA that is not bound to antibody or beads. Methylated DNA fragments are then eluted from the beads and isolated through DNA extraction. This purified DNA is now suitable for quantifying methylation levels using qPCR or sequencing. On the right is the ChIP procedure. The first step in this procedure is to cross-link the protein-DNA complexes using formaldehyde. These complexes are then fractionated using ultrasonication. An antibody that recognizes histone modifications is then added followed by beads used for immunoprecipitation. After several washes the protein-DNA complexes are eluted from the beads. These protein-DNA complexes are then uncross-linked, the protein is digested, and what is left is the DNA that was bound to histone proteins with the modification recognized by the antibody. This DNA is suitable for qPCR or sequencing quantification

4.3 Chromatin Immunoprecipitation Procedures

A ChIP method was first described by Gilmour and Lis in 1984 using UV to cross-link DNA. The difficulty with their approach was in uncross-linking, the protein-DNA complexes in order to localize the genes that the protein they were studying were targeting [78]. Not long after the development of this method, it was refined and improved with the discovery that formaldehyde could be used to cross-link protein and DNA and that this reaction could be easily reversed with heat [79].

The basic method of ChIP involves cross-linking proteins to DNA, fragmenting the chromatin into smaller pieces, using an antibody and beads to immunoprecipitate protein-DNA complexes, eluting the complexes from the beads, uncross-linking, and finally protein digestion and DNA extraction (Fig. 2). Subsequently, the locations in which the protein-DNA interactions occurred can be measured using Southern blotting, qPCR, microarray, or sequencing. Some of the steps in this basic ChIP method can vary based on the interest of the investigator. For example, in native ChIP there is no cross-linking step. This can be advantageous because cross-linking with formaldehyde can cause proteins that do not directly interact with DNA to be immunoprecipitated as well. Another advantage of native ChIP is that antibodies better recognize proteins that have not had their conformation distorted by cross-linking. One disadvantage is that native ChIP is limited by the fact that only those proteins that directly and tightly adhere to DNA can be immunoprecipitated. Because cross-linking is not performed in native ChIP, the chromatin cannot be sheared using ultrasonication as the sound waves are too intense for this technique and may dislodge proteins from DNA. Rather, in native ChIP an endonuclease is used to fragment the chromatin. This endonuclease digestion may be more biased to digesting open chromatin areas rather than sonication which will shear open and closed chromatin alike. Choosing between traditional ChIP and native ChIP is dependent on the scientific question being investigated.

The first step of our ChIP method differs slightly whether we are using tissue or cell samples. For tissue we homogenize 50 mg of tissue in 500 μl of RPMI Media 1640 (Thermo, 11875093). To this we will add methanol-free formaldehyde (Thermo, 28908) to a final concentration of 1.4%. If ChIP is to be performed on cells in culture, we will add methanol-free formaldehyde directly to the media such that the final concentration in the cell culture well or plate is 1.4%. The cells or tissue is then incubated with the formaldehyde at 37 °C for 5 minutes. After 5 minutes quench the reaction using 1 M glycine solution (glycine in water) such that the final concentration of glycine is 125 mM. Centrifuge the tissue/cells at 1000 g for 10 minutes at 4 °C. All the remaining steps of the ChIP protocol should be conducted on ice.

After centrifugation remove the supernatant, and add 500 μl of PBS with protease inhibitors (Sigma 539134; 1:100 dilution).

Resuspend the tissue/cells in PBS with protease inhibitors by pipetting up and down. Centrifuge the tissue/cells again at 800 g for 10 minutes at 4 °C. After centrifugation remove the supernatant, and add SDS lysis buffer (see *MeDIP procedures* section for formulation) with protease inhibitors (Sigma 539134; 1:100 dilution). We add 250 μl of SDS lysis buffer because that is the volume of the tubes used for our ultrasonicator. Sonicate samples such that the chromatin is 100–300 base pairs.

After sonication add ChIP dilution buffer (see above) with protease inhibitors (Sigma 539134; 1:1000 dilution). We generally add enough ChIP dilution buffer to samples to make the final volume 2 ml. To three new 1.7 ml tubes, add 500 μl to two of them (one to be used for antibody pull-down and one for immunoglobulin negative control) and 166.7 μl to the third (this will be the input) of the now sheared and diluted chromatin. The tube with the 166.7 μl amount and the remainder of the sheared chromatin should be stored at this time at −80 °C.

To the tubes with 500 μl of sheared chromatin, add 20 μl Protein A/G PLUS-Agarose beads (Santa Cruz, sc-2003) for 2 hours. This is a preclearing step aimed at removing any nonspecific binding of chromatin to agarose beads. Place the tube on an end-over-end mixer for 2 hours at 4 °C. After 2 hours centrifuge the samples at 1200 RPM for 1 minute at 4 °C. Place the supernatant in a new 1.7 ml tube and discard the remaining beads.

Now that the samples have been precleared, they are ready for the addition of the antibody or immunoglobulin negative control. The quality and volume of the antibody to be used is quite variable and best optimized using knockout mice or some other similar method as a negative control. One good resource for information regarding the quality of histone modification antibodies is a paper by Egelhofer et al. [80]. The volume of antibody to be added is also quite variable and dependent on the affinity and specificity of the antibody. Therefore, this step may require optimization depending on the protein to be immunoprecipitated. Generally, we incubate chromatin with antibodies on an end-over-end mixer overnight at 4 °C.

The following day add 40 μl Protein A/G PLUS-Agarose beads (Santa Cruz, sc-2003), and place the tube on an end-over-end mixer for 3 hours at 4 °C. After 3 hours perform the same wash steps as above for MeDIP: low salt wash, high salt wash, LiCl wash, and TE buffer twice. Additionally, the elution step for ChIP is identical as for MeDIP as above. However, unlike MeDIP following the elution step, the chromatin needs to be uncross-linked. Remove the tubes with the input samples from the freezer and add 333.3 μl water. To uncross-link the chromatin from immunoprecipitated samples, negative controls, and inputs, add 20 μl of 5 M NaCl. After adding the NaCl, leave the samples on a dry bath set for 65 °C overnight. The remaining steps of the protocol describing protein digestion and DNA extraction are identical to MeDIP.

5 Conclusions

In conclusion, cell cultures have been extensively used in the cancer field to explore epigenetic mechanisms modulating proliferation. We propose that brain cell cultures could be very valuable in exploring the involvement of epigenetic mechanism during normal and pathological brain development as well as in the mature brain. It is now apparent that neurons and astrocytes act in partnership to allow the brain to function properly. It is also apparent that these two cell types have very different roles in the brain and therefore the investigation of them separately in enriched in vitro cultures allows for the elucidation of the relative role played by each cell type. DNA methylation and histone acetylation are probably the most studied epigenetic modifications associated with inhibition (the first one) and induction (the latter) of gene transcription. The investigation of changes in these two epigenetic marks induced in neurons and astrocytes by xenobiotics will dramatically advance the understanding to pathological mechanisms occurring in the brain.

This chapter describes the methods for the preparation of cortical neuron and astrocyte primary cultures and the methods to measure DNA methylation (the MeDIP method) and histone acetylation (the ChIP method). These methods can be applied to numerous scientific questions relative to the physiology and pathology of the brain.

Acknowledgments

This work was supported by the Department of Veterans Affairs Merit Review Awards BX001819 (M.G.) and BX004091 (D.P.G.) and National Institutes of Health (R01AA021468), (R01AA022948) (M.G.), (R01AA025035) (D.P.G.). The contents do not represent the views of the US Department of Veterans Affairs or the US Government.

References

1. Rudenko A, Tsai LH (2014) Epigenetic modifications in the nervous system and their impact upon cognitive impairments. Neuropharmacology 80:70–82
2. Itzhak Y, Liddie S, Anderson KL (2013) Sodium butyrate-induced histone acetylation strengthens the expression of cocaine-associated contextual memory. Neurobiol Learn Mem 102:34–42
3. Gupta-Agarwal S, Franklin AV, Deramus T, Wheelock M, Davis RL, McMahon LL, Lubin FD (2012) G9a/GLP histone lysine dimethyltransferase complex activity in the hippocampus and the entorhinal cortex is required for gene activation and silencing during memory consolidation. J Neurosci 32:5440–5453
4. Chatterjee S, Mizar P, Cassel R, Neidl R, Selvi BR, Mohankrishna DV, Vedamurthy BM, Schneider A, Bousiges O, Mathis C et al (2013) A novel activator of CBP/p300 acetyltransferases promotes neurogenesis and extends memory duration in adult mice. J Neurosci 33:10698–10712
5. Hanson JE, La H, Plise E, Chen YH, Ding X, Hanania T, Sabath EV, Alexandrov V, Brunner

D, Leahy E et al (2013) SAHA enhances synaptic function and plasticity in vitro but has limited brain availability in vivo and does not impact cognition. PLoS One 8:e69964

6. Kaas GA, Zhong C, Eason DE, Ross DL, Vachhani RV, Ming GL, King JR, Song H, Sweatt JD (2013) TET1 controls CNS 5-methylcytosine hydroxylation, active DNA demethylation, gene transcription, and memory formation. Neuron 79:1086–1093
7. Rudenko A, Dawlaty MM, Seo J, Cheng AW, Meng J, Le T, Faull KF, Jaenisch R, Tsai LH (2013) Tet1 is critical for neuronal activity-regulated gene expression and memory extinction. Neuron 79:1109–1122
8. Nagy C, Suderman M, Yang J, Szyf M, Mechawar N, Ernst C, Turecki G (2015) Astrocytic abnormalities and global DNA methylation patterns in depression and suicide. Mol Psychiatry 20:320–328
9. Fuchikami M, Morinobu S, Segawa M, Okamoto Y, Yamawaki S, Ozaki N, Inoue T, Kusumi I, Koyama T, Tsuchiyama K, Terao T (2011) DNA methylation profiles of the brain-derived neurotrophic factor (BDNF) gene as a potent diagnostic biomarker in major depression. PLoS One 6:e23881
10. Sales AJ, Biojone C, Terceti MS, Guimaraes FS, Gomes MV, Joca SR (2011) Antidepressant-like effect induced by systemic and intra-hippocampal administration of DNA methylation inhibitors. Br J Pharmacol 164:1711–1721
11. Tulisiak CT, Harris RA, Ponomarev I (2017) DNA modifications in models of alcohol use disorders. Alcohol 60:19–30
12. Nestler EJ (2014) Epigenetic mechanisms of drug addiction. Neuropharmacology 76(Pt B): 259–268
13. Gavin DP, Floreani C (2014) Epigenetics of schizophrenia: an open and shut case. Int Rev Neurobiol 115:155–201
14. von Bartheld CS, Bahney J, Herculano-Houzel S (2016) The search for true numbers of neurons and glial cells in the human brain: a review of 150 years of cell counting. J Comp Neurol 524:3865–3895
15. Wang DD, Bordey A (2008) The astrocyte odyssey. Prog Neurobiol 86:342–367
16. Mulder M (2009) Sterols in the central nervous system. Curr Opin Clin Nutr Metab Care 12:152–158
17. Kaczor P, Rakus D, Mozrzymas JW (2015) Neuron-astrocyte interaction enhance GABAergic synaptic transmission in a manner dependent on key metabolic enzymes. Front Cell Neurosci 9:120
18. Soueid J, Nokkari A, Makoukji J (2015) Techniques and methods of animal brain surgery: perfusion, brain removal, and histological techniques. In: Kobeissy FH (ed) Brain neurotrauma: molecular, neuropsychological, and rehabilitation aspects. Taylor & Francis Group, LLC, Boca Raton
19. Neal M, Richardson JR (2018) Epigenetic regulation of astrocyte function in neuroinflammation and neurodegeneration. Biochim Biophys Acta 1864:432–443
20. Zorec R, Parpura V, Verkhratsky A (2018) Astroglial vesicular network: evolutionary trends, physiology and pathophysiology. Acta Physiol (Oxf) 222. https://doi.org/10.1111/apha.12915. Epub 2017 Aug 3
21. Ullian EM, Sapperstein SK, Christopherson KS, Barres BA (2001) Control of synapse number by glia. Science 291:657–661
22. Song H, Stevens CF, Gage FH (2002) Astroglia induce neurogenesis from adult neural stem cells. Nature 417:39–44
23. McGann JC, Oyer JA, Garg S, Yao H, Liu J, Feng X, Liao L, Yates JR 3rd, Mandel G (2014) Polycomb- and REST-associated histone deacetylases are independent pathways toward a mature neuronal phenotype. Elife 3:e04235
24. Bird A (2007) Perceptions of epigenetics. Nature 447:396–398
25. Goldberg AD, Allis CD, Bernstein E (2007) Epigenetics: a landscape takes shape. Cell 128:635–638
26. Yao B, Christian KM, He C, Jin P, Ming GL, Song H (2016) Epigenetic mechanisms in neurogenesis. Nat Rev Neurosci 17:537–549
27. Luger K, Mader AW, Richmond RK, Sargent DF, Richmond TJ (1997) Crystal structure of the nucleosome core particle at 2.8 A resolution. Nature 389:251–260
28. Henikoff S, Shilatifard A (2011) Histone modification: cause or cog? Trends Genet 27:389–396
29. Berger SL (2002) Histone modifications in transcriptional regulation. Curr Opin Genet Dev 12:142–148
30. Zhang Y, Reinberg D (2001) Transcription regulation by histone methylation: interplay between different covalent modifications of the core histone tails. Genes Dev 15:2343–2360
31. Lachner M, O'Sullivan RJ, Jenuwein T (2003) An epigenetic road map for histone lysine methylation. J Cell Sci 116:2117–2124
32. Santos-Rosa H, Schneider R, Bannister AJ, Sherriff J, Bernstein BE, Emre NC, Schreiber SL, Mellor J, Kouzarides T (2002) Active genes are tri-methylated at K4 of histone H3. Nature 419:407–411
33. Lachner M, O'Carroll D, Rea S, Mechtler K, Jenuwein T (2001) Methylation of histone H3

lysine 9 creates a binding site for HP1 proteins. Nature 410:116–120
34. Fischle W, Wang Y, Jacobs SA, Kim Y, Allis CD, Khorasanizadeh S (2003) Molecular basis for the discrimination of repressive methyl-lysine marks in histone H3 by Polycomb and HP1 chromodomains. Genes Dev 17:1870–1881
35. Sharma RP, Gavin DP, Grayson DR (2010) CpG methylation in neurons: message, memory, or mask? Neuropsychopharmacology 35:2009–2020
36. Robertson KD (2005) DNA methylation and human disease. Nat Rev Genet 6:597–610
37. Gackowski D, Zarakowska E, Starczak M, Modrzejewska M, Olinski R (2015) Tissue-specific differences in DNA modifications (5-Hydroxymethylcytosine, 5-Formylcytosine, 5-Carboxylcytosine and 5-Hydroxymethyluracil) and their interrelationships. PLoS One 10: e0144859
38. Hahn MA, Qiu R, Wu X, Li AX, Zhang H, Wang J, Jui J, Jin SG, Jiang Y, Pfeifer GP, Lu Q (2013) Dynamics of 5-hydroxymethylcytosine and chromatin marks in mammalian neurogenesis. Cell Rep 3:291–300
39. Szulwach KE, Li X, Li Y, Song CX, Wu H, Dai Q, Irier H, Upadhyay AK, Gearing M, Levey AI et al (2011) 5-hmC-mediated epigenetic dynamics during postnatal neurodevelopment and aging. Nat Neurosci 14:1607–1616
40. Bachman M, Uribe-Lewis S, Yang X, Burgess HE, Iurlaro M, Reik W, Murrell A, Balasubramanian S (2015) 5-Formylcytosine can be a stable DNA modification in mammals. Nat Chem Biol 11:555–557
41. Lister R, Mukamel EA, Nery JR, Urich M, Puddifoot CA, Johnson ND, Lucero J, Huang Y, Dwork AJ, Schultz MD et al (2013) Global epigenomic reconfiguration during mammalian brain development. Science 341:1237905
42. Gavin DP, Chase KA, Sharma RP (2013) Active DNA demethylation in post-mitotic neurons: a reason for optimism. Neuropharmacology 75:233–245
43. Gavin DP, Kusumo H, Sharma RP, Guizzetti M, Guidotti A, Pandey SC (2015) Gadd45b and N-methyl-d-aspartate induced DNA demethylation in postmitotic neurons. Epigenomics 7:567–579
44. Chen WG, Chang Q, Lin Y, Meissner A, West AE, Griffith EC, Jaenisch R, Greenberg ME (2003) Depression of BDNF transcription involves calcium-dependent phosphorylation of MeCP2. Science 302:885–889
45. Martinowich K, Hattori D, Wu H, Fouse S, He F, Hu Y, Fan G, Sun YE (2003) DNA methylation-related chromatin remodeling in activity-dependent BDNF gene regulation. Science 302:890–893
46. Jones PL, Veenstra GJ, Wade PA, Vermaak D, Kass SU, Landsberger N, Strouboulis J, Wolffe AP (1998) Methylated DNA and MeCP2 recruit histone deacetylase to repress transcription. Nat Genet 19:187–191
47. Nan X, Ng HH, Johnson CA, Laherty CD, Turner BM, Eisenman RN, Bird A (1998) Transcriptional repression by the methyl-CpG-binding protein MeCP2 involves a histone deacetylase complex. Nature 393:386–389
48. Fuks F, Hurd PJ, Wolf D, Nan X, Bird AP, Kouzarides T (2003) The methyl-CpG-binding protein MeCP2 links DNA methylation to histone methylation. J Biol Chem 278:4035–4040
49. Takizawa T, Nakashima K, Namihira M, Ochiai W, Uemura A, Yanagisawa M, Fujita N, Nakao M, Taga T (2001) DNA methylation is a critical cell-intrinsic determinant of astrocyte differentiation in the fetal brain. Dev Cell 1:749–758
50. Hatada I, Namihira M, Morita S, Kimura M, Horii T, Nakashima K (2008) Astrocyte-specific genes are generally demethylated in neural precursor cells prior to astrocytic differentiation. PLoS One 3:e3189
51. Cheng PY, Lin YP, Chen YL, Lee YC, Tai CC, Wang YT, Chen YJ, Kao CF, Yu J (2011) Interplay between SIN3A and STAT3 mediates chromatin conformational changes and GFAP expression during cellular differentiation. PLoS One 6:e22018
52. Zhang L, He X, Liu L, Jiang M, Zhao C, Wang H, He D, Zheng T, Zhou X, Hassan A et al (2016) Hdac3 interaction with p300 histone acetyltransferase regulates the oligodendrocyte and astrocyte lineage fate switch. Dev Cell 37:582
53. Coppieters N, Dieriks BV, Lill C, Faull RL, Curtis MA, Dragunow M (2014) Global changes in DNA methylation and hydroxymethylation in Alzheimer's disease human brain. Neurobiol Aging 35:1334–1344
54. Phipps AJ, Vickers JC, Taberlay PC, Woodhouse A (2016) Neurofilament-labeled pyramidal neurons and astrocytes are deficient in DNA methylation marks in Alzheimer's disease. Neurobiol Aging 45:30–42
55. Bailey ZS, Grinter MB, VandeVord PJ (2016) Astrocyte reactivity following blast exposure involves aberrant histone acetylation. Front Mol Neurosci 9:64
56. Cantuti-Castelvetri L, Fitzner D, Bosch-Queralt M, Weil MT, Su M, Sen P, Ruhwedel T, Mitkovski M, Trendelenburg G, Lutjohann D et al (2018) Defective cholesterol clearance limits remyelination in the aged central nervous system. Science 359:684–688
57. Dresselhaus E, Duerr JM, Vincent F, Sylvain EK, Beyna M, Lanyon LF, LaChapelle E,

Pettersson M, Bales KR, Ramaswamy G (2018) Class I HDAC inhibition is a novel pathway for regulating astrocytic apoE secretion. PLoS One 13:e0194661

58. Wu X, Chen PS, Dallas S, Wilson B, Block ML, Wang CC, Kinyamu H, Lu N, Gao X, Leng Y et al (2008) Histone deacetylase inhibitors up-regulate astrocyte GDNF and BDNF gene transcription and protect dopaminergic neurons. Int J Neuropsychopharmacol 11:1123–1134
59. Lioy DT, Garg SK, Monaghan CE, Raber J, Foust KD, Kaspar BK, Hirrlinger PG, Kirchhoff F, Bissonnette JM, Ballas N, Mandel G (2011) A role for glia in the progression of Rett's syndrome. Nature 475:497–500
60. Zhang X, Bhattacharyya S, Kusumo H, Goodlett CR, Tobacman JK, Guizzetti M (2014) Arylsulfatase B modulates neurite outgrowth via astrocyte chondroitin-4-sulfate: dysregulation by ethanol. Glia 62:259–271
61. Banker GA, Cowan WM (1979) Further observations on hippocampal neurons in dispersed cell culture. J Comp Neurol 187:469–493
62. Banker GA, Cowan WM (1977) Rat hippocampal neurons in dispersed cell culture. Brain Res 126:397–342
63. James CD, Davis R, Meyer M, Turner A, Turner S, Withers G, Kam L, Banker G, Craighead H, Isaacson M et al (2000) Aligned microcontact printing of micrometer-scale poly-L-lysine structures for controlled growth of cultured neurons on planar microelectrode arrays. IEEE Trans Biomed Eng 47:17–21
64. Harnett EM, Alderman J, Wood T (2007) The surface energy of various biomaterials coated with adhesion molecules used in cell culture. Colloids Surf B Biointerfaces 55:90–97
65. Kaech S, Banker G (2006) Culturing hippocampal neurons. Nat Protoc 1:2406–2415
66. Shimizu S, Abt A, Meucci O (2011) Bilaminar co-culture of primary rat cortical neurons and glia. J Vis Exp 57:3257. https://doi.org/10.3791/3257
67. Brewer GJ, Torricelli JR, Evege EK, Price PJ (1993) Optimized survival of hippocampal neurons in B27-supplemented Neurobasal, a new serum-free medium combination. J Neurosci Res 35:567–576
68. McCarthy KD, de Vellis J (1980) Preparation of separate astroglial and oligodendroglial cell cultures from rat cerebral tissue. J Cell Biol 85:890–902
69. Guizzetti M, Costa LG (1996) Inhibition of muscarinic receptor-stimulated glial cell proliferation by ethanol. J Neurochem 67: 2236–2245
70. Chen J, Zhang X, Kusumo H, Costa LG, Guizzetti M (2013) Cholesterol efflux is differentially regulated in neurons and astrocytes: implications for brain cholesterol homeostasis. Biochim Biophys Acta 1831:263–275
71. Bird AP, Southern EM (1978) Use of restriction enzymes to study eukaryotic DNA methylation: I. The methylation pattern in ribosomal DNA from *Xenopus laevis.* J Mol Biol 118:27–47
72. Cedar H, Solage A, Glaser G, Razin A (1979) Direct detection of methylated cytosine in DNA by use of the restriction enzyme MspI. Nucleic Acids Res 6:2125–2132
73. Hayatsu H, Wataya Y, Kai K, Iida S (1970) Reaction of sodium bisulfite with uracil, cytosine, and their derivatives. Biochemistry 9: 2858–2865
74. Wang RY, Gehrke CW, Ehrlich M (1980) Comparison of bisulfite modification of 5-methyldeoxycytidine and deoxycytidine residues. Nucleic Acids Res 8:4777–4790
75. Harris RA, Wang T, Coarfa C, Nagarajan RP, Hong C, Downey SL, Johnson BE, Fouse SD, Delaney A, Zhao Y et al (2010) Comparison of sequencing-based methods to profile DNA methylation and identification of monoallelic epigenetic modifications. Nat Biotechnol 28:1097–1105
76. Beck S, Rakyan VK (2008) The methylome: approaches for global DNA methylation profiling. Trends Genet 24:231–237
77. Eckhardt F, Lewin J, Cortese R, Rakyan VK, Attwood J, Burger M, Burton J, Cox TV, Davies R, Down TA et al (2006) DNA methylation profiling of human chromosomes 6, 20 and 22. Nat Genet 38:1378–1385
78. Gilmour DS, Lis JT (1984) Detecting protein-DNA interactions in vivo: distribution of RNA polymerase on specific bacterial genes. Proc Natl Acad Sci USA 81:4275–4279
79. Solomon MJ, Varshavsky A (1985) Formaldehyde-mediated DNA-protein crosslinking: a probe for in vivo chromatin structures. Proc Natl Acad Sci USA 82:6470–6474
80. Egelhofer TA, Minoda A, Klugman S, Lee K, Kolasinska-Zwierz P, Alekseyenko AA, Cheung MS, Day DS, Gadel S, Gorchakov AA et al (2011) An assessment of histone-modification antibody quality. Nat Struct Mol Biol 18:91–93
81. Frommer M, McDonald LE, Millar DS, Collis CM, Watt F, Grigg GW, Molloy PL, Paul CL (1992) A genomic sequencing protocol that yields a positive display of 5-methylcytosine residues in individual DNA strands. PNAS 89:1827–1831. https://doi.org/10.1073/pnas.89.5.1827

Chapter 8

The Neurosphere Assay as an In Vitro Method for Developmental Neurotoxicity (DNT) Evaluation

Laura Nimtz, Jördis Klose, Stefan Masjosthusmann, Marta Barenys, and Ellen Fritsche

Abstract

The human developing central nervous system is more vulnerable to the adverse effects of chemical agents than the adult brain. At present, due to the lack of available data on human neurodevelopmental toxicants, there is an urgent need for testing and subsequently regulating chemicals for their potential to interfere with nervous system development. Alternative testing strategies might fill that gap as they allow fast and resource-efficient compound screenings for a variety of neurodevelopmental endpoints. Nervous system development is complex calling for a battery of tests that cover early and late developmental stages and a variety of neurodevelopmental processes. One of such assays is the "neurosphere assay," an in vitro 3D model for developmental neurotoxicity (DNT) evaluation based on human neural progenitor cells. With this assay, one can identify compounds that disturb basic neurodevelopmental processes, such as NPC proliferation, migration, neuronal- and oligodendrocyte differentiation, as well as thyroid hormone (TH)-dependent oligodendrocyte maturation. By including viability and cytotoxicity assays into the workflow, the assays allow the distinction of specific DNT from general cytotoxicity. This chapter explains how the different test methods of the "neurosphere assay," i.e., NPC1–6, are performed and how some of them can be multiplexed in a time- and cost-efficient manner.

Key words Developmental neurotoxicity, Neural progenitor cells, In vitro, Brain development, Neurodevelopment, DNT, Human, Alternative methods, New approach method, NAM

1 General Introduction

Within the last two decades, there has been considerable concern that exposure toward chemicals might be a contributing factor to the increasing incidence of neurodevelopmental disorders in children [1–4]. For few compounds the evidence for causing developmental neurotoxicity (DNT) is clear, whereas for the majority of chemicals this concern hampers a solid scientific basis because they have not been evaluated for their neurodevelopmental toxicity [5, 6].

Laura Nimtz, Jördis Klose, and Stefan Masjosthusmann are contributed equally to this manuscript.

Michael Aschner and Lucio Costa (eds.), *Cell Culture Techniques*, Neuromethods, vol. 145,
https://doi.org/10.1007/978-1-4939-9228-7_8, © Springer Science+Business Media, LLC, part of Springer Nature 2019

The main reason for this data gap lies in the resource intensity of the current guideline studies: EPA 870.6300 developmental neurotoxicity (DNT) guideline [7] and the draft OECD 426 guideline [8]. These guidelines are highly demanding with regard to time, money, and animals [5, 9] and are therefore not suited for testing large number of chemicals. Therefore, international researchers have been establishing alternative methods for faster and cheaper DNT evaluation based on in vitro methods. In addition, concepts on how to use and interpret such methods with the final goal of regulatory application have been developed [9–16]. For alternative DNT evaluation, the complex procedure of brain development is disassembled into spatiotemporal neurodevelopmental processes that are necessary for forming a functional brain and can be tested for adverse effects of compounds in in vitro assays [10, 11, 16, 17]. Here, human-based systems are preferred because species differences in toxicokinetics, e.g., due to developmental timing, and/or toxicodynamics might affect responses to compounds [18–24]. In this chapter we describe one of the methods suitable for DNT evaluation, the "neurosphere assay." This assay consists of six individual test methods (NPC1–6) measuring different endpoints, some of which can be multiplexed (Fig. 1) [11]. In the following

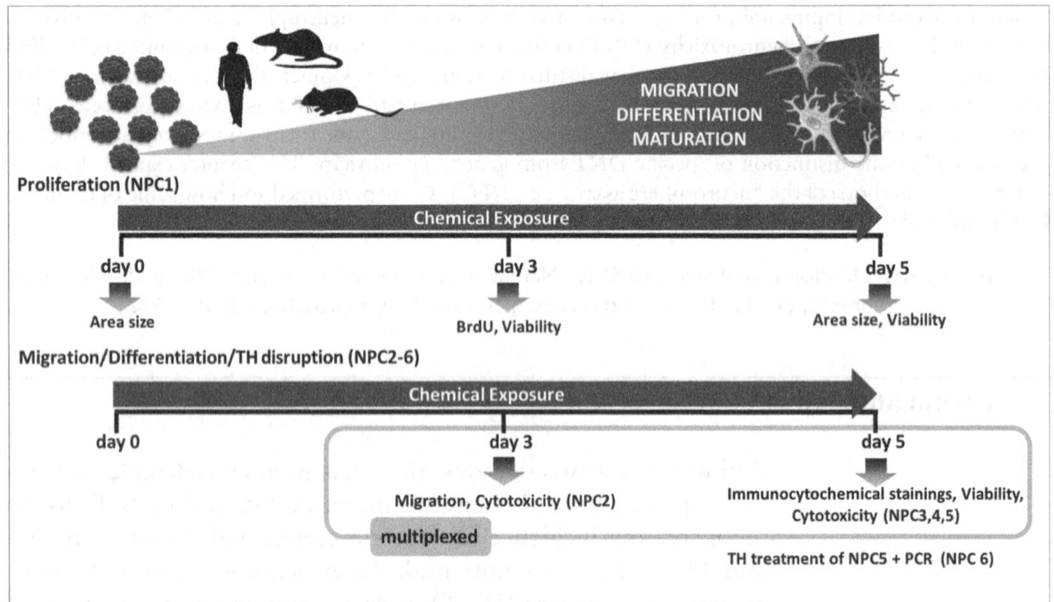

Fig. 1 Experimental setup of the "neurosphere assay". (Adapted from Masjosthusmann et al. [23]). NPC were generated from fetal human brain (Lonza, Verviers, Belgium) or postnatal day (PND)1 mouse and rat brains and cultivated as free-floating neurospheres. The "neurosphere assay" consists of six individual assays: proliferation assay (NPC1), migration assay (NPC2), as well as neuronal differentiation assay (NPC3), neuronal morphology assay (NPC4), oligodendrocyte differentiation assay (NPC5), and thyroid hormone (TH)-dependent oligodendrocyte maturation assay (NPC6) for measuring different endpoints under chemical exposure. For saving time and resources, the assays NPC2-5 can be multiplexed. Gene expression analyses in the TH disruption assay (NPC6) quantify the oligodendrocyte-specific maturation genes *MBP*, *MOBP*, and *MOG* [18]

paragraphs, a detailed description of the individual endpoint evaluations will be given. These endpoints can also be evaluated with neurospheres generated from rodents, which goes beyond the scope of this chapter [18, 19, 25, 26].

1.1 Cell Culture

Human neurospheres gain size by cultivation in the presence of growth factors. By mechanical passaging with a razor blade (chopping), neurospheres are cut into small cubes, which round up to smaller neurospheres of a uniform size within 1 day in proliferation medium and subsequently again continue to grow in size. By using this method, neurospheres are expanded and cultivated over several months without losing their proliferative capacity.

1.2 Materials

Proliferation medium:

DMEM (Gibco GlutaMAX, Life Technologies GmbH) and Hams F12 (Gibco GlutaMAX, Life Technologies GmbH) 3:1 supplemented with 2% B27 (Life Technologies GmbH). Epidermal growth factor (EGF, Life Technologies GmbH) and recombinant human fibroblast growth factor (FGF, R&D Systems) are dissolved at 10 μg/mL in sterile PBS containing 0.1% BSA and 1 mM DTT and stored at −20 °C. EGF and FGF are diluted in the medium at a final concentration of 20 ng/mL each for human neurospheres. Antibiotics (100× penicillin/streptomycin) are added to a 1× final concentration. Prepared medium containing supplements, growth factors, and antibiotics can be stored up to 2 weeks at 4 °C.

1.3 Methods

All steps need to be performed under sterile conditions.

1.3.1 Culturing of Human Neurospheres

Feed human neurospheres every 2–3 days by replacing half of the medium with fresh proliferation medium.

1.3.2 Expand the Human Neurospheres by Mechanical Chopping

1. To increase growth and survival, neurospheres are chopped the latest when they reach a diameter of 0.7 mm, alternatively once a week.
2. For chopping use a McIlwain tissue chopper, and clean it with terralin.
3. Soak a double-edged razor blade in 100% acetone, and sterilize the sliding table and the chopping arm with terralin.
4. Carefully secure the blade onto the chopping arm. Make sure the blade is parallel to the chopping surface.
5. Check and if necessary set the chopper settings (*see Note 1*).
6. Prepare 10-cm petri dishes for the newly chopped neurospheres by filling them with 20 mL proliferation medium each (*see Note 2*).

7. Transfer the neurospheres with as little medium as possible from the 10-cm petri dish into the middle of an inverted lid of a 6-cm petri dish.
8. Remove the remaining medium with a pipet in order to prevent the neurospheres from moving during the chopping process.
9. Place the dish lid on the chopper, and move the sliding table to the starting position.
10. Turn on the power, and press "reset."
11. When all neurospheres on the lid are chopped, stop and raise the chopping arm, and reposition the table in the starting position.
12. Rotate the dish lid 90°, and repeat steps 10 and 11.
13. When the neurospheres are chopped in the second direction, remove the dish lid from the chopper, and add about 1 mL proliferation medium to the cells.
14. Resuspend the chopped neurospheres by gently pipetting them up and down, and then equally distribute the cell suspension into the prepared new petri dishes.
15. Put the cells back into the cell culture incubator (*see Note 3*).
16. After chopping is complete, clean the chopper with terralin, and eventually discard the razor blade (usually one blade can be used three times each side) (*see Note 4*).
17. Feed human neurospheres every 2–3 days by replacing half of the medium with fresh B27 proliferation medium with growth factors.
18. When the chopped neurospheres gain a sphere diameter of 0.3 mm (usually 2–3 days after chopping), they are used for experiments.

1.4 Notes

Note 1: The blade force should be set straight up, and the chop distance should be set between 0.15 and 0.25 mm depending on the sphere size you wish to recover.

Note 2: Usually two to three new petri dishes are gained from one petri dish of chopped neurospheres.

Note 3: We keep the dishes in the incubator until we need them the next time, either for feeding or for plating an experiment.

Note 4: Make sure the neurospheres are well distributed in the petri dish to avoid aggregation.

2 Proliferating Neurospheres

2.1 The Proliferation Assay (NPC1)

A fundamental neurodevelopmental key event (KE) is cell proliferation. Disturbance of proliferation during brain development can lead to smaller brains as an adverse outcome in vivo [27]. The NPC1 assay measures neural proliferation by using primary hNPC of fetal origin (Lonza, Verviers, Belgium) grown as neurospheres in 3D. Proliferation can be studied by measuring the increase in sphere diameter (by area) over 5 days in vitro (DIV) using phase-contrast microscopy [19, 25, 28–32] and/or by measuring BrdU incorporation after 3 DIV using a luminescence-based BrdU assay (Roche) and a luminometer (Fig. 2) [25, 28]. Sphere diameter is measured with ImageJ and change in diameter monitored for each individual sphere. The same setup is used for the BrdU assay, where BrdU incorporation into the DNA of hNPC is measured by using a luminometer. The endpoint-specific control for this assay is withdrawal of growth factors significantly reducing hNPC proliferation. Because measuring the area in individual spheres is quick, easy, and inexpensive, it can be used as a first tier for a first-hit identification. Yet, the direct proliferation assay, quantifying BrdU incorporation into the DNA, is subsequently used for final assessment of disturbance of NPC proliferation by substances.

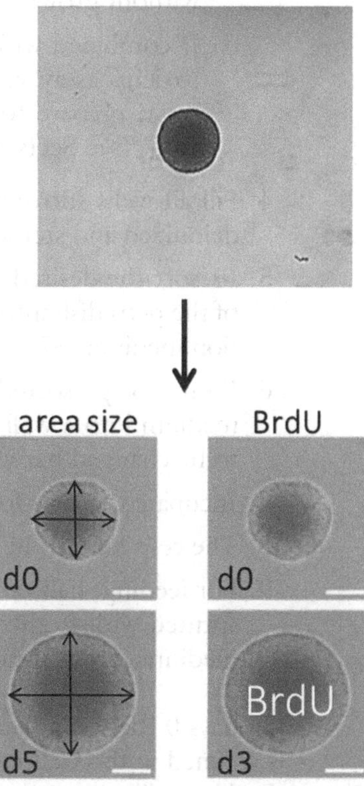

Fig. 2 Proliferation assay (NPC1). Human neural progenitor cells (NPC) are grown as neurospheres under proliferating conditions. Proliferation is assessed by performing a BrdU Cell Proliferation ELISA (Roche©) on day3 (d3) and/or measuring the increase in sphere area size on day 5 (d5). The latter is a quick and inexpensive test that can be used as a first tier screen assay. Scale bars = 0.15 mm

3 Proliferation Assay (NPC1): Readout – Area Size

3.1 Materials

Proliferation medium (*see above*)

H_2O, deionized and sterile

U-bottom 96-well plate (clear)

3.2 Methods

All steps need to be performed under sterile conditions.

3.2.1 Plating Spheres for NPC1: Readout – Area Size

1. Chop spheres 1–3 days prior to the experiment (*see Note 1*).
2. Equilibrate proliferation medium with and without growth factors to 37 °C, 5% CO_2 for 60 min.
3. Prepare all treatment and control solutions, and add 100 μL to each well of a U-bottom 96-well plate.
   Controls:
   (a) Media control (MC): only proliferation medium.
   (b) Solvent control (SC): proliferation medium with respective solvent.
   (c) Endpoint-specific control (PC): proliferation medium without growth factors.
   (d) If combined with the lactase dehydrogenase (LDH) cytotoxicity assay and/or the CellTiter-Blue (CTB) viability assay, prepare lysis control (LC) and background control (BG) (see Sects. 9 and/or 10).
4. Fill all wells surrounding wells with plated cells with 100 μL deionized and sterile water.
5. Presort the desired amount of spheres (0.3 ± 0.025 mm) out of the petri dish into a new petri dish with 3–5 mL of proliferation medium (37 °C).
6. From the presorted spheres, transfer 1 sphere in 2,5 μL of medium into a well in a U-bottom 96-well plate. The tip has to be changed between different conditions.
7. Incubate the cells for 5 days at 37 °C, 5% CO_2.
8. The cells have to be fed on day 2 or 3.
9. For feeding, half of the medium (50 μL) is removed and substituted with freshly prepared control/treatment proliferation medium (pre-warmed to 37 °C).

3.2.2 Taking Images of Individual Spheres

On days 0 (between 0 and 2 h after plating), 1, 2, and 5, the plate is scanned in the Cellomics ArrayScan VTI (Thermo Fisher) using the bioapplication colocalization V4 of the vHCS-Scan software (V6.6.0; build 8153). This software takes a phase-contrast image of each sphere in its individual well and directly analyzes the sphere area. Alternatively, individual phase-contrast images can be taken with a regular microscope equipped with a camera. Images can be analyzed for sphere area size using ImageJ.

All results in fold of control are pooled from independent experiments, SD and SEM are calculated, and statistical analyses are performed. Data are analyzed with the software GraphPad Prism 6.0 using OneWay ANOVA and Bonferroni's post hoc test. Significance threshold is established at $p < 0.05$.

3.3 Notes

There are no relevant notes for this protocol.

4 Proliferation Assay (NPC1): Readout – BrdU Incorporation

4.1 Materials

Proliferation medium with and without growth factors (*see above*)

H_2O, deionized and sterile

Cell Proliferation ELISA, BrdU (#11669915001, Roche/Sigma, +2 to +8 °C)

1 (Red flip-up cap) BrdU labeling reagent

2 (Red) Fix Denat

3 (Blue) Anti-BrdU-POD

4 (Blue) Antibody dilution solution

5 (Green) Washing buffer

6 (Black) Substrate component A

7 (Yellow) Substrate component B

Accutase (#A1110501; Life Technologies)

U-bottom 96-well plate (clear)

Flat-bottom 96-well plate (black)

4.2 Methods

All steps need to be performed under sterile conditions.

Follow the instructions from NPC1 with the readout "area size" for preparing and plating of neurospheres.

4.2.1 Preparation for Cell Proliferation ELISA

1. Dissolve anti-BrdU-POD (3, blue) in 1.1 mL double dist. water for 10 min and mix thoroughly. Solution can be stored at +2 to +8 °C for several months.
2. Dilute BrdU labeling reagent 1:100 in proliferation medium without growth factors (shortly before use).
3. Pre-warm an aliquot of Accutase at 37 °C for 15 min.
4. Dilute anti-BrdU-POD stock solution (3, blue) 1:100 in antibody dilution solution (4, blue) to prepare anti-BrdU working solution (shortly before use).
5. Dilute washing buffer (5, green) 1:10 in deionized water to prepare washing solution.

6. Dilute substrate component B (7, yellow) 1:100 in substrate component A (6, black) to prepare substrate solution. Prepare substrate solution before washing (see Sect. 4.2.2, *step 15*), and incubate for 5–10 min at 37 °C (protect from light).

4.2.2 Procedure of the Cell Proliferation ELISA

1. Eighteen hours before termination of the experiment, add 10 μL of BrdU labeling dilution (see Sect. 4.2.1, *2*) to each well except for the background for BrdU.
2. Incubate the spheres for additional 18 h at 37 °C, 5% CO_2.
3. After 16 h of incubation, the LDH and the CTB Assays are performed (see Sects. 9 and 10).
4. For sphere dissociation needed for BrdU evaluation, pipette 25 μL pre-warmed Accutase to the edge of each well of a new, black 96-well plate, and incubate for 10 min at 37 °C.
5. Transfer each sphere with 10 μL medium into the Accutase drop of the new plate. Make sure that each sphere is pipetted into the same well position of the new plate.
6. Incubate 10 min at 37 °C.
7. Singularize cells by pipetting up and down (15–20 times) with a 100 μL tip (set the pipette to 50 μL, and use a new set of tips for each treatment).
8. Spread the drop well across the well.
9. Heat the plate with a hairdryer until it is completely dry (there should be at least 10 cm space between the hairdryer and the plate).
10. Add 200 μL Fix Denat solution to each well, and incubate for 45 min at room temperature.
11. Remove fixation solution by flicking off into the sink and tapping the plate on a paper towel.
12. Add 100 μL anti-BrdU-POD working solution (see Sect. 4.2.1, *step 4*) to each well, and incubate for 1.5 h at room temperature.
13. Prepare substrate solution (see Sect. 4.2.1 *bullet 5*) and washing solution (see Sect. 4.2.1, *step 6*) 15 min before the end of the incubation period.
14. Remove solution by flicking off into the sink and tapping the plate on a paper towel.
15. Wash wells three times with 200 μL/well washing solution (see Sect. 4.2.1, *step 6*) for 1 min, and shake the plate horizontally for 10–20 s.
16. Remove washing solution by flicking off into the sink and tapping the plate on a paper towel.

17. Add 100 µL substrate solution (see Sect. 4.2.1, *step 6*) to each well, and incubate 5 min at room temperature. During this time, protect the plate from light.
18. Place the plate without lid into the plate reader for measuring luminescence.
19. Open a protocol (*see below Specification of plate reader protocol for TECAN fluorescence plate reader Infinite M200Pro*) for reading the BrdU assay.
20. Select the wells that you want to measure, and start the measurement.

4.2.3 Specification of TECAN Plate Reader Protocol

Mode	Luminescence
Attenuation	NONE
Integration time	1000 ms
Settle time	0 ms
Plate Thermo Fisher Scientific-Nunclon 96 Flat Bottom Black Polystyrene Cat. No.: 437869/437958 [NUN96fb_LumiNunc FluoroNunc.pdfx]	

4.2.4 Data Export and Evaluation

1. Safe the Excel file with raw data.
2. Copy the raw data into your evaluation sheet.
3. Calculate MEAN of all technical replicate measurements.
4. Subtract MEAN of BrdU from the MEAN of all conditions.
5. Calculate BG subtracted MEAN of each condition as percent of SC.
6. All results in fold of control are pooled from independent experiments, SD and SEM are calculated, and statistical analyses are performed. Data are analyzed with the software GraphPad Prism 6.0 using OneWay ANOVA and Bonferroni's post hoc test. Significance threshold is established at $p < 0.05$.

4.3 Notes

There are no relevant notes for this protocol.

5 Differentiating Neurospheres

5.1 Migration Assay (NPC2)

During cortex development glial cells form a scaffold for neurons to migrate from proliferating niches to their final cortical destination. This migration is a fundamental neurodevelopmental KE since its disturbance leads to alterations in cortex formation, as it occurs, e.g., in methylmercury-exposed children [33].

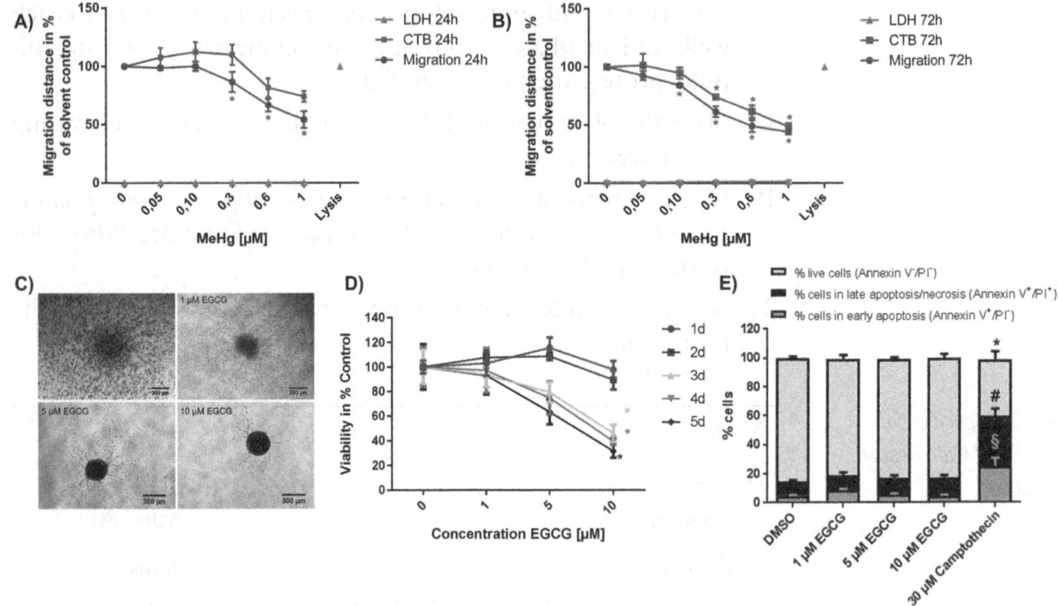

Fig. 3 Migration assay (NPC2) with exemplary migration effect of EGCG treatment. (Taken from Fritsche et al. [36] with permission of *Toxicological Science*). hNPCs, growing as neurospheres in proliferation culture, were plated for migration analyses onto poly-D-lysine/laminin-coated glass slides in presence and absence of MeHgCl. After (**a**) 24 h and (**b**) 72 h, migration distance was measured from the outer sphere rim to the furthest migrated cells at four opposite positions. CellTiter-Blue (CTB) and Lactate Dehydrogenase (LDH) Assays were performed as described previously [25]. (**c**), Neurospheres were plated as described in (**a**) in presence and absence of epigallocatechin gallate (EGCG). After 24 h the migration area was analyzed visually by phase-contrast microscopy, and for better visualization images were subjected to a black/white filter. (**d**), Viability analyses using the CTB Assay were performed on each day up to 5 DIV. (**e**), On day 5, FACS analyses of dissociated hNPC were performed after annexin V/PI staining. As a positive control, spheres were treated with the topoisomerase I inhibitor camptothecin. (**a**, **b**, **d**) $*p \leq 0.05$; (**e**) § $p \leq 0.05$ of annexin^{+} /PI^{-} cells; # $p \leq 0$ 0.05 of annexin^{+} /PI^{+} cells; $*p \leq 0$ 0.05 of live cells

Plating primary human neurospheres of fetal origin on a poly-D-lysine/laminin-coated surface in medium devoid of growth factors initiates radial NPC migration out of the sphere core, thus allowing the study of this essential KE in a plate, i.e., a 96-well plate. In this culture, first migrating cells are nestin- and SOX-2-positive and show radial glia morphology [25, 30, 34, 35]. Moreover, their migration is laminin-integrin dependent (Fig. 3) [36, 37]. In this protocol we present a method to evaluate the ability of hNPCs to migrate out of the sphere core by measuring the maximum covered distance over the coated surface after 3 days in vitro. The migration distance can also be determined at any day up to 5 days by using high-content image analyses (HCA) and the Omnisphero program [37, 38] as described in Sect. 6.2.6.

5.2 Materials

Proliferation medium (see Sect. 1.1)

Flat-bottom 96-well plates (clear)

Poly-D-lysine: Dissolve poly-D-lysine (#P0899, Sigma-Aldrich) at 0.1 mg/mL in sterile water. Store up to 1 month at −20 °C.

Sterile water

One milligram per milliliter laminin (#L2020, Sigma-Aldrich)

Phosphate-buffered saline with $Ca2^+$ and $Mg2^+$ (PBS; Gibco, Life Technologies GmbH), sterile

Differentiation medium: DMEM (Gibco GlutaMAX, Life Technologies GmbH), Hams F12 (Gibco GlutaMAX, Life Technologies GmbH) 3:1 supplemented with 1% of N2 (Life Technologies GmbH), and antibiotic solution (100× penicillin/streptomycin) to 1× final. Store for up to 2 weeks at 4 °C.

PP2 (#P0042-5MG, Sigma-Aldrich)

5.3 Methods

All steps need to be performed under sterile conditions.

Follow the instructions in "Expand the Human Neurospheres by Mechanical Chopping" to prepare neurospheres for plating.

5.3.1 Coating of 24- or 96-Well Plates

1. Fill each well with 300 μL (24-well plate) or 50 μL (96-well plate) poly-D-lysine (PDL), and incubate for 1 h at 37 °C.
2. Wash each well with 500 μL (24-well plate) or 100 μL (96-well plate) sterile deionized water.
3. Afterward, fill every well with 300 μL (24-well plate) or 50 μL (96-well plate) laminin (1:80 dilution of laminin in sterile deionized water), and incubate for 1 h at 37 °C (*see Note 1*).
4. Wash each well with 500 μL (24-well plate) or 100 μL (96-well plate) sterile deionized water.
5. Wash each well with 500 μL (24-well plate) or 100 μL (96-well plate) sterile PBS.

5.3.2 Preparation of Exposure Solutions and Endpoint-Specific Controls

1. Prepare the desired dilution series of the test compound from a stock solution in differentiation medium. The final concentration of the solvent should be the same in all dilutions and not exceed 0.3% (v/v) for DMSO and 1% (v/v) for deionized water or PBS. The maximum solvent concentration for other solvents needs to be tested separately.
2. Prepare the following controls:

   Controls:

   (a) Media control (MC): only differentiation medium.

   (b) Solvent control (SC): differentiation medium with respective solvent.

   (c) Positive control (PC): 10 μM PP2 in differentiation medium. PP2 is a src-kinase inhibitor and therefore serves as an endpoint-specific control for cell migration [34].

(d) If combined with LDH Assay and/or CTB Assay, prepare lysis control (LC) and background control (BG) (see Sects. 9 and/or 10) (*see Note 2*).

3. Fill the needed wells of a coated 96-well plate with 100 μL of the exposure solutions (*see Note 3*).
4. Fill all wells surrounding cell-containing wells with 100 μL deionized and sterile water.

5.3.3 Plating Neurospheres for NPC2

1. Fill a 6-cm petri dish with 5 mL differentiation medium (*see Note 4*).
2. Select neurospheres with a diameter of 0.3 ± 0.025 mm with a 100-μL tip using a binocular microscope, and transfer them into the 6-cm petri dish with N2 differentiation medium (*see Note 5*).
3. Transfer each neurosphere in 2.5 μL medium into the middle of a well of the 96-well plate filled with the experimental solutions. The tip should be changed between different conditions (*see Note 6*).
4. Incubate neurospheres at 37 °C, 5% CO_2.

5.3.4 Cultivation and Image Acquisition

1. After a total incubation time of 3 days at 37 °C, 5% CO_2, take a picture of each sphere including the whole migration area with a phase-contrast microscope.
2. The experiment finishes with the assessment of cell viability and cytotoxicity using the CTB Assay and/or the LDH Assay (see Sects. 9 and/or 10 for description) (*see Note 2*).
3. In case you want to multiplex NPC2 with NPC3, 4, 5, or 6, you should solely perform the LDH Assay, and not the CTB Assay, because the CTB Assay might disturb the differentiation process, while the LDH Assay uses medium that can be collected during the necessary medium change on day 3 (*see descriptions of NPC3–6*).
4. When multiplexed with NPC3–6, then a second measure of migration will automatically be done when the sphere is imaged after 5 days of differentiation by analyzing positions of Hoechst-stained nuclei (*see description of NPC3–5*).

5.3.5 Image Analysis

1. Use ImageJ for measuring four radii of the migration area in perpendicular angles from the edge of the neurosphere to the furthest migrated cells in each phase-contrast image.
2. Calculate the mean of the four radii to obtain the mean migrated distance of each neurosphere.
3. Calculate the mean migrated distance of all neurospheres from each exposure condition.

4. In addition, Omnisphero will calculate migration distance using positions of Hoechst stained nuclei automatically across the whole migration area of each sphere (*see description of NPC3–5*).

All results in fold of control are pooled from independent experiments, SD and SEM are calculated, and statistical analyses are performed. Data are analyzed with the software GraphPad Prism 6.0 using OneWay ANOVA and a Bonferroni's post hoc test. Significance threshold is established at $p < 0.05$.

5.4 Notes

Note 1: Plates can be stored with laminin at 4 °C for up to 7 days.

Note 2: The distinction of compound-specific effects on NPC migration with the "neurosphere assay" and secondary migration effects due to cytotoxicity is an important issue. Recent data reveal that the migration distance and/or the pattern of the migration area defines the dimension of signal of viability assays like the CellTiter-Blue Assay (CTB Assay; Promega), due to cell number relation. Specific effects of DNT compounds, e.g., methylmercury (MeHgCl), on migration without producing cell death can be measured by combining to different cytotoxicity/viability assays at a time. The combination of a cell viability assay and the additional readout of a LDH Assay, which is not directly cell number dependent, indicates compound-specific cytotoxicity effects at two different time points (Fig. 3a, b) [36].

Similarly, epigallocatechin gallate (EGCG) inhibits the migration and adhesion of hNPC on an extracellular matrix by changing the migration pattern and area (Fig. 3c) [36, 37, 39]. The migration in presence of EGCG and the CTB Assay suggests that EGCG reduces cell viability after 3 days (Fig. 8.3d) [36, 37]. However, annexin V-/PI-positive cells, identified with FACS analysis, clearly demonstrate that EGCG does not cause cell death, but reduces the cell area with access to the CTB substrate (Fig. 3e) [36].

Note 3: The plate should be equilibrated at 37 °C and 5% CO_2 until the neurospheres are plated.

Note 4: This step is important to wash out remaining growth factors from the proliferation medium.

Note 5: Make sure the neurospheres for the experiment are round and uniform in size.

Note 6: Make sure that the sphere is placed in the middle of the well.

6 Assays for Differentiation of NPC into Neurons (NPC3/4) and Oligodendrocytes (NPC5)

Differentiation of NPC into different effector cells (neurons, oligodendrocytes, and astrocytes) and the subsequent outgrowth of neurites are major processes of brain development and a prerequisite for the formation of a functional brain. In the "neurosphere assay," the test methods NPC3, 4, and 5 represent the crucial processes of neuronal differentiation, neuronal morphology, and oligodendrocyte differentiation, respectively, and are applied to assess chemical actions on these processes. Therefore, hNPCs are plated as neurospheres on a poly-D-lysine/laminin matrix in a 96-well plate format as described under NPC2. Those cells migrate radially out of the sphere core (NPC2) and differentiate into radial glia, neurons, and oligodendrocytes [25, 26, 35, 40]. In the NPC3 assay, neuronal cells in the migration area are identified by immunocytochemical staining for β(III)tubulin (Fig. 4), while in the NPC4 assay, these β(III)tubulin^{+} neurons are analyzed for their morphological parameters (neurite length, number of branching points, number of neurites). Oligodendrocytes are identified in the migration area by immunocytochemical stainings for O4 (Fig. 8.4) in the NPC5 assay. The test methods NPC2–5 in combination with an assessment of cell viability and cytotoxicity can be multiplexed in one experimental setup.

6.1 Materials

Poly-D-lysine: Dissolve poly-D-lysine (#P0899, Sigma-Aldrich) at 0.1 mg/mL in sterile water. Store up to 1 month at −20 °C.

Sterile water

One milligram per milliliter laminin (#L2020, Sigma-Aldrich)

Phosphate-buffered saline with Ca2^{+} and Mg2^{+} (PBS; Gibco, Life Technologies GmbH), Proliferation medium (see Sect. 1.1)

Differentiation medium (see Sect. 4.1)

Epidermal growth factor (EGF, Life Technologies GmbH)

Bone morphogenetic protein 7 (BMP7, provided by Prof. Pamela Lein, UC Davis; [28])

Paraformaldehyde (12%): Dissolve 12 g PFA in 100 mL phosphate-buffered saline (PBS) and add 5 drops of 1 N NaOH. Heat the solution carefully to 70°–80 °C in a fume hood, and cool to room temperature. Divide into 1-mL aliquots. Store aliquots up to 1 year at −80 °C, and use them freshly.

Goat Serum (#G9023, Sigma-Aldrich)

Triton X-100 (#T8787, Sigma-Aldrich)

bisBenzimide, Hoechst 33258 (#B1155, Sigma-Aldrich)

Mouse IgM anti-O4 (R&D Systems)

Rabbit IgG anti-β(III)tubulin (#T2200, Sigma-Aldrich)

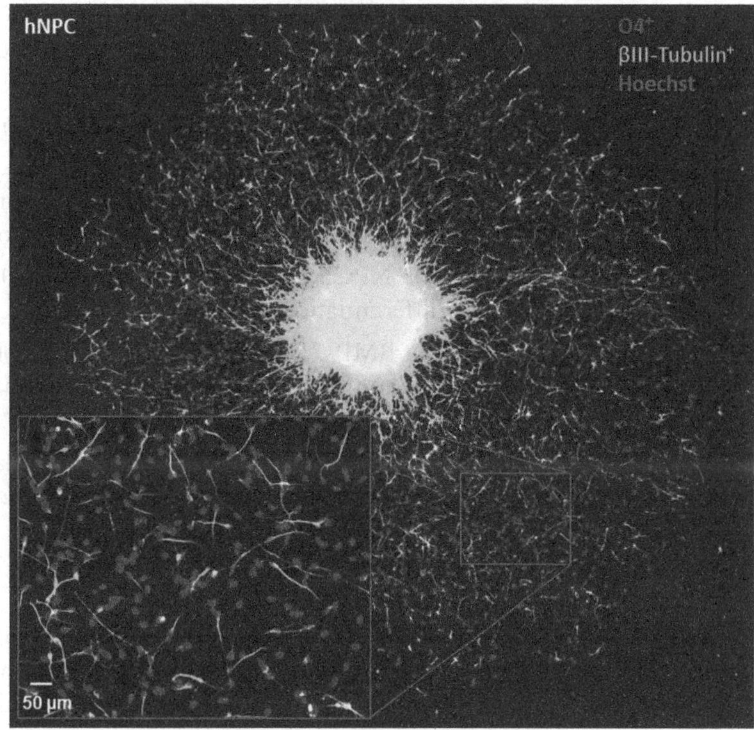

Fig. 4 Immunocytochemical double stainings of neurons and oligodendrocytes differentiated in the migration area of hNPCs. One neurosphere was plated in each well of a 96-well plate in differentiation medium. After 5 days of differentiation, spheres were fixed, and neuronal as well as oligodendrocyte differentiation was assessed by immunocytochemical stainings of migrated and simultaneously differentiated hNPCs for the neuronal marker β(III)-Tubulin+ (green), the oligodendrocyte marker O4+ (pink), and Hoechst+ (blue) for nuclei. One hundred images were taken using the Cellomics Array Scan VTI HCS Reader from Thermo Fisher Scientific to cover the whole sphere. The self-written algorithm Omnisphero (www.omnisphero.com; [38]) assembled the 100 images into one sphere depiction. The inlay is an enlargement of the stained sphere migration area. Scale bar = 50 µm

Alexa 546 IgG (#A11010, Invitrogen)

Alexa 488 IgM (#A11001, Invitrogen)

Flat-bottom 96-well plates (clear)

6.2 Methods

6.2.1 Preparation of Exposure Solutions

1. Prepare the desired dilution series of the test compound from a stock solution in differentiation medium.
   (a) The final concentration of the solvent should be the same in all dilutions and not exceed 0.3% (v/v) for DMSO and 1% (v/v) for deionized water or PBS. The maximum solvent concentration for other solvents needs to be tested separately.

2. Prepare a solvent control by adding solvent to the differentiation medium in the same concentration as used for the exposure solutions.
3. For NPC3 prepare a solution of 20 ng/mL EGF in N2 differentiation medium.
   (a) EGF inhibits the formation of neurons and therefore serves as an endpoint-specific control for neuronal differentiation.
4. For NPC5 prepare a solution of 100 ng/mL BMP-7 in N2 differentiation medium.
   (a) BMP-7 inhibits the formation of $O4^+$ oligodendrocytes and therefore serves as an endpoint-specific control for this endpoint.
5. Fill each well of a coated 96-well plate with 100 μL of the respective exposure solutions (*see Note 2*).

6.2.2 Plating of Neurospheres

1. Fill a 6-cm petri dish with 5 mL N2 differentiation medium (*see Note 3*).
2. Select neurospheres with a diameter of 0.3±0.025 mm with a 100-μL tip using a binocular microscope, and transfer them into the 6-cm petri dish with N2 differentiation medium (*see Note 4*).
3. Transfer each neurosphere in 2.5 μL medium into the middle of a well of the 96-well plate filled with the experimental solutions (*see Note 5*).

6.2.3 Cultivation and Feeding of the Experiments

1. The 96-well plates containing the plated neurospheres are incubated for 3 days at 37 °C and 5% CO_2.
2. On day 3 the cells are fed by removal of 50 μL of the exposure solution and addition of 50 μL freshly prepared exposure solution.
   (a) The removed solution can be used to determine cytotoxicity using the LDH Assay (see Sect. 10).
   (b) The exposure solutions are prepared as described in points 1–4 (Sect. 6.2.1).
3. The neurospheres are incubated for another 2 days until day 5 of differentiation.
4. After a total incubation time of 5 days, the experiment is ended with the assessment of cell viability and cytotoxicity using the CTB Assay and the LDH Assay, respectively (see Sects. 9 and/or 10 for description).

6.2.4 Immunocytochemical (ICC) Stainings of Differentiated Neurospheres

1. Add 12% PFA to each well for achieving a final concentration of 4% PFA/well.
   (a) In case you performed no viability assessment, the final volume of your experiment did not change, and you have to add 50 μL 12% PFA to each well.

(b) In case you performed a viability assessment using the CTB Assay (see Sect. 9 for description), the final volume of your experiment increased, so you have to add 66,6 μL 12% PFA to each well.

(c) In case you performed a cytotoxicity assessment using the LDH Assay (see Sect. 10 for description), the volume of the well decreased, so you have to add 25 μL 12% PFA to each well.

(d) In case you performed a viability and a cytotoxicity assessment, you have to add 33,3 μL 12% PFA to each well.

2. Incubate 30 min at 37 °C.
3. Remove the PFA solution with a residual of 50 μL/well, and discard it into the PFA waste (*see Note 6*).
4. Wash 6 × 3 min by addition and removal of 100 μL PBS (discard to PFA waste).
   (a) After the last washing step, remove 110 μL with 40 μL remaining in the well.
5. Add 10 μL blocking solution to each well.
   (a) Blocking solution: PBS with 50% (v/v) GS.
6. Incubate 15 min at 37 °C.
7. Remove 10 μL from each well.
8. Add 10 μL O4 first antibody solution to each well.
   (a) O4 first antibody solution: 1:40 dilution of mouse IgM anti-O4 in PBS with 10% (v/v) GS.
9. Incubate overnight at 4 °C.
10. Wash 6 × 3 min by addition and removal of 100 μL PBS.
    (a) After the last washing step, remove 110 μL with 40 μL remaining in the well.
11. Add 10 μL Alexa 488 2nd antibody solution to each well.
    (a) Alexa 488 2nd antibody solution: 1:50 dilution of Alexa 488 IgM in PBS with 10% (v/v) GS and 5% (v/v) Hoechst33258.
12. Incubate 30 min at 37 °C.
13. Wash 6 × 3 min by addition and removal of 100 μL PBS each.
    (a) After the last washing step, remove 116 μL with 34 μL remaining in the well.
14. Repeat fixation by addition of 16 μL 12% PFA to a final concentration of 4%.
15. Incubate 30 min at 37 °C.
16. Wash 6 × 3 min by addition and removal of 100 μL PBS each, and discard to PFA waste.

(a) After the last washing step, remove 110 μL with 40 μL remaining in the well.

17. Add 10 μL blocking solution to each well.
    (a) Blocking solution: PBS-T (0.5% (v/v) Triton X-100 in PBS) with 50% (v/v) GS.
18. Incubate 15 min at 37 °C.
19. Add 10 μL β(III)tubulin first antibody solution to each well.
    (a) β(III)tubulin first antibody solution: 1:40 dilution of rabbit IgG anti-β(III)tubulin in PBS-T (0.1% (v/v) Triton X-100 in PBS) with 10% (v/v) GS.
20. Incubate 60 min at 37 °C.
21. Wash 6 × 3 min by addition and removal of 100 μL PBS each.
    (a) After the last washing step, remove 110 μL with 40 μL remaining in the well.
22. Add 10 μL Alexa 546 2nd antibody solution to each well.
    (a) Alexa 546 2nd antibody solution: 1:20 dilution of Alexa 546 IgG in PBS with 10% (v/v) GS and 5% (v/v) Hoechst33258.
23. Incubate 30 min at 37 °C.
24. Wash 6 × 3 min by addition and removal of 100 μL PBS each.
    (a) After the last washing step, add 150 μL PBS to a final volume of 200 μL (*see Note 7*).

6.2.5 Image Acquisition and Analysis

1. Images of each well are acquired with the ArrayScan VTI (Thermo Fisher).
    (a) Objective: LD Plan Neofluar 20×/0.4 (Zeiss).
    (b) Resolution: 552 × 552 pixel (1 pixel = 0.88 μm).
    (c) Channels:
    (i) BGRFR-386-23-filter for Hoechst-stained nuclei
    (ii) BGRFR-549-15-filter for β(III)tubulin-stained neurons
    (iii) BGRFR-488-20-filter for O4-stained oligodendrocytes
    (d) One well is imaged with a total of 100 images per channel.
2. Nuclei are quantified by rescanning all images using the Spot Detector bioapplication colocalization V4 of the vHCS-Scan software (V6.6.0; build 8153).
3. Nuclei coordinates are exported form the vHCS-View software into a comma-separated value file.
4. All images are exported as 16-bit images from the vHCS-View software.
5. High-content image analysis (HCA) is performed using the Omnisphero Platform [38].

6.2.6 Omnisphero

The Omnisphero platform (https://www.omnisphero.com/; [38]) is a freely available high-content image analysis (HCA) tool that was specifically developed for the automated analysis of immunocytochemically stained, heterogeneous cell cultures with varying cell densities as experienced in the "neurosphere assay." Omnisphero is able to identify, count, skeletonize, and position β(III)tubulin^{+} neurons and identify and quantify O4^{+} oligodendrocytes. Thereby, Omnisphero allows an automated assessment of neuronal and oligodendrocyte numbers, cell distributions (neuronal and oligodendrocyte density distribution), and neuronal morphology (neurite length, number of branching points, number of neurites). Other novel endpoints that can be assessed with the Omnisphero software are cell type-specific migration (radial glia, neuronal, and oligodendrocyte migration) as well as migration patterning.

All results in fold of control are pooled from independent experiments, SD and SEM are calculated, and statistical analyses are performed. Data are analyzed with the software GraphPad Prism 6.0 using OneWay ANOVA and a Bonferroni's post hoc test. Significance threshold is established at $p < 0.05$.

6.3 Notes

Note 1: Plates can be stored with laminin at 4 °C for up to 7 days.

Note 2: The plate should be equilibrated at 37 °C and 5% CO_2 until the neurospheres are plated.

Note 3: This step is important to wash out remaining growth factors from the proliferation medium.

Note 4: Make sure the neurospheres for the experiment are round and uniform in size.

Note 5: Make sure that the sphere is placed in the middle of the well.

Note 6: To avoid washing off of cells, the neurospheres should always be covered with at least 50 μL of liquid.

Note 7: ICC images should be taken within the next week. Store plates at 4–8 °C and protected from light.

7 The NPC Thyroid Hormone (TH) Disruption Assay (NPC6)

Maturation of oligodendrocytes plays an important role during brain development. By quantifying mRNA expression of myelin basic protein (*MBP*), myelin-associated oligodendrocytic basic *protein* (*MOBP*), *or* myelin oligodendrocyte glycoprotein (*MOG*) divided by the percentage of O4^{+} cells differentiated after 5 days in the neurosphere migration area (assessed in NPC5), the maturation of O4^{+} oligodendrocytes can be examined. This *MBP/MOBP/MOG*-oligodendrocyte ratio is defined as the oligodendrocyte maturation quotient (Q_M). During NPC development Q_M increases when cultures are exposed to the thyroid hormone (TH) triiodothyronine (T3, [18]). Cellular TH disruption of developing hNPC can be tested by assessment of Q_M with test compounds. Here, cellular TH signaling is disturbed when Q_M (TH + compound) < Q_M (TH) when Q_M is not disturbed by the compound alone in absence of TH. In case the compound affects oligodendrocyte differentiation (NPC5) without lowering Q_M, the compound is an oligodendrocyte toxin rather than a TH disruptor.

7.1 Materials

Flat-bottom 24-well plate (clear), precoated (coating protocol see Sect. 5.1)

Flat-bottom 96-well plate (clear), precoated (coating protocol see Sect. 5.1)

Differentiation medium (see Sect. 5.1)

H_2O, deionized and sterile

Phosphate-buffered saline with $CaCL^2$ and $MgCl^2$ (PBS; Gibco, Life Technologies)

3,3′,5-Triiodo-L-thyronine (T3; Sigma Aldrich, #T2877)

Ninety-six percent ethanol (EtOH)

Hydrogen chloride (HCl)

7.2 Methods

All steps need to be performed under sterile conditions.

7.2.1 Preparation for Cell Plating

1. Chop primary human neurospheres (hNPCs) 3 days before plating (chopping protocol see Sect. 1.3.2).
2. Use primary human neurospheres passage 1–3.

7.2.2 Plating Spheres for NPC6

1. Equilibrate differentiation media to 37 °C, 5% CO_2 for 60 min.
2. Prepare all treatment, co-treatment, and control solutions. Add 5 × 100 μL to each well of a pre-coated (see Sect. 5.1) flat-bottom 96-well plate (for later O4 staining) and 3 × 1 mL to each well of a pre-coated flat-bottom 24-well plate (for later polymerase chain reaction (PCR)) (*see Note 1*).

   Controls:

   (a) Media control (MC): only differentiation media.

   (b) Solvent control (SC): differentiation media with respective solvent.

   Solvent concentrations in the experiments contain % of respective chemical solvent, 0.01% EtOH/HCl and 3 nM T3. For 0.01% EtOH/HCl and T3 with a solvent concentration of 0.01% EtOH/HCl in the stock solutions the following components are required:

   – 100% EtOH/HCl in differentiation medium 1:10 (10% EtOH/HCl)

- Three hundred micrometer T3 in EtOH/HCl 1:10 (30 μM T3 in 100% EtOH/HCl)
- Thirty micrometer T3 in differentiation medium 1:10 (3 μM T3 in 10% EtOH/HCl) (*see Note 2*).

(c) Positive control (PC): BMP-7100 ng/mL.

(d) If combined with LDH Assay and/or CTB Assay, prepare lysis control (LC) and background control (BG) (see Sects. 9 and/or 10).

3. Fill all wells surrounding cell-containing wells with 100 μL (96-well plate) and 1 mL (24-well plate) deionized and sterile water, respectively.
4. Presort 0.3 mm neurospheres in differentiation medium: for each condition (treatment) 35 × 0.3 mm neurospheres are plated (5 × 1 sphere in a 96-well plate for O4 staining and 3 × 10 spheres in a 24-well plate for PCR analysis); furthermore 10 additional neurospheres are needed for controls (BMP-7 as the endpoint-specific control for oligodendrocyte differentiation and Triton X-100 as a lysis control for the cytotoxicity assay).
5. Take up 1 neurosphere in 2 μL differentiation medium from the presorted spheres, and place it into the middle of a well of a 96-well plate.
6. Take up 10 neurospheres in 20 μL differentiation medium from the presorted spheres, and place them into a well of a 24-well plate (*see Note 3*).
7. Incubate these differentiating spheres for 5 days at 37 °C, 5% CO_2.
8. Feed spheres on day 3.
9. For feeding, half of the media (96-well 50 μL, 24-well 500 μL) is removed and substituted with freshly prepared control respective treatment solution (pre-warmed to 37 °C).

7.2.3 Viability Assay and Fixation

1. After 5 days of differentiation, measure viability with the CTB Assay, and afterward fix the spheres in the 96-well plates for later immunocytochemical stainings.
2. For viability/cytotoxicity assays and fixation of spheres, see Sects. 9 and/or 10.
3. After fixation the plate can be stored in the fridge until staining is performed (for a maximum of 1 week, see Sect. 6).

7.2.4 Immunocytochemical Staining and Counting of $O4^+$ Cells and Fluorescence Microscopy

1. For the 96-well oligodendrocyte staining with the O4 antibody, see protocol Sect. 6.2.4.
2. For plate scanning using fluorescence microscopy with the Cellomics ArrayScan, see protocols Sects. 6.2.5 and 6.2.6.

7.2.5 Harvesting of RNA

1. After 5 days of differentiation, remove medium from the wells of the 24-well plate.
2. Add 350 mL RLT buffer (RNeasy Mini Kit, Qiagen).
3. Scratch the cells.
4. Pipette lysates up and down thrice, and fill them into 1.5 mL Eppendorf cups.
5. Store RNA lysates at −80 °C until you purify the RNA.
6. For RNA purification use the RNeasy Mini Kit from Qiagen according to the manufacturer's protocol.
7. The RNA can be stored at −80 °C (*see Note 4*).
8. For reverse transcription the Quantitect Reverse Transcription Kit from Qiagen is used.
9. The real time RT-PCR is carried out with QuantiFast® SYBR® Green PCR (*see Note 5*).
10. Gene product-specific copy numbers were determined as published earlier [18].

7.2.6 Normalization of Gene Expression to the Percentage of Oligodendrocytes

1. Perform both, O4 staining and gene expression experiments in parallel, three times independently.
2. Gene expression (copy numbers *MBP/MOBP/MOG/*10.000 *ßACTIN*) of each treatment condition (pooled technical replicates) of each experiment is divided by its respective percentage of oligodendrocytes from the parallel immunostaining experiment.
3. Gene expression/oligodendrocytes (Q_M) of the solvent control is set to 1 for each experiment, and the Q_M for the TH treatment as the positive control is expressed in fold of solvent control for each experiment.
4. After pooling Q_M from independent experiments, SD and SEM are calculated, and statistical analyses are performed. Data are analyzed with the software GraphPad Prism 6.0 using TwoWay ANOVA and a Tukey's post hoc test. Significance threshold is established at $p < 0.05$.
5. Assess the Q_M of a chemical by itself and in co-treatment with 3 nM T3. If the Q_M of the chemical by itself is not lower than that of the solvent control, but is significantly different from the T3 treatment, the compound is a TH disruptor with regard to TH-induced oligodendrocyte maturation in vitro.

7.3 Notes

Note 1: Plate 96-well plates for O4 staining and 24-well plates for PCR of oligodendrocyte maturation markers on the same day using the same neurosphere stock cultures. Dilution series for the 96-well and the 24-well plates are prepared at once.

Note 2: Always use 10% EtOH/HCl, T3 in 10% EtOH/HCl and the chemical 1:1000 as the highest concentration, a predilution of chemical in solvent might be required.

Note 3: Shake the 24-well plate gently to distribute neurospheres within the well. They need enough space from each other to migrate correctly.

Note 4: RNA of 10 NA0.3 mm neurospheres per sample is used. The RNA content is not measured before each reverse transcription but lies between 10 and 30 ng/μL for ten human NPCs (depending on size and migration area).

Note 5: Primer sequences for the oligodendrocyte maturation gene *MBP/MOBP/MOG* and the housekeeping gene *ßACTIN* are:

Species	Gene	Forward primer 5′–3′	Reverse primer 5′–3′
Human	*MBP*	CAGAGCGTCCGACTATAAATCG	GGTGGGTTTTCAGCGTCTA
	MOBP	ACCCATCTGCCCTCAGACTTA	GCATCTGTAGTTGTTACATCAGC
	MOG	CAATTACCGGAGTGGAGGCA	GTGCATGTCCCCTTACTGCT
	ßACTIN	CAGGAAGTCCCTTGCCATCC	ACCAAAAGCCTTCATACATCTCA

8 Assessment of Cell Viability and Cytotoxicity

The six individual test methods (NPC1–6) described above assess the effect of a chemical on processes that are specific for brain development. The unspecific effect of a compound on general cell viability however will ultimately affect DNT-specific endpoints like NPC proliferation, migration, or differentiation. To distinguish between effects on general cell viability and specific effects on DNT endpoints, cell viability and/or cytotoxicity needs to be determined in the same assay.

Cell viability can be measured with the CTB Assay. This assay provides a fluorometric measure for viable cells as living cells reduce the indicator dye resazurin to the fluorescent metabolite resorufin thus reflecting their metabolic (cellular reductase) capacity [39]. This assay is made for determining viability in a plate reader format. As the cells are incubated with the indicator dye, the experiment needs to be terminated shortly after incubation, i.e., fixed with PFA for subsequent immunocytochemical stainings (NPC3–5).

Cytotoxicity can be assessed with the LDH Assay. This assay is a fluorescent measure of the release of lactate dehydrogenase (LDH) from cells with a damaged membrane. LDH released into the culture medium is proportional to the number of damaged/dead cells and can be measured with an enzymatic reaction that also results in the conversion of resazurin to the fluorescent resorufin [41, 42]. As the LDH Assay uses the supernatant of cultured cells, the assay does not disturb cell cultivation and is thus the method of choice when cells need to be cultivated further like with the migration assay (NPC2) when multiplexed with the differentiation assays (NPC3–5) or the TH disruption assay (NPC6).

9 CellTiter-Blue (CTB) Assay

9.1 Materials

CellTiter-Blue cell viability assay kit (#G8081, Promega; −20 °C)

CellTiter-Blue reagent (protect from light)

Triton X-100 (#T8787, Sigma-Aldrich)

Differentiation medium (see Sect. 5.1)

Proliferation medium (without growth factors; see Sect. 1.1)

Flat-bottom 96-well plate (clear)

9.2 Methods

For the assessment of cell viability using the CTB Assay, a lysis control (LC) and a background control (BG; culture medium without cells) are needed in addition to the endpoint-specific controls of each assay NPC1–6.

1. Culture cells for the desired period and with the desired compound exposure.
2. Thaw and equilibrate CellTiter-Blue reagent to RT for 2 h protected from light.
3. Dilute 10% (v/v) Triton X-100 solution 1:5 in each well of lysis control (2 μL in each well; final concentration 2%).
4. Fill empty wells just with medium.
5. Preincubate 20 min at 37 °C and 5% CO_2.

9.2.1 With Removal of Medium for LDH Assay in the Differentiation Assays

1. Dilute CellTiter-Blue reagent 1:7.5 in differentiation medium. For one well dilute 11 μL of CellTiter-Blue reagent in 72 μL of differentiation medium.
2. Add 83 μL to each well, and incubate at 37 °C and 5% CO_2 for 2 h.

9.2.2 Without Removal of Medium in the Differentiation Assays

1. Dilute CellTiter-Blue reagent 1:3 in differentiation medium. For one well dilute 11 μL of CellTiter-Blue reagent in 22 μL of differentiation medium.
2. Add 33 μL to each well, and incubate at 37 °C and 5% CO_2 for 2 h.

9.2.3 In the BrdU Assay

1. Dilute CellTiter-Blue reagent 1:3 in proliferation medium without growth factors. For one well dilute 12 μL of CellTiter-Blue reagent in 24 μL of proliferation medium.
2. Add 36 μL to each well, and incubate at 37 °C and 5% CO_2 for 2 h.

9.2.4 Fluorescence Measurement and Data Evaluation

1. Measure fluorescence at an excitation wavelength of 540 nm and emission wavelength of 590 nm.
2. Calculate the MEAN of all replicate measurements.
3. Subtract the MEAN of BG from the MEAN of all conditions.
4. Calculate BG subtracted MEAN of each condition as percent of SC.
5. All results in fold of control are pooled from independent experiments, SD and SEM are calculated, and statistical analyses are performed. Data are analyzed with the software GraphPad Prism 6.0 using OneWay ANOVA and Bonferroni's post hoc test. Significance threshold is established at $p < 0.05$.

9.3 Notes

There are no relevant notes for this protocol.

10 Lactate Dehydrogenase (LDH) Assay

10.1 Materials

CytoTox-ONE Homogeneous Membrane Integrity Assay Kit (#G7891, Promega; −20 °C)

Substrate mix
Assay buffer

Triton X-100 (#T8787, Sigma-Aldrich)

Differentiation medium (see Sect. 5.1)

Proliferation medium (without growth factors; see Sect. 1.1)

Flat-bottom 96-well plate (clear)

10.2 Methods

For the assessment of cytotoxicity using the LDH Assay, the same controls as for the CTB Assay are needed (*see above*).

1. Culture cells for the desired period and with the desired compound exposure.
2. Prepare CytoTox-ONE Reagent as indicated in the supplier's instructions.
3. Dilute 10% (v/v) Triton X-100 solution 1:5 in each well of lysis control (2 µL in each well; final concentration 2%).
4. Incubate 20 min at 37 °C and 5% CO_2.
5. Remove 50 µL of medium, and transfer into a new 96-well plate.
6. Equilibrate medium in new plate to RT for 10 min.
7. Add 50 µL of CytoTox-ONE Reagent (RT) to each well.
8. Incubate at RT for 2 h.

10.2.1 Fluorescence Measurement and Data Evaluation

1. Measure fluorescence at an excitation wavelength of 540 nm and emission wavelength of 590 nm.
2. Calculate the MEAN of all replicate measurements.
3. Subtract the MEAN of BG from the MEAN of all conditions.
4. Calculate BG subtracted MEAN of each condition as percent of LC.
5. All results in percent of LC are pooled from independent experiments, SD and SEM are calculated, and statistical analyses are performed. Data are analyzed with the software GraphPad Prism 6.0 using OneWay ANOVA and Bonferroni's post hoc test. Significance threshold is established at $p < 0.05$.

10.3 Notes

There are no relevant notes for this protocol.

References

1. Bennett D, Bellinger D, Health LB-E et al (2016) Project TENDR: targeting environmental neuro-developmental risks. the TENDR consensus statement. Environ Health Perspect 124:A118–A122
2. Grandjean P, Landrigan P (2006) Developmental neurotoxicity of industrial chemicals. Lancet 368:2167–2178
3. Grandjean P, Landrigan PJ (2014) Neurobehavioural effects of developmental toxicity. Lancet Neurol 13:330–338
4. Schettler T (2001) Toxic threats to neurologic development of children. Environ Health Perspect 109:813–816
5. Crofton KM, Mundy WR, Shafer TJ (2012) Developmental neurotoxicity testing: a path forward. Congenit Anom (Kyoto) 52: 140–146
6. Goldman LR, Koduru S (2000) Chemicals in the environment and developmental toxicity to children: a public health and policy perspective. Environ Health Perspect 108(Suppl 3): 443–448
7. US-EPA (1998) Health effects test guidelines OPPTS 870.6300. Dev Neurotox Study EPA 712-C-98-239
8. OECD (2007) OECD guidelines for the testing of chemicals/section 4: Health effects. Test no. 426: developmental neurotoxicity study. Dev Neurotox study P.26
9. Lein P, Silbergeld E, Locke P et al (2005) In vitro and other alternative approaches to developmental neurotoxicity testing (DNT). Environ Toxicol Pharmacol 19:735–744
10. Bal-Price A, Crofton KM, Leist M et al (2015) International STakeholder NETwork (ISTNET): creating a developmental neurotoxicity (DNT) testing road map for regulatory purposes. Arch Toxicol 89:269–287
11. Bal-Price A, Hogberg HT, Crofton KM et al (2018) Recommendation on test readiness criteria for new approach methods in toxicology: exemplified for developmental neurotoxicity. Altex 35:306–352
12. Bal-Price AK, Coecke S, Costa L et al (2012) Conference report: advancing the science of developmental neurotoxicity (DNT): testing for better safety evaluation. ALTEX 29: 202–215
13. Crofton KM, Mundy WR, Lein PJ et al (2011) Developmental neurotoxicity testing: recommendations for developing alternative methods for the screening and prioritization of chemicals. ALTEX 28:9–15
14. Fritsche E, Crofton KM, Hernandez AF et al (2017) OECD/EFSA workshop on developmental neurotoxicity (DNT): the use of non-animal test methods for regulatory purposes. ALTEX 34:311–315
15. Fritsche E, Grandjean P, Crofton KM et al (2018) Consensus statement on the need for innovation, transition and implementation of developmental neurotoxicity (DNT) testing for regulatory purposes. Toxicol Appl Pharmacol 354:3–6
16. Lein P, Locke P, Goldberg A (2007) Meeting report: alternatives for developmental neurotoxicity testing. Environ Health Perspect 115:764–768
17. Fritsche E (2016) Report on integrated testing strategies for the identification and evaluation of chemical hazards associated with the devel-

opmental neurotoxicity (DNT). In: Report of the OECD/EFSA workshop on developmental neurotoxicity (DNT): the use of non-animal test. OECD Environ Heal Saf Publications Ser Test Assess 242

18. Dach K, Bendt F, Huebenthal U et al (2017) BDE-99 impairs differentiation of human and mouse NPCs into the oligodendroglial lineage by species-specific modes of action. Sci Rep 7:1–11
19. Gassmann K, Abel J, Bothe H et al (2010) Species-specific differential AhR expression protects human neural progenitor cells against developmental neurotoxicity of PAHs. Environ Health Perspect 118:1571–1577
20. Gold LS, Manley NB, Slone TH et al (2005) Supplement to the carcinogenic potency database (CPDB): results of animal bioassays published in the general literature through 1997 and by the national toxicology program in 1997–1998. Toxicol Sci 85:747–808
21. Knight A (2008) Systematic reviews of animal experiments demonstrate poor contributions toward human healthcare. Rev Recent Clin Trials 3:89–96
22. Leist M, Hartung T (2013) Reprint: inflammatory findings on species extrapolations: humans are definitely no 70-kg micel. ALTEX 30: 227–230
23. Masjosthusmann S, Becker D, Petzuch B et al (2018) A transcriptome comparison of time-matched developing human, mouse and rat neural progenitor cells reveals human uniqueness. Toxicol Appl Pharmacol 354:40–55
24. Seok J, Warren HS, Cuenca AG et al (2013) Genomic responses in mouse models poorly mimic human inflammatory diseases. Proc Natl Acad Sci U S A 110:3507–3512
25. Baumann J, Barenys M, Gassmann K, Fritsche E (2014) Comparative human and rat "neurosphere assay" for developmental neurotoxicity testing. In: Costa LG, Davila JC, Lawrence DA, Reed DJ (eds) Current protocols in toxicology, vol 59. Wiley, pp 12.21.1–12.21.24. https://doi.org/10.1002/0471140856.tx1221s59
26. Baumann J, Gassmann K, Masjosthusmann S, DeBoer D, Bendt F, Giersiefer S, Fritsche E (2016) Comparative human and rat neurospheres reveal species differences in chemical effects on neurodevelopmental key events. Arch Toxicol 90:1415–1427
27. Tang H, Hammack C, Ogden SC et al (2016) Zika virus infects human cortical neural progenitors and attenuates their growth. Cell Stem Cell 18:587–590
28. Baumann J, Dach K, Barenys M, Giersiefer S, Goniwiecha J, Lein PJ, Fritsche E (2015) Application of the Neurosphere assay for DNT Hazard assessment: challenges and limitations. Humana Press, Totowa, pp 1–29
29. Gassmann K, Baumann J, Giersiefer S, Schuwald J, Schreiber T, Merk HF, Fritsche E (2012) Automated neurosphere sorting and plating by the COPAS large particle sorter is a suitable method for high-throughput 3D in vitro applications. Toxicol In Vitro 26:993–1000
30. Moors M, Rockel TD, Abel J, Cline JE, Gassmann K, Schreiber T, Schuwald J, Weinmann N, Fritsche E (2009) Human neurospheres as three-dimensional cellular systems for developmental neurotoxicity testing. Environ Health Perspect 117:1131–1138
31. Schreiber T, Gassmann K, Götz C et al (2010) Polybrominated diphenyl ethers induce developmental neurotoxicity in a human in vitro model: evidence for endocrine disruption. Environ Health Perspect 118:572–578
32. Tofighi R, Wan Ibrahim WN, Rebellato P et al (2011) Non-dioxin-like polychlorinated biphenyls interfere with neuronal differentiation of embryonic neural stem cells. Toxicol Sci 124:192–201
33. Choi BH (1989) The effects of methylmercury on the developing brain. Prog Neurobiol 32:447–470
34. Moors M, Cline JE, Abel J, Fritsche E (2007) ERK-dependent and -independent pathways trigger human neural progenitor cell migration. Toxicol Appl Pharmacol 221:57–67
35. Edoff K, Raciti M, Moors M et al (2017) Gestational age and sex influence the susceptibility of human neural progenitor cells to low levels of MeHg. Neurotox Res 32:683–693
36. Fritsche E, Barenys M, Klose J, Masjosthusmann S, Nimtz L, Schmuck M, Wuttke S, Tigges J (2018) Current availability of stem cell-based in vitro methods for developmental neurotoxicity (DNT) testing. Toxicol Sci 165:21–30
37. Barenys M, Gassmann K, Baksmeier C, Heinz S, Reverte I, Schmuck M, Temme T, Bendt F, Zschauer TC, Rockel TD (2017) Epigallocatechin gallate (EGCG) inhibits adhesion and migration of neural progenitor cells in vitro. Arch Toxicol 91:827–837
38. Schmuck MR, Temme T, Dach K et al (2017) Omnisphero: a high-content image analysis (HCA) approach for phenotypic developmental neurotoxicity (DNT) screenings of organoid neurosphere cultures in vitro. Arch Toxicol 91:2017–2028
39. Bal-Price A, Lein PJ, Keil KP, Sethi S, Shafer T, Barenys M, Fritsche E, Sachana M, Meek MEB (2017) Developing and applying the adverse outcome pathway concept for understanding

and predicting neurotoxicity. Neurotoxicology 59:240–255

40. Moors M, Bose R, Johansson-Hague K et al (2012) Dickkopf 1 mediates glucocorticoid-induced changes in human neural progenitor cell proliferation and differentiation. Toxicol Sci 125:488–495

41. TECHNICAL BULLETIN, CellTiter-Blue® cell viability assay, revised 3/16, TB317, Promega, USA

42. TECHNICAL BULLETIN, CytoTox-ONE™ homogeneous membrane integrity assay, revised 5/09, TB306, Promega, USA

Chapter 9

Zebrafish as a Tool to Assess Developmental Neurotoxicity

Keturah G. Kiper and Jennifer L. Freeman

Abstract

The zebrafish (*Danio rerio*) is an emerging biological model system in toxicological studies. The zebrafish is used to fill the gap between various in vitro and mammalian models currently being used to identify mechanisms of developmental neurotoxicity. The high-throughput characteristics that contribute to the strength of this small animal model combined with novel and/or standardized high-throughput technology can be used by researchers to conduct a robust volume of developmental embryonic and larval neurotoxicity assays. Using analytical toxicological methods, dose-response-time relationships can be established for comparison to other research animal model systems and translation to humans. Toxicogenomic and targeted molecular evaluations of xenobiotics can be used to identify pathways of toxicity and linked with phenotypic and behavioral insults to define mechanisms of developmental neurotoxicity using wild-type or transgenic zebrafish. This chapter describes analytical approaches for examining toxicokinetics of xenobiotics in the developing zebrafish, imaging techniques being used to identify phenotypic neurological abnormalities, behavioral assays in embryonic and larval zebrafish, and targeted and -omic approaches to identify molecular targets and pathways of neurotoxicity for an integrated approach to investigate developmental neurotoxicity using the zebrafish model system.

Key words Behavior, Development, Morphology, Neurotoxicology, Transcriptomics, Zebrafish

1 Introduction

Explosion of the chemical engineering industry, nanotechnology industry, and many other industries that produce consumer and industrial products has increased the list of chemicals being used in commerce and has outpaced toxicity assessments of these chemicals [1]. As such, there is a need for high-throughput biological model systems for chemical toxicity testing. Various alternative and complementary model systems are being explored to fulfill this need, and over the past two decades, we have seen establishment of the zebrafish (*Danio rerio*) as a complementary animal model system to mammals for toxicity testing [2].

Zebrafish are complex vertebrates with highly conserved organ systems and metabolic pathways that can be used to assess a wide range of toxicological outcomes, spanning acute toxicity to detailed

Michael Aschner and Lucio Costa (eds.), *Cell Culture Techniques*, Neuromethods, vol. 145,
https://doi.org/10.1007/978-1-4939-9228-7_9,

mechanistic studies [3]. The zebrafish as a biological model has many strengths including being economical, smaller in size relative to other animal models, and having simple husbandry. Furthermore, rapid ex vivo embryonic development with well-defined developmental stages makes this fish an attractive biological model system for developmental toxicology. The optically translucent chorion of the zebrafish embryo is practical for monitoring morphological endpoints at even the earliest embryonic stages, and the rapid embryogenesis contributes to the model's high-throughput capacity. In addition, evaluation of an embryo's bioconcentration and xenobiotic transportation capacity can provide the opportunity to produce more exact estimates of toxicokinetic parameters including rate of uptake, metabolism, and excretion [4].

One of the most notable strengths of the zebrafish is a sequenced genome. The zebrafish genome has over 70% gene homology with humans and over 80% match of conserved pathways related to human diseases [5]. When compared to other mammalian models (e.g., mice, rats, and rabbits) currently used for evaluating developmental toxicity, the zebrafish has a concordance of 55–100%, supporting the use of zebrafish in place of these mammalian models [6]. Beyond the shared homology of their coding sequences, zebrafish also display conserved gene regulatory sequences with mammals [7], including those sequences related to developmental regulatory genes [8].

The zebrafish, contrary to the stark external anatomical differences, shares a high degree of internal anatomical and physiological homology with higher-order vertebrates used in the laboratory. For example, zebrafish have similar cellular structure, cellular signaling pathways, cognitive behavior, visual and olfaction systems, and balance and hearing systems to mammals [9]. There is also evidence identifying similarities in zebrafish developmental brain structural counterparts to the developing mammalian brain [10, 11]. Studies assessing behavior observe strong associations with structural regions of the brain (e.g., the amygdala and habenula) in zebrafish, which can be compared to identical parallels drawn from human neurobehavioral studies [12]. The concordance in brain structure and phenotypic behavioral endpoints between zebrafish and mammals can be paralleled in the development of the neuroectoderm with neurogenesis in the zebrafish, neural plate formation, and neurulation being similar to the general processes observed in other vertebrates [13]. Furthermore, the zebrafish brain completes neurogenesis by 3 days after fertilization, presenting a short time interval for developmental neurotoxicity assessments. The morphology, size, and shape of an embryonic zebrafish's telencephalon, diencephalon, tectum, tegmentum, midbrain-hindbrain boundary, hindbrain, and brain ventricles can be inspected for malformations and determination of significant morphological comparisons to gross anatomy of the brain, eyes, and/or head [14]. Thus, although there are some structural differences between the zebrafish and human brain, multiple striking similarities are shared when it comes to the functional areas for modeling disease states. For example, a past study used dopaminergic nerve terminals to confirm that neuronal degeneration in the striatum of the human brain can be modeled by the zebrafish's ventral telencephalon [14].

Shared gene homology between mammalian models with the zebrafish central nervous system supports newly found evidence of transcriptional changes in response to neurotoxicants. For example, similar expression of basic helix-loop-helix (bHLH) genes such as *zash1a (ascl1a)* and *Mash1*, the respective zebrafish and mouse homologs of the mammalian gene *achaete-scute*, support that there are strong similarities in forebrain cell differentiation and organization [15]. The discovery of similar gene expression patterns and homology supports the notion that similar genetic pathways are involved in neurogenesis of the zebrafish and mammalian brain [16]. This same pattern of similarities has been observed in the early formation of the posterior forebrain and has been documented in a variety of focal ortholog and paralog gene studies [15]. The prominent similarities in gene expression patterns in the zebrafish telencephalon are promising; however, the library of evidence on the functionality of these orthologs is still growing. This striking agreement between zebrafish and mammalian models leads us to infer that recorded observations of the zebrafish's response to neurodevelopmental toxicants can be comparable with the observations of other mammalian models and translated to humans. The similar brain structures and gene expression of homologs, orthologs, and paralogs suggest that both qualitative and quantitative molecular methods can be used to investigate the relationship between phenotypic neurological alterations and genetic mechanisms responsible for the observed toxicity.

As this model's advantages have continued to demonstrate its robust strength of evidence, the application of the zebrafish in chemical toxicity screening and developmental neurotoxicity testing has expanded [3]. Epidemiological studies have highlighted an abundance of causal relationships between developmental exposures to elemental metals and other neurotoxic xenobiotics and potential significantly different phenotypic endpoints; these relationships require further scientific investigation [2, 6, 9]. Current findings conjecture that the zebrafish model is both relevant and useful for investigating developmental neurotoxicity hypotheses. Developmental neurotoxicity studies with elemental metals such as lead are great examples of the power of the zebrafish model [13, 16]. Examples of gene/molecular targets involved in neuronal development include protein kinase C (PKC), the N-methyl-D-aspartate (NMDA) subtype of glutamate receptor, and cell adhesion molecules (CAMs) [such as L1 (humans) or L1.2 paralog—L1.2

ortholog (zebrafish)] [17–19]. Research shows that both acute and chronic exposure to lead can affect both PKC and NMDA subtypes of glutamate receptor; these are widely important to neuronal development in learning and cognitive functions [17]. L1.1 and L1.2, the paralog and ortholog of L1, are major biomarkers of several mechanisms involved in neural development and spinal cord regeneration [18]. Modes of action including differential adhesion, signal transduction, and physical/mechanical effects are all potential targets for developmental neurotoxicants; the presence of this CAM mediates adhesion, neurite extension, neuronal migration, and axon fasciculation. In rodent studies, the lack of expression of L1 produced evidence supporting a causal relationship with extensive neuropathological and abnormal behavioral endpoints [18].

Several attributes of the zebrafish have been suggested as traits that will boost the identification of chemical toxicity mechanisms. High-throughput dynamic biological models, the automation of screening tests, and many other advancements have broadened the application of investigating behavioral, morphological, and transcriptional endpoints either as phenotypic or genotypic measures of toxicity [20]. Integrating these assessments with analytical chemistry methods has enabled the determination of dose-response relationships. Here, analytical methods for establishing a dose-response relationship are described. Next, the imaging, behavioral, and molecular-based methods used to investigate developmental neurotoxicity are summarized.

2 Materials and Methods

2.1 Animal Husbandry and Treatment

There are a number of wild-type zebrafish (*Danio rerio*) strains available (https://zfin.org/action/feature/wildtype-list). Some of the most commonly used wild-type strains include AB, Tupel long fin (TL), Tuebingen (TU), and WIK. The TU line was used by the Wellcome Sanger Institute for the zebrafish sequencing project. In addition, there are numerous transgenic strains available for mechanistic studies of xenobiotics (www.zfin.org). The small low-maintenance biological model has a simple diet that changes based on life stage. The zebrafish can be bred year-round and, under consistent laboratory conditions, have high fecundity. On average the sample count of embryos that can be produced per breeding pair, 100–200, also contributes to the desirability of this animal model for high-throughput toxicity screening studies. The zebrafish strain(s) used in the laboratory are kept separated in aquatic habitats suited to imitate the ratio of light to dark hours, pH, salinity, and temperature of their natural habitat. Zebrafish accrue in rice paddies in Northeast India, Bangladesh, and Nepal and are also known to come from freshwater streams. Based on the studies of pioneering zebrafish laboratories, standards such as time ratios for light and dark periods, water and room temperatures, pH ranges, and conductivity have been set accordingly [21]. The maintenance of these conditions is essential to the successful fecundity of this model; more importantly successful integration of consistent aquatic conditions can assist in controlling random error caused by the normal variability of a clutch or strain. Embryonic health is linked to both maternal and paternal quality of health. Most commonly females are used for breeding once per week, and males are bred one to two times a week for optimal husbandry. Many laboratories have different cycles, but it is generally understood that breeding less than once per month will not result in the production of high quality or quantities. The tank used for mating depends on the sexual maturity of the zebrafish [21]. In addition, some laboratories are now utilizing mass spawning tanks for daily collection of embryos instead of the traditional weekly approach [22]. Fertilization occurs after the eggs are laid (~1 hour after the lights turn on). Once the fertilized embryos are collected and rinsed with filtered fish water, they can be used for experimentation based on study design or raised to maturity to maintain the proper breeding stocks and strain diversity in the laboratory.

2.2 Analytical Assessment of Tissue Dose

Most commonly embryos are submerged in a solution at a chosen treatment concentration during the desired developmental period. Concentration of the treatment solution should be confirmed using a suitable analytical method. In addition, it is important to understand the relationship between the concentration of the xenobiotic treatment solution and the absorbed tissue dose [20]. The toxicokinetics in embryos and larvae are especially crucial to understand, because this is the life stage at which a vast majority of neurogenesis and development of organ systems take place including the establishment of metabolism. Following the xenobiotic exposure period, embryos or larvae are collected and digested for analytical assessment. Uptake, biotransformation, and elimination can be studied using various spectrometry methods, depending on the xenobiotic of interest and the sample nature. Commonly used methods include inductively coupled plasma-mass spectrometry (ICP-MS), atomic absorption spectrometry (AAS), gas chromatography-mass spectrometry (GC-MS), or liquid chromatography-mass spectrometry (LC-MS).

ICP-MS is a hyphenated analytical method that uses temperature as a separation property by heating samples with an ICP torch. Depending on the preceding treatment of the sample and desired analyte, the ICP torch will heat to various element's temperatures of vaporizations. The vaporization of these elements produces ions with an ultimate fate in the mass spectrometer of separation based on mass-to-charge (m/z) ratio, detection, and a report of the counted ions using an electron multiplier [23, 24]. ICP-MS is especially useful for quantifying the dose of heavy metals and isotopes of the same element [25]. Quantitative tissue dose can be calculated when known standards are included in the analysis. ICP-MS analysis includes (1) introduction of the sample into the apparatus initial ionization, (2) ion focusing for the analyte of interest, (3) separation, (4) detection, and (5) collection and illustration of the apparatus. ICP-MS can be completed with treated groups at different concentrations to characterize and confirm uptake and tissue dose. The resulting concentration can then be included in subsequent analysis to see if the response or endpoint expression increased as the concentration of a toxicant measured in the tissue increased [26, 27].

Similar to ICP-MS, metal concentrations can also be measured by AAS from digested tissues collected from whole embryo or larvae or specific tissues to determine where the metal accumulates in the body. For example, cadmium has been determined to accumulate in the gills, liver, skeletal muscle, and brain of zebrafish after initial exposure [28]. In AAS, the analyte must be converted into free atoms by stripping away the solvent, volatilizing the analyte, and dissociating the analyte. Most commonly, atomization is completed by flame atomization or electrothermal atomization. AAS detects the loss of light transmitted through the vaporized sample, and this loss of light is proportional to the number of atoms in the selected assaying system (flame or electrothermal). Ultimately, the experimental question should drive whether ICP-MS or AAS is used as there are differences between the two techniques. For example, ICP-MS usually has a better detection limit and is more economical when considering analysis of many metals in a single sample. In addition, ICP-MS can detect nonmetals such as sulfur and nitrogen.

Mass spectrometry (MS) in tandem with gas chromatography (GC) or liquid chromatography (LC) produces powerful analytical tools for the analysis of a wide range of xenobiotics and their metabolites. GC-MS and LC-MS are based on differential migration as a form of separation of target molecules [29]. Mass spectrometry is used in toxicological analyses to determine the mass-to-charge (m/z) ratio of xenobiotics and their metabolites. The ion sources, which are a part of the MS apparatus, are identified using specific techniques unique to the nature of the sample. More important is the efficiency of the ionizer in regard to sample ionization, which influences the determination of the MS instrument's true analytical sensitivity [30, 31]. For instance, in GC-MS electron ionization (EI), specifically used for volatile and thermodynamically stable compounds, compiles interlaboratory libraries of reproducible fragmentation patterns; when accurately done the likelihood that an unknown compound will be able to be identified is increased [1]. LC-MS uses electrospray ionization (ESI), which is a commonly used ionizer for nonvolatile and thermodynamically liable compounds. ESI provides a softer ionization and no production of fragmented ions. While the use of EI in GC-MS produces an interlaboratory library, the ESI in LC-MS can only contribute to "in-house" spectral libraries [32, 33]. Another important component of MS is the analyzer, which determines the instrument's resolution, accuracy, and range. Typical mass analyzers used in toxicology investigations include quadrupole (2D and 3D fields), ion traps (orbit traps, time of flight (TOF)), and sector (magnetic and electric) [30, 34, 35]. The selection of the analyzer can be based on several factors, from the affordability and robustness to the required resolution and mass range for analyte analysis. The components of a MS technique listed above, along with carefully selected detectors, all contribute to this apparatus' analytical capacity. Owing to the continued improvement of these components (ion sources, mass analyzers, and detectors) and relatively relentless consistency of the MS instrument itself, the combination of sensitivity and selectivity needed for unknown and known analyte analysis can be satisfied by the robust capacity of MS in toxicology [1]. The diverse analytical power of MS comes from its seamless capacity to be integrated into analytical techniques following purification/separation techniques such as GC and LC. Due to the innate analytical versatility of MS and the improved specificity and selectivity, along with several other improved factors that increase the ability for high- and low-throughput screens of unrelated compounds, the applications in toxicology are centralized [1].

GC is used as a separation or purification technique prior to the analysis and detection. GC can incorporate either liquid or polymer stationary phases in conjunction with the gas phase to separate the analyte molecules based on column retention time, which are then ionized and detected by adjoined MS instruments [30]. GC–MS is typically used for analysis of analytes that are volatile and thermodynamically stable compounds. GC–MS can also be coupled with specific ionization techniques such as EI–GC–MS to produce a tool capable of detecting and determining untargeted molecules [30, 33, 36]. The advantages of using GC–MS include efficient separation by GC and resulting high resolution and peak capacity. The disadvantage of using GC–MS is the limitation of the types of samples that can be analyzed (volatile and heat stable), which elicits the need to use chemical derivatization of the sample [33, 37, 38]. While GC–MS is still used as a tool to quantify and/or identify unknown volatile compounds in biological samples, most toxicology laboratories have replaced protocols involving GC–MS with LC–MS, because of the targeted approach for identifying and quantifying unknown compounds [1].

LC–MS is used for the analysis of compounds that are nonvolatile and inherently not thermodynamically stable [30, 37–39].

Due to the switch from GC to LC as a separation technique, non-volatile and heat-labile compounds can be analyzed. Atmospheric pressure ionization (API)–ESI is a source of ionization that makes it possible to purify samples in the condensed phase and then inject the analyte ions directly into the mass spectrometer for analysis [30, 31]. API–ESI is a soft ionization technique and does not include fragmentation but can contribute to an in-house library of previously unknown compounds that are now identifiable [1, 30, 31]. While the exact mechanism for the ionization of samples is not fully understood, this is an efficient mechanism used to follow the purification of a sample in the MS instrument [1]. Currently, there are no known limits for detection in regard to molecular size when using LC–MS [31]. The limitations of LC–MS are linked to the flexibility of the ionizer used in the MS apparatus [33, 37]. The ionizer is capable of multiple protonation and deprotonation events, which translate to a yield of multiple m/z peaks from a single compound. This advantage can improve the precision of the observations during analysis [31]. The switch from GC–MS to LC–MS was an integral change for metabolic studies on xenobiotics that undergo biotransformation by way of phase I and phase II metabolic reactions [30, 40]. LC–MS allows investigators to begin to explore the relationship between xenobiotics that enter the body and the ADME characteristics with numerous studies, demonstrating that the use of LC–MS greatly increases the ability to detect a xenobiotic and its metabolites in biological samples [41].

Regardless of the analytical methods included, it may take some preliminary studies to determine the number of embryos or larvae that will be needed to be pooled in each sample for adequate detection. It is recommended to ensure your sample will be within the limit of detection of the equipment being used. Furthermore, inclusion of a standard curve will enable quantitative dose concentration in the tissue. One potential limitation is the small size of the embryo and larval zebrafish for targeted organ analysis, such as the brain for developmental neurotoxicity studies. Instead, laboratories will commonly rely on whole embryos or larvae. While in the past, only a few laboratories have dissected brain tissue for targeted assessments in larval zebrafish, newer technologies are now making this an achievable option. For example, matrix-assisted laser desorption ionization mass spectrometry imaging (MALDI-MSI) has been used to determine cocaine distribution [20]. These approaches are becoming more refined (such as including laser capture microdissection) and are expected to enhance dose analysis in the developing zebrafish.

2.3 Morphology: Embryo and Larval Imaging

Rapid development in zebrafish embryos and their translucent chorion becomes advantageous characteristics of the model for imaging gross and fine morphological alterations in developmental neurotoxicology studies. A standard developmental neurotoxicity assay may initially include a general assessment of malformations. Typical measurements for morphological responses include yolk sac edema, bent body axis, eye, snout, jaw, pericardial edema, somites, pectoral fin, caudal fin, circulation, pigmentation, swim bladder, and trunk or whole body length. In addition, the translucent chorion permits assessment of brain morphological alterations. For these assessments, zebrafish are collected following the xenobiotic exposure period, rinsed, and anesthetized. Common developmental time points assessed include 24, 72, 120, or 144 hours postfertilization (hpf). Zebrafish are positioned for microscopic imaging, oriented laterally or dorsally (Fig. 1). Brain length is measured from the limit of the hindbrain and spinal cord to the rostral brain. Brain length can be normalized with head length for comparison of relative size among xenobiotic treatments [42]. Visible phenotypic fluctuations are considered morphological responses to xenobiotic exposure. It is recommended to image multiple samples in each xenobiotic treatment (considered as subsamples) in each biological replicate (generally blocked by clutch).

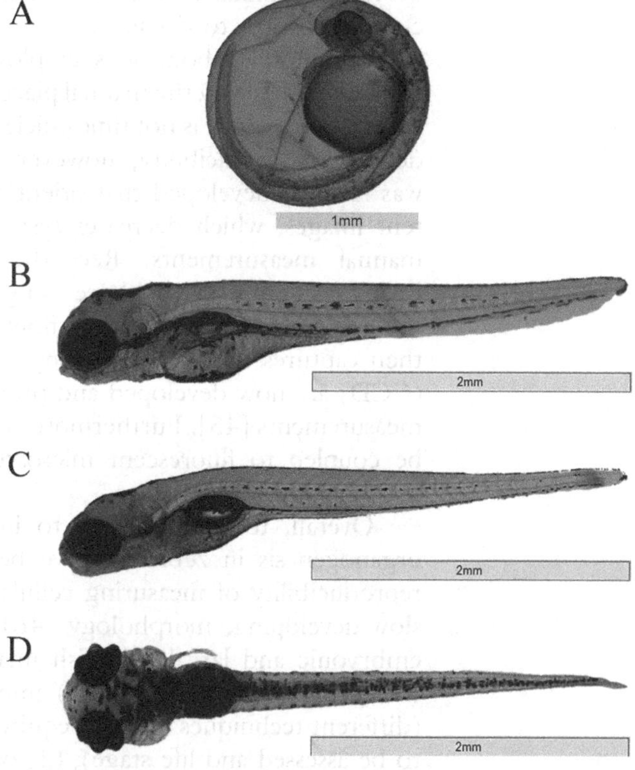

Fig. 1 Zebrafish developmental stages for assessment of abnormal morphological endpoints. (**a**) Wild-type AB zebrafish embryos at 24 hours postfertilization (hpf). (**b**) Wild-type AB larvae at 72 hpf. (**c**) Lateral and (**d**) dorsal view of wild-type AB zebrafish larvae at 120 hpf. Scale bars represent 1 mm (**a**) or 2 mm (**b–d**)

Moreover, specific proteins used as biomarkers of neuronal and other major organ development can be labeled fluorescently and precisely imaged for a more detailed assessment of neurotoxicity. Furthermore, there are a number of transgenic zebrafish lines available to assess specific structures. Transgenic zebrafish lines most commonly contain a green fluorescent protein (GFP) and/or other fluorescent proteins in specific organs and tissues. The known presence of a GFP warrants real-time visualization of the development of specified organs and tissues [43]. The power of transgenic fish that express specific fluorescent proteins from identified classes of neurons can be used in developmental and functional studies to investigate developmental neurotoxicity. For example, the transgenic zebrafish, Tg[isl1: GFP], expresses GFP in specific central nervous system neurons. A study investigating *alx*-expressing neurons illustrated the power of using this specific transgenic model to measure the response of single cell morphology of *alx*-neurons, which are homologs of mammalian CEH-10 homeodomain-containing homolog (*CHX10*) expressed in a subset of developmental spinal neurons [44]. Imaging and analysis of the transgenic zebrafish is similar to as described above but incorporates fluorescent microscopy to visualize.

Commonly laboratories employ a low-throughput imaging approach, involving the manual placement of single or a few zebrafish at a time, which is not time efficient, can increase variability, and decrease reproducibility; however, high-throughput equipment was recently developed that orient the zebrafish and take consistent images, which decreases systematic error due to unprecise manual measurements. Recently, developed high-throughput imaging systems such as those using a sheet of light that irradiates the zebrafish embryos as they flow through a capillary tube and then captures the embryos using a linear charge-coupled device (CCD) are now developed and provide efficiency in collection of measurements [45]. Furthermore, some of these technologies can be coupled to fluorescent microscopes for use with transgenic zebrafish lines.

Overall, techniques used to image brain development and organogenesis in zebrafish have been modified to improve the reproducibility of measuring cellular mechanics and related, but slow-developing, morphology [46]. Factors to consider during embryonic and larval zebrafish imaging for both wild-type and transgenic models include (1) immobilization of the specimen (different techniques may be required depending on the endpoint to be assessed and life stage), (2) orientation (maladjusted specimens can lead to indistinct imaging results and most likely inaccurate and/or incorrect elucidations of the results), and (3) laser irradiation (the effects of the irradiation can subject the specimen to further cellular toxicity by way of excessive z-stack slice acquisition) [47]. Investigations on gene mutations that are suspected or known to affect neurogenesis and brain morphology aligned with morphological assays that measure and observe deviation from the norm can both lead and refine the hypothesized mechanism of neurotoxicity [48].

2.4 Behavioral Tests

Neurobehavioral assessment of zebrafish can include sensory, motor control, or attention endpoints. The analysis of the zebrafish neurobehavior can be used to model functional changes in the brain [49]. The behavior of a zebrafish larva is simplified into spontaneous swimming, startle response to stimuli, and learned behavior [50]. In toxicology studies, behavior assays are used as measurable units of phenotypic expression of physiological deviations from normal. While there are a variety of behavioral tests for adults, larvae, and embryos, this section will focus on the photomotor response (PMR) test most commonly used during embryogenesis and the visual motor response (VMR) test most commonly used for larvae, for which there are well-characterized behavioral responses [51, 52]. These modified behavioral tests are fitted to test conserved behavioral endpoints of the zebrafish larvae including thigmotaxis (wall-hugging), scototaxis (light/dark preference), geotaxis (diving preference), exploration, habituation, and stress- and anxiety-related parameters [53]. These assays can be beneficial when the displayed deviation in normal behavior becomes a phenotypic biomarker of the molecular responses, resulting from xenobiotic toxicity. The behavior analysis provides qualitative data that can support information provided about the xenobiotic mechanism of action [54]. Variability of normal behavior and inter-strain differences should both be considered as practical challenges in the execution of the behavioral assays [3]. A key to having confidence in discriminating against these factors is the accurate selection of quasi-static developmental time periods (Fig. 2).

At 24 hpf, zebrafish eyes have not yet fully developed; however, use of non-visual photosensation permits animals to sense light without the sense of sight. This sense allows researchers to understand nonvisual behaviors elicited by visual wavelengths of light. An animal's nonvisual photosensation response is linked to quickly develop motor circuits. The developed motor circuits involved in this response are directly related to the hindbrain of the zebrafish. Researchers have identified a distinct neuronal set found in the hindbrain. When this set of neurons is inhibited, photomotor response (PMR) behaviors are decreased, suggesting that opsin-based photoreceptors control nonvisual photosensation behaviors [55]. Based on these behaviors, a PMR test was developed to assess four phases associated with embryonic photo-motor responses: (1) pre-stimulus background phase, (2) latency phase, (3) excitation phase, and (4) a refractory phase, which are all characterized by embryo movement. The PMR test is most commonly performed at 24 hpf (Fig. 2).

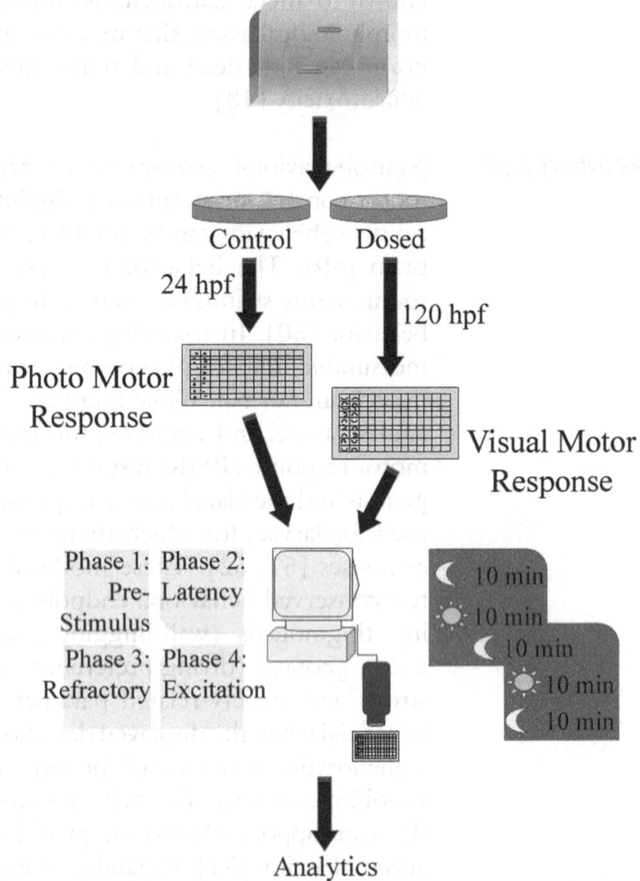

Fig. 2 Developmental behavioral assessments with the zebrafish. Standard methods are developed for assessing the photo-motor response (PMR). The PMR test is usually administered at 24 hpf to evaluate the development of the central nervous system by incorporating pre-stimulus, latency, refractory, and excitation phases. Similarly, standard methods are also developed for the visual motor response (VMR) assay, which uses a white light routine. The VMR test is most commonly assessed at 120 hpf in larval zebrafish. Laboratories can develop their own systems or purchase commercial systems. Commercial systems usually provide software packages that incorporate statistical processing of raw output data obtained from the behavioral tests

During the larval developmental stage, the zebrafish displays vigorous locomotor responses that are well characterized. These responses are stimulated through a visually controlled reflex. During a light/dark routine, larval zebrafish will possess different behavior in light and dark with increased activity during light periods. Larvae will display increasingly inconsistent behavior with a sudden change to light following a dark period [56]. A visual motor response (VMR) test was developed for the phenotype endpoints of larvae behavior ranging from distance and time moved and angular velocity and is most commonly performed at 120 hpf (Fig. 2). These endpoints can be used to draw a relationship between xenobiotic exposures and behavior outcomes such as anxiety, which can be predictive of developmental neurotoxicity. The VMR test has been used for toxicity assessments and validated using predictive rodent behavioral models [53, 57]. For example, a review of several behavioral assessments, posttreatment of polybrominated diphenyl ethers, concluded that early life exposure, 0–144 hpf, altered larval swimming behavior and has lasting effects on zebrafish behavior later in life at 45 days of age [58, 59].

In addition to tracking equipment developed by specific laboratories, there are various zebrafish behavioral equipment available from commercial suppliers (e.g., the Noldus DanioVision or the Viewpoint ZebraBox). These systems often include an imaging platform for multi-well plates to image multiple fish at one time, a camera for video tracking the fish, and temperature control units to maintain consistent temperatures during testing. In addition, software programs enable tracking and analysis of a variety of endpoints.

For the behavior analysis, zebrafish are rinsed following xenobiotic exposure and placed in multi-well plates containing aquarium water or embryo media. It is recommended to randomize placement of fish within the plate to alleviate any location-specific artifacts and to include multiple fish from each xenobiotic treatment (considered as subsamples) with the well-plate serving as the replicate unit. Temperature of the apparatus should be the same at which the embryos/larvae were developing (i.e., usually 26–28 °C) and is pre-warmed before adding the fish. A calibration distance for the software gives the "real-world distance" of, in this case, one well in a multi-well plate. Commercial tracking systems may come with pre-programmed behavior test routines. For example, the Noldus system uses the white light routine for the VMR test. Typically, it is a good idea to add a 10-min step before starting the track to allow the larvae to acclimate to their environment. After the completion of a behavioral trial, track visualization data can be used to collect the larval paths and analyzed in a concentration-response curve. Different software packages provide additional analysis such as the heat map visualization and integrated visualization, which allows viewing of the video of the larvae and their paths, in Noldus EthoVision XT software. While behavioral tests are useful phenotypic endpoints to observe after a toxic insult, both PMR and VMR protocols and implementation of methods can vary in each lab. This lack of overall standardization can affect the quality of statistically significant results. It is also critical to consider variation due to circadian rhythm at different times of the day when planning a time to conduct a behavioral assay. It is important to note that studies to assess the reliability and validity of the developmental behavioral assays are completed and report there are optimal times to conduct these assays to produce results that minimize the amount of error that can come from spontaneous variation in "normal" behavior. It is recommended to review [51] for a full description of factors that can influence outcomes in zebrafish behavioral assays.

2.5 Molecular Mechanisms of Developmental Neurotoxicity

Taking advantage of the sequenced and well-described genome, the zebrafish can be used to link phenotypic and behavioral alterations with molecular mechanisms, driving the observed developmental neurotoxicity (Fig. 3). The zebrafish can be used to further assess mechanisms of previously identified developmental neurotoxicants (e.g., previous studies with metals [13, 60–62]) or of xenobiotics that are not yet identified to be developmental neurotoxicants. Ultimately, the experimental approach will depend on the question being asked, but it is common to evaluate changes in gene expression either using an *-omics* or targeted assessment of mRNA transcripts. mRNA is in constant production by DNA and participates in many dynamic processes including regulatory protein reactions and translation into significant proteins or posttranscriptional alterations, especially during development. In this analysis, the relative abundance of mRNA transcripts in a sample has a metric unit of change known as gene expression.

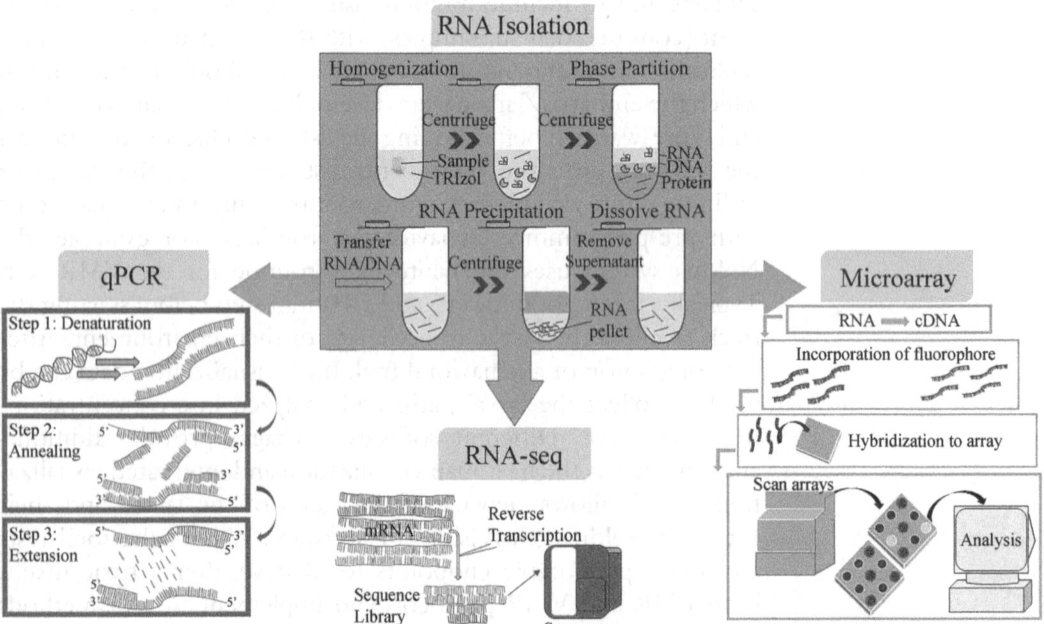

Fig. 3 Methods to assess differential expression of transcripts associated with exposure to developmental neurotoxicants. RNA is isolated from whole zebrafish embryos/larvae or specific tissues through a series of homogenization, partitioning, and precipitation steps. High-quality RNA can then subsequently be used in various targeted (e.g., quantitative PCR (qPCR)) or transcriptome (e.g., RNA-Seq or microarrays) approaches to determine differential expression of transcripts following exposure to neurotoxicants

Gene expression versus mRNA abundance is preferred, due to the dynamic rate of gene transcription that mathematically depends on events where the transcription or turnover is altered (e.g., after a xenobiotic exposure). The resulting inferences that can be made from evaluation of differential gene expression support mechanisms of action of the xenobiotic (the toxicodynamics). Approaches being employed with zebrafish and other animal models include targeted approaches such as quantitative PCR (qPCR) or toxicogenomic approaches (e.g., transcriptomics) to guide future hypotheses on developmental neurotoxicity using microarrays or next-generation sequencing (Fig. 3). The first step for most of these techniques is to isolate high-quality RNA. For zebrafish, this may include isolating RNA from whole embryos or larvae or specific organs, tissues, or cells. It is important to note the storage demands of RNA samples and more specifically mRNA that must be considered when designing experiments. Some transcripts have a half-life of minutes, while others can be stable for over a day [63].

2.5.1 RNA Isolation

RNA molecules are fragile, and that fragility must be considered during the execution of any total RNA isolation protocol. RNAse-free water is used throughout. The risk of contamination and RNA degradation is decreased in most laboratories that typically perform total RNA isolation by frequent cleaning with removal products like *RNAse Away*, designation of a part of the lab for RNA procedures only, by handling RNA samples with gloves at all times, and by keeping the samples on ice during the isolation procedure [64].

In complex procedures the value of high quality, intact RNA, is extremely important. The quality of subsequent outputs is set by many factors, with RNA quality holding more weight [64]. To isolate total RNA, commercially available RNA isolation kits can be used. The commercial kits typically include a chemical denaturant and cleanup that can be used to remove DNA residue and other impurities. As mentioned, RNA is unstable, especially in the presence of RNAses. RNA can be converted to cDNA, a more structurally stable molecule, to assay gene expression in some approaches (explained in detail below). Depending on the endpoint desired for analysis, cDNA synthesis may not be necessary; however, synthesis of high-purity cDNA can be used for microarray analysis, qPCR, or long-term storage.

Following xenobiotic exposure, fish are collected and rinsed. RNA isolation may be completed on a single embryo/larva, pooled samples of multiple embryos/larvae, or dissected tissues/cells dependent on experimental question. The samples are homogenized, and cells lysed in a small volume of a phenol guanidine-based solution (e.g., TRIzol (Invitrogen), QIAzol (Qiagen), or similar product) to rapidly separate fractions of RNA, DNA, and proteins from cells and tissues. For the small samples, a microfuge tube coupled with a pestle works well for homogenization (e.g., pellet pestle, Kimble Kontes). It is recommended to follow the manufacturer recommendations of the phenol guanidine-based solution for total RNA extraction, which generally includes addition of chloroform and centrifugation to separate molecular components, transfer of the upper aqueous layer that contains the RNA into a separate tube, followed by precipitation of the RNA pellet with isopropanol, dehydration with ethanol, and resuspension of the RNA in RNAse-free water. Depending on the quality of RNA that is needed for subsequent analysis, further cleanup of RNA may be needed (e.g., using the Qiagen RNEasy Kit). Due to the nature of the original sample, collection methods, and preparation of the sample, RNA cleanup is needed to remove contaminants and impurities that will interfere with downstream analysis. RNA quality can be assessed using spectrometry, gel electrophoresis, or an Agilent 2100 Bioanalyzer.

An optional consecutive step to RNA isolation is the synthesis of cDNA. Due to the vulnerability of RNA during preparation and analysis and susceptibility to further degradation during storage, cDNA, directly synthesized from the isolated RNA, is an ideal option as an input for many of the gene expression protocols. Commercial kits are available to synthesize single- or double-stranded cDNA, depending on the stability and input needed for downstream applications. Briefly, for single-stranded synthesis using the Invitrogen SuperScript First-Strand Synthesis System, a RNA/primer mixture containing total RNA, dNTP mix, primer, and water is prepared and incubated at 65 °C for 5 min. The primer used in the reaction is dependent on the nature of the sample and relevant experimental conditions. Random hexamer primers are used for nonspecific priming methods and are typically only used when RNA is difficult to copy. In order to maximize the size of cDNA fragments, an empirically calculated ratio of random hexamers to RNA should be determined with a general ratio recommendation for cDNA synthesis being 50 ng: 5 μg. A more specific primer used to make cDNA from total RNA or mRNA is oligo (dT). The oligo (dT) primer provides higher and more complex yields of cDNA when compared to random hexamer primers. For the greatest specificity, gene-specific primers can be used for reverse transcription of mRNA to form cDNA. Gene-specific primers hybridize to the nearest 3′ terminus of the mRNA with its increasing yield of specific cDNA when compared to the oligo (dT) and random hexamer primers. The Invitrogen SuperScript First-Strand Synthesis System uses oligo (dT) as the primer.

After quick cooling of the RNA/primer reaction, a master mix containing $MgCl_2$, DTT, RNAseOUT, and RT buffer is added to the sample and the sample incubated at 42 °C for 2 min. Superscript II RT is then added and incubated at 42 °C for 1 hour.

The reaction is terminated at 70 °C for 5 min, the sample chilled to 4 °C, and then RNAse H is added to each tube and incubated at 37 °C for 20 min. The addition of RNAse H to the sample will degrade the RNA template and leave only the single-stranded cDNA product. cDNA is then isolated and precipitated using a phenol:chloroform:isoamyl alcohol procedure. Optionally, glycogen can be included to assist in precipitation. To increase the stability of the synthesized cDNA, double-stranded cDNA synthesis can be performed instead. For more details and troubleshooting, please refer to [64]. While the referenced protocol used commercially available reagents and kits, there is a wide variety of additional reagents and RNA isolation kits available on the market that will yield high-quality RNA.

2.5.2 Quantitative Polymerase Chain Reaction (qPCR)

Quantitative polymerase chain reaction (qPCR) is a gene expression assay used to quantify target genes relative to a reference gene or genes. Input may either be RNA or cDNA, depending upon the approach. qPCR uses a panel to select a few transcriptional targets. For example, to understand the response of zebrafish neurogenesis to a xenobiotic exposure, the relationship between developing tissue and specialized endothelial cells must be monitored. Neuro-panels and target genes such as vascular endothelial growth factor a (*vegfa*) and fms-related tyrosine kinase 1 (*flt1*) are used to assess changes in embryonic neurogenesis. In the case of the aforementioned targets, these produce changes in spinal cord neurons [65].

The two most common qPCR approaches utilize DNA-binding dyes (e.g., SYBR green) or hydrolysis probes (Taqman). SYBR green binds to double-stranded DNA to detect PCR product, while Taqman uses a fluorogenic probe (e.g., FAM) to detect the PCR product of a specific target gene using the 5′ nuclease activity of Taq DNA polymerase. Taqman is based on degradation of special Taqman probes that contain a reporter and quencher dye. Taqman requires a specific probe, which is targeted to the amplicon. During the elongation phase, the probe is degraded by the 5′–3′ exonuclease activity of the Taq enzyme. The probe separates the reporter from the quencher, which increases fluorescence. This fluorescent signal is then detected by the qPCR machine. The specificity and reproducibility of Taqman are higher than SYBR green, but Taqman requires the synthesis of different probes for different sequences with probes being much more costly than primer pairs used for SYBR green analysis. Taqman probes are usually purchased pre-optimized from commercial suppliers with confirmation to work at or near 100% efficiency, but are species-specific, which for zebrafish can sometimes be a limitation.

SYBR green reaction kits are available through multiple manufacturers and can be used with any forward and reverse primer sets. With the SYBR green reaction, the amount of sample doubles with each cycle and will increase until the reaction is limited by one of the reagents. The amplification of each cycle can be measured by attaining reads after the extension phase. The threshold cycle value (Cq) is the main variable measured in SYBR green reactions, which mathematically reflects the point at which enough target amplicons are significantly accumulated compared to baseline points. The Cq value is less sensitive to amplification differences when observed with post-PCR endpoint-based methods, because the Cq value is determined early in the PCR reaction since this is the point at which the amplification efficiency is highest. One of the most common methods to calculate relative expression uses the ΔΔCq method for quantification. In this method, the difference between the gene of interest and reference gene Cq is calculated (ΔCq). Two samples are then compared by calculating the difference between each of their ΔCq values (ΔΔCq). This method is only accurate when reaction efficiencies are around 100% and assumes consistent expression among all samples for the chosen reference gene. Ultimately, the output of relative quantity will be determined and can be statistically compared among samples to determine differences. It is not recommended to divide all samples by the control treatment or to set the value of every control treatment to "1" as this will then not account for any variation in the control samples in the analysis.

If using SYBR green technologies, primer design is then extremely important to ensure optimization of the PCR reaction, minimizing false positive signals since SYBR binds any double-stranded DNA. Literature reviews and a plethora of computer resources including computational online sites (e.g., http://bioinfo.ut.ee/primer3 or https://www.ncbi.nlm.nih.gov/tools/primer-blast/index.cgi?LINK_LOC=BlastHome) and computer software can be used to guide primer design and permit designs based on user-specific requirements (e.g., primer length, melting temperature, product size ranges, among other conditions). Upon completion, it is suggested to validate the primer pairs in situ to confirm amplification in the correct genome location with the region of interest (e.g., http://rohsdb.cmb.usc.edu/GBshape/cgi-bin/hgPcr). Once primer design is complete, primer pairs can be ordered commercially through various suppliers (e.g., www.idtdna.com or https://www.thermofisher.com/au/en/home/life-science/oligonucleotides-primers-probes-genes/custom-dna-oligos.html). Primers should be validated for PCR amplification, and the product can be visualized using gel electrophoresis to confirm product size.

Since qPCR is based on relative expression, choice of the reference gene(s) (also known as housekeeping genes) is very important. Ideally, a panel of reference genes is used to determine if changes are occurring in the gene target of interest following xenobiotic exposure but is usually limited due to costs. Several commonly used reference genes include glyceraldehyde 3-phosphate dehydrogenase (*gapdh*), *β-actin*, or *mhcI* mRNA, or *28S* or *18S* rRNA. An optimal reference gene will be expressed at a constant level in all cells at all stages of development during all conditions (i.e., not changed by exposure to the xenobiotic of interest).

To assess optimization of the SYBR green reaction, additional considerations such as inclusion of a no template control (i.e., a reaction mixture with only primers, master mix, and water) should be included to ensure there are no signs of nonspecific amplification or contamination. In addition, a standard curve is recommended to be included to determine amplification efficiency of the reaction for use of the ΔΔCq method for quantification. Analysis of the melting curve following the final cycle will allow assessment of the presence of any primer dimers, multiple peaks, and no template amplification. All wells with the same primer pair should exhibit a single sharp peak with no template controls having no observable peaks. Primer dimers are usually observed as a broad peak that appears at low temperature. Additional optimization of the reaction may be needed. Specificity can be increased by decreasing the primer concentration or increasing the annealing temperature. Efficiency can be increased by increasing the primer concentration or decreasing the annealing temperature. If a primer pair is found to behave oddly, it is often easiest to redesign and try out a different set of primers for the gene of interest. Regardless of the desired application of qPCR, acceptable standards for publication of results are to follow the minimum information for publication of quantitative (real-time) PCR experiment (MIQE) guidelines [66].

2.5.3 Microarray Gene Expression Analysis

A commonly used transcriptomic technique is the microarray. The high-throughput capacity of microarrays increases the volume of transcriptomic evaluations of many xenobiotics. Microarrays detect nucleic acids in a collected sample using hybridized probes manufactured on microchips. The biggest advantage of this method is that it can evaluate the transcriptome of experimental treatment groups to illustrate any differential expression of genes [67].

A microarray is composed of short nucleotide sequences of known gene loci spotted onto a slide. Using exclusively designed probes, microarrays can hybridize with cDNA or RNA, depending on the system. The synthesis of fluorescently labelled cDNA/RNA from experimentally treated zebrafish tissue is completed using a fluorescent dye. For a single-color array, all samples are labeled with the same fluorophore, and only one sample is hybridized onto the array. For a two-color array, a different fluorophore is used for the experimental control group and the exposed group with two samples being mixed and co-hybridized to the array. Cy3 and Cy5 are typical fluorophores incorporated into the samples and are sensitive to photodegradation. The reagents used for labeling commonly include a random primer buffer, dNTP mix, and nuclease-free water. A separate labeling reaction is completed for each sample following manufacturer recommendations. Dye incorporation into the sample can be quantified with a spectrophotometer for quality control assessment and to inform subsequent steps. The labeled sample is then hybridized onto the microarray. After hybridization, the microarray is washed and scanned. The microarray should be scanned immediately after the washing step due to the sensitivity of the dye to light and ozone. Following scanning, a multistep analysis approach including image and data processing, data visualization, and statistical comparison of chemically exposed and unexposed groups is completed. Appropriate corrections (e.g., correcting for background fluorescence) and normalization should be applied following manufacturer recommendations of the platform being used. Overall, the level of relative fluorescence provides information about the relative gene expression including upregulation, downregulation, or constitutive expression when a chemically treated sample is compared to the control treatment. It is recommended that calling parameters include both calculated *p*-values and the quantified magnitude of the fold differentiation in gene expression from the control. Most manufacturers of microarray platforms have detailed guidelines on data processing now available, increasing standardization and decreasing interlaboratory variability of experiments. For minimal standards and information on data analysis practices for microarray technology, it is recommended to review the Microarray Quality Control (MAQC) project and the Minimum Information About a Microarray Experiment (MIAME) [67]. Commercial and custom array platforms for zebrafish transcriptome targeting are currently available through Agilent Technologies, ThermoFisher/Affymetrix, and other companies.

2.5.4 RNA-Sequencing

RNA-sequencing (RNA-Seq), a tool used to identify and determine the quantity of RNA in a sample, capitalizes on next-generation sequencing (NGS) methods by way of direct sequencing. A small amount of RNA is required, which is useful when dealing with zebrafish embryos, larvae, and tissue. RNA is isolated from a zebrafish sample and fractioned into mRNA, tRNA, and rRNA to remove the rRNA. Following the typical reverse transcription and fragmentation of the product, ligation of the terminal ends of the cDNA fragments with sequencing adaptors is completed. To retain directional information, the strands of cDNA may be chemically modified. NGS kits sequence the cDNA fragments based on the specific sequencing equipment being used, and the reassembled output is delivered as a list of transcripts. RNA-Seq is being applied in many zebrafish studies to compare transcriptomes of exposed and non-exposed samples. While, historically, RNA-Seq was not as economical as microarrays, costs continue to decrease, making this approach viable for routine application [67]. In addition, while standardization is mainstream for microarrays, standards for RNA-Seq are still developing, especially when considering the expertise required for analysis [68–72].

2.5.5 Bioinformatics: Microarray and RNA-Seq Analysis

The microarray and RNA-Seq transcriptomic approaches both produce large amounts of informative data but can also present difficulties in data processing and analysis. The field of bioinformatics incorporates multiple areas of expertise (biology, mathematics, and computer science) to enable analysis of these -omic data sets. There are a number of software programs currently available for this analysis with standards still in development. Briefly, once gene lists are finalized using either a microarray or RNA-Seq approach, a variety of available software packages (e.g., KEGG, Ingenuity Pathway Analysis, among others) can be utilized to further understand and mine ontology, functions, networks, and pathways in which the genes of interest are involved and associated. The tools available vary among the different software programs but ultimately allow a more informed understanding of the connections and relationships of the genes determined to be differentially expressed in the toxicity test. Overall, this analysis can predict adverse neurotoxicological outcomes and/or identify molecular pathways involved in previously observed neurotoxic insults guiding future mechanistic studies.

3 Discussion/Conclusion

The zebrafish is a strong biological model for studying morphological, behavioral, and molecular endpoints after embryonic, larval, and adult xenobiotic exposure. With each morphological, behavioral, and molecular analyses, a more refined approach can be implemented to further understand the neurotoxic mechanisms behind the chemical insult. Depending on the xenobiotic and the nature of the sample, combinations of the reviewed assessments can be integrated to investigate the toxicokinetics, toxicodynamics, and time-dose-response relationships, while considering the strengths, limitations, and key quality controls of each zebrafish assay to provide informed outcomes of developmental neurotoxicity (Fig. 4).

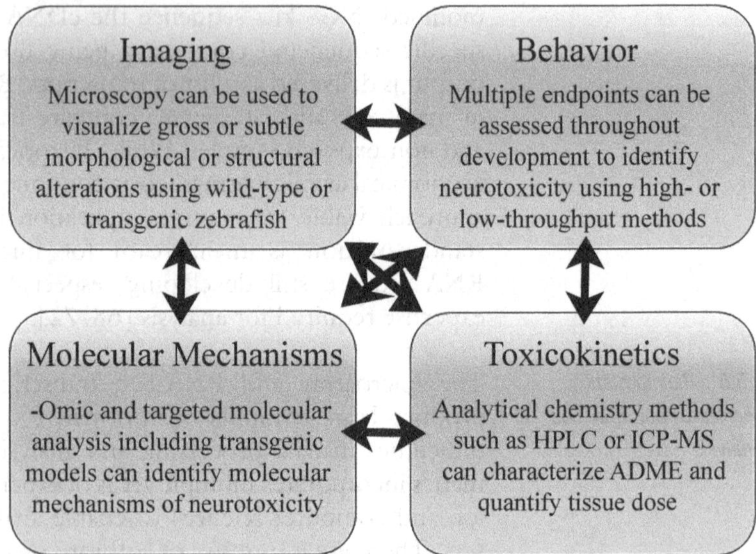

Fig. 4 Tools available for the zebrafish enable integration of multiple endpoints to assess and define mechanisms of developmental neurotoxicity. Low- and high-throughput imaging systems take advantage of ex vivo embryonic development, smaller-sized larvae, and transgenic lines for morphological assessments. Imaging observations can be integrated with behavioral outcomes and molecular analysis to link phenotypic and behavioral insults with molecular mechanisms of developmental neurotoxicity. Furthermore, use of analytical chemistry methods can define ADME in the zebrafish for comparison to other animal research models and humans for translation

References

1. Mbughuni MM, Jannetto PJ, Langman LJ (2016) Mass spectrometry applications for toxicology. eJIFCC 27(4):272–287
2. Giordano G, Costa LG (2012) Developmental neurotoxicity: some old and new issues. ISRN Toxicol 2012, Article ID 814795, 12 pages.
3. Horzmann KA, Freeman JL (2018) Making waves: new developments in toxicology with the zebrafish. Toxicol Sci 163(1):5–12
4. Scholz S, Fischer S, Gündel U, Küster E, Luckenbach T, Voelker D (2008) The zebrafish embryo model in environmental risk assessment—applications beyond acute toxicity testing. Environ Sci Pollut Res 15(5):394–404
5. Howe K et al (2013) The zebrafish reference genome sequence and its relationship to the human genome. Nature 496(7446):498–503
6. Selderslaghs IWT, Van Rompay AR, De Coen W, Witters HE (2009) Development of a screening assay to identify teratogenic and embryotoxic chemicals using the zebrafish embryo. Reprod Toxicol 28(3):308–320
7. Müller F, Blader P, Strähle U (2002) Search for enhancers: teleost models in comparative genomic and transgenic analysis of *cis* regulatory elements. BioEssays 24(6):564–572
8. Kikuta H et al (2007) Genomic regulatory blocks encompass multiple neighboring genes and maintain conserved synteny in vertebrates. Genome Res 17(5):545–555
9. McCollum CW, Ducharme NA, Bondesson M, Gustafsson J-A (2011) Developmental toxicity screening in zebrafish. Birth Defects Res Part C Embryo Today Rev 93(2):67–114
10. Wullimann MF, Mueller T (2004) Erratum: Teleostean and mammalian forebrains contrasted: evidence from genes to behavior. (Journal of Comparative Neurology (2004) 475 (143–162)). J Comp Neurol 478(4): 427–428
11. Wullimann MF (2009) Secondary neurogenesis and telencephalic organization in zebrafish and mice: a brief review. Integr Zool 4(1): 123–133

12. Kalueff AV, Stewart AM, Gerlai R (2014) Zebrafish as an emerging model for studying complex brain disorders. Trends Pharmacol Sci 35(2):63–75

13. Lee J, Freeman JL (2014) Zebrafish as a model for investigating developmental lead (Pb) neurotoxicity as a risk factor in adult neurodegenerative disease: a mini-review. Neurotoxicology 43:57–64

14. Rink E, Rink E, Wullimann MF, Wullimann MF (2002) Development of the catecholaminergic system in the early zebrafish brain: an immunohistochemical study. Dev Brain Res 137:89–100

15. Mueller T, Wullimann MF (2016) Atlas of early Zebrafish brain development, 2nd edn. ELSEVIER B.V, San Diego

16. Wirbisky SE, Weber GJ, Lee JW, Cannon JR, Freeman JL (2014) Novel dose-dependent alterations in excitatory GABA during embryonic development associated with lead (Pb) neurotoxicity. Toxicol Lett 229(1):1–8

17. Marchetti C (2003) Molecular targets of lead in brain neurotoxicity. Neurotox Res 5(3):221–235

18. Bearer CF (2001) L1 cell adhesion molecule signal cascades: targets for ethanol developmental neurotoxicity. Neurotoxicology 22:625–633

19. Chen T, Yu Y, Hu C, Schachner M (2016) L1.2, the zebrafish paralog of L1.1 and ortholog of the mammalian cell adhesion molecule L1 contributes to spinal cord regeneration in adult zebrafish. Restor Neurol Neurosci 34(2):325–335

20. Hamadeh HK, Afshari CA (2004) Toxicogenomics: principles and applications. Wiley-Liss Hoboken, NJ

21. Westerfield M (2007) THE Zebrafish book a guide for the laboratory use of Zebrafish Danio* (Brachydanio) Rerio, 5th Edition (2007): Monte Westerfield: Amazon.com: Books, 5th edn. University of Oregon Press, Eugene

22. Sessa AK, White R, Houvras Y, Burke C, Pugach E, Baker B, Gilbert R, Thomas Look A, Zon LI (2008) The effect of a depth gradient on the mating behavior, oviposition site preference, and embryo production in the zebrafish, Danio rerio. Zebrafish 5(4): 335–339

23. Wilson CA, Bacon JR, Cresser MS, Davidson DA (2006) Lead isotope ratios as a means of sourcing anthropogenic lead in archaeological soils: a pilot study of an abandoned SHETLAND croft*. Archaeometry 48(3): 501–509

24. Nelms SM (ed) (2005) Inductively coupled plasma mass spectrometry handbook. Blackwell Publishing Ltd., Oxford

25. al-Saleh IA, Fellows C, Delves T, Taylor A (1993) Identification of sources of lead exposure among children in Arar, Saudi Arabia. Ann Clin Biochem 30(Pt 2):142–145

26. Heitland P, Köster HD (2006) Biomonitoring of 30 trace elements in urine of children and adults by ICP-MS. Clin Chim Acta 365(1–2):310–318

27. Heitland P, Köster HD (2004) Fast, simple and reliable routine determination of 23 elements in urine by ICP-MS. J Anal At Spectrom 19(12):1552–1558

28. Gonzalez P, Baudrimont M, Boudou A, Bourdineaud J-P (2006) Comparative effects of direct cadmium contamination on gene expression in gills, liver, skeletal muscles and brain of the zebrafish (*Danio rerio*). Biometals 19(3):225–235

29. Ettre LS, Sakodynskii KI (1993) Tswett, M. S. and the discovery of chromatography II: completion of the development of chromatography (1903–1910). Chromatographia 35(5–6):329–338

30. Wen B, Zhu M (2015) Applications of mass spectrometry in drug metabolism: 50 years of progress. Drug Metab Rev 47(1):71–87

31. Glish GL, Vachet RW (2003) The basics of mass spectrometry in the twenty-first century. Nat Rev Drug Discov 2(2):140–150

32. Vogeser M (2003) Liquid chromatography-tandem mass spectrometry – application in the clinical laboratory. Clin Chem Lab Med 41(2):117–126

33. Lynch K, Breaud A (2010) Performance evaluation of three liquid chromatography mass spectrometry methods for broad spectrum drug screening. Clin Chim Acta 411:1474–1481

34. Grebe SKG, Singh RJ (2011) LC-MS/MS in the clinical laboratory – where to from here? Clin Biochem Rev 32(1):5–31

35. Maurer HH, Meyer MR (2016) High-resolution mass spectrometry in toxicology: current status and future perspectives. Arch Toxicol 90(9):2161–2172

36. Chauhan A (2014) GC-MS technique and its analytical applications in science and technology. J Anal Bioanal Tech 5:6

37. Viette V, Hochstrasser D, Fathi M (2012) LC-MS (/MS) in clinical toxicology screening methods. Chim Int J Chem 66(5):339–342

38. Garg U, Zhang YV (2016) Clinical applications of mass spectrometry in drug analysis, methods

and protocols, vol 1383. Humana Press, New York

39. Nair H, Woo F, Hoofnagle AN, Baird GS (2013) Clinical validation of a highly sensitive GC-MS platform for routine urine drug screening and real-time reporting of up to 212 drugs. J Toxicol 2013:329407.
40. Caldwell GW, Yan Z, Tang W, Dasgupta M, Hasting B (2009) Drug, ADME optimization and toxicity assessment in early- and late-phase discovery. Curr Top Med Chem 9(11): 965–980
41. Yuan C, Chen D, Wang S (2015) Drug confirmation by mass spectrometry: identification criteria and complicating factors. Clin Chim Acta 438:119–125
42. Collery RF, Veth KN, Dubis AM, Carroll J, Link BA (2014) Rapid, accurate, and non-invasive measurement of zebrafish axial length and other eye dimensions using SD-OCT allows longitudinal analysis of myopia and Emmetropization. PLoS One 9(10):e110699
43. Kamei M, Weinstein BM (2005) Long-term time-lapse fluorescence imaging of developing zebrafish. Zebrafish 2(2):113–123
44. Higashijima S-I (2008) Transgenic zebrafish expressing fluorescent proteins in central nervous system neurons. Develop Growth Differ 50:407–413
45. Liu L et al (2017) High-throughput imaging of zebrafish embryos using a linear-CCD-based flow imaging system. Biomed Opt Express 8(12):5651–5662
46. Langenberg T, Brand M, Cooper MS (2003) Imaging brain development and organogenesis in zebrafish using immobilized embryonic explants. Dev Dyn 228(3):464–474
47. Renaud O, Herbomel P, Kissa K (2011) Studying cell behavior in whole zebrafish embryos by confocal live imaging: application to hematopoietic stem cells. Nat Protoc 6(12):1897–1904
48. Schier AF et al (1996) Mutations affecting the development of the embryonic zebrafish brain. Development 123:165–178
49. Saverino C, Gerlai R (2008) The social zebrafish: behavioral responses to conspecific, heterospecific, and computer animated fish. Behav Brain Res 191(1):77–87
50. Tierney KB (2011) Behavioural assessments of neurotoxic effects and neurodegeneration in zebrafish. Biochim Biophys Acta Mol basis Dis 1812(3):381–389
51. Kalueff AV (2017) Illustrated zebrafish neurobehavioral glossary. In: The rights and wrongs of zebrafish: behavioral phenotyping of zebrafish. Springer International Publishing, Cham, pp 291–317
52. Kalueff AV et al (2013) Towards a comprehensive catalog of zebrafish behavior 1.0 and beyond. Zebrafish 10(1):70–86
53. Ahmad F, Noldus LPJJ, Tegelenbosch RAJ, Richardson MK (2012) Zebrafish embryos and larvae in behavioural assays. Behaviour 149(10–12):1241–1281
54. Reif DM, Truong L, Mandrell D, Marvel S, Zhang G, Tanguay RL (2016) High-throughput characterization of chemical-associated embryonic behavioral changes predicts teratogenic outcomes. Arch Toxicol 90(6):1459–1470
55. Kokel D et al (2013) Identification of nonvisual photomotor response cells in the vertebrate hindbrain. J Neurosci 33(9):3834–3843
56. MacPhail RC, Brooks J, Hunter DL, Padnos B, Irons TD, Padilla S (2009) Locomotion in larval zebrafish: influence of time of day, lighting and ethanol. Neurotoxicology 30(1):52–58
57. MacPhail RC, Hunter DL, Irons TD, Padilla S (2011) Locomotion and behavioral toxicity in larval zebrafish: background, methods, and data. In: Zebrafish. Wiley, Hoboken, pp 151–164
58. Chou C-T, Hsiao Y-C, Ko F-C, Cheng J-O, Cheng Y-M, Chen T-H (2010) Chronic exposure of 2,2′,4,4′-tetrabromodiphenyl ether (PBDE-47) alters locomotion behavior in juvenile zebrafish (*Danio rerio*). Aquat Toxicol 98(4):388–395
59. Macaulay LJ, Bailey JM, Levin ED, Stapleton HM (2015) Persisting effects of a PBDE metabolite, 6-OH-BDE-47, on larval and juvenile zebrafish swimming behavior. Neurotoxicol Teratol 52(Pt B):119–126
60. Tilton FA, Tilton SC, Bammler TK, Beyer RP, Stapleton PL, Scholz NL, Gallagher EP (2011) Transcriptional impact of organophosphate and metal mixtures on olfaction: copper dominates the chlorpyrifos-induced response in adult zebrafish. Aquat Toxicol 102:205–215
61. Ling XP, Lu YH, Huang HQ (2012) Differential protein profile in zebrafish (Danio rerio) brain under the joint exposure of methyl parathion and cadmium. Environ Sci Pollut Res 19(9):3925–3941
62. Green AJ, Planchart A (2018) The neurological toxicity of heavy metals: A fish perspective. Comp Biochem Physiol C Toxicol Pharmacol 208:12–19
63. Schwanhäusser B et al (2011) Global quantification of mammalian gene expression control. Nature 473(7347):337–342

64. Peterson SM, Freeman JL (2009) RNA isolation from embryonic zebrafish and cDNA synthesis for gene expression analysis. J Vis Exp 30:1–5
65. Wild R et al (2017) Neuronal sFlt1 and Vegfaa determine venous sprouting and spinal cord vascularization. Nat Commun 8:13991
66. Bustin SA, Wittwer CT (2017) MIQE: a step toward more robust and reproducible quantitative PCR. Clin Chem 63(9):1537–1538
67. Horzmann KA, Freeman JL (2017) Toxicogenomic evaluation using the zebrafish model system. In: Encyclopedia of analytical chemistry. Wiley, Chichester, pp 1–19
68. Mantione KJ et al (2014) Comparing bioinformatic gene expression profiling methods: microarray and RNA-Seq. Med Sci Monit Basic Res 20:138–142
69. Wang Z, Gerstein M, Snyder M (2009) RNA-Seq: a revolutionary tool for transcriptomics. Nat Rev Genet 10(1):57–63
70. Trapnell C et al (2012) Differential gene and transcript expression analysis of RNA-seq experiments with TopHat and Cufflinks. Nat Protoc 7(3):562–578
71. Drewe P et al (2013) Accurate detection of differential RNA processing. Nucleic Acids Res 41(10):5189–5198
72. Zhang Z, Wang W (2014) RNA-skim: a rapid method for RNA-Seq quantification at transcript level. Bioinformatics 30(12): i283–i292

Chapter 10

Rat Brain Slices: An Optimum Biological Preparation for Acute Neurotoxicological Studies

Gabriela Aguilera-Portillo, Aline Colonnello-Montero, Marisol Maya-López, Edgar Rangel-López, and Abel Santamaría

Abstract

The search for alternative biological preparations comprising the structural and functional complexity of the whole brain is needed for the accurate development of acute and chronic pharmacological and toxicological studies. A good, easy-to-obtain and easy-to-use preparation is the isolated cortical slices from rodents. This preparation maintains all the cytoarchitecture of the brain cortex, preserving the complexity of this region and keeping functional aspects intact for a short period of time. If appropriately incubated and/or cultured, it represents an accurate model of the synaptic circuitry and combines the functions all cell types that are found in an in vivo system. Indeed, neurochemical, neurophysiological, and neurotoxicological studies using incubated and/or cultured slices represent a practical interphase between in vitro studies with cultured brain cells and in vivo studies in living animals. Here we describe several advantages of this preparation for simple acute neurotoxicological studies and some of the experimental approaches that, in our experience, can reveal, in a rapid manner, neurotoxic and neuroprotective profiles of several molecules in different acute toxic models. We also review and describe basic elements for the establishment of culturing conditions for this preparation to be employed in chronic toxic models. Through these review processes, we aim to provide simple basis for extending the use of this preparation in neuroscience research.

Key words Rat brain cortical slices, Synaptic circuitry preservation, Structural/functional conservation, Cell viability, Oxidative stress, Cell damage

1 Introduction

The three basic principles (3Rs) in animal research remain dictating the design and administration of experiments elsewhere. Whenever possible, *replacement* of "upper" animals (rodents and nonhuman primates) by "lower" living models (lower vertebrates

Gabriela Aguilera-Portillo, Aline Colonnello-Montero, and Marisol Maya López are authors who equally contributed to this review; therefore, they should be considered as first authors. Edgar Rangel and Abel Santamaría should both be considered as corresponding authors.

Michael Aschner and Lucio Costa (eds.), *Cell Culture Techniques*, Neuromethods, vol. 145,
https://doi.org/10.1007/978-1-4939-9228-7_10, © Springer Science+Business Media, LLC, part of Springer Nature 2019

or invertebrates), *reduction* in the number of animals used to obtain the same amount and quality of information, and *refinement* of experimental procedures to improve animal welfare and reduce distress are all key premises to follow nowadays, especially when ethical issues are stricter every day. Therefore, the search for alternative models and platforms to improve the quality of research obeying these principles is accompanied by criteria such as the achievement of solid inferences similar to those collected from "upper" vertebrates.

Despite that it is not the intention of this review to describe and discuss these principles and the issues behind them since this topic has been recently and nicely discussed in previous reports [1], basic concepts of this preparation are needed to better understand this work: the brain cortical slices represent a good ex vivo platform where a fragment of the brain is explanted and preserved either for short- or long-term studies. This preparation preserves the cellular diversity of the region (including neuronal, glial, and endothelial cells) and the many functional interactions within; therefore, brain slices are useful for both physiological and pharmacological studies. Moreover, since the slices can be preserved at the structural and functional levels for a short period of time (ranging 6–8 h) with the appropriate culture conditions [1], electrophysiological, neurochemical, and pharmacological studies carried out under these conditions represent an acute phase of study where the cytoarchitecture of the region, the complexity of cellular interactions, and the synaptic circuits remain all well preserved. Taking advantage of these properties, it is possible to conduct acute or short-term toxicological studies; in particular, for chronic or long-term studies, the organotypic cultures represent a good alternative when culture conditions (oxygen supply, pH, nutrients, etc.) are appropriate.

This review provides a simple and accurate description of methods to start and perform experiments with rat cortical slices aimed to characterize endpoints of acute toxicity at biochemical and cellular levels representing the changes that can be also found in experiments with cultured cells, invertebrates, or even in in vivo experiments with rodents. These assays include simple experimental approaches to characterize loss of cell viability/mitochondrial function, cell damage, and oxidative stress using different toxic models. We also include a brief description of methods oriented to establish the conditions for organotypic cultures as we are currently starting this useful protocol in our laboratory. Through all this information, our intention is to provide basic elements and tools to start these simple protocols for neurotoxicological studies.

2 Materials

2.1 Krebs Buffer Preparation

To prepare 1 L of Krebs buffer solution for the incubation of rat cortical slices, 73.8 g of sucrose (215 mM) are dissolved in 100 ml of distilled water, and then 5.4 g of glucose (30 mM) + 2.1 g of sodium bicarbonate (25 mM) + 0.2236 g of potassium chloride (3 mM) and 0.952 g magnesium chloride (10 mM) are added. When all components are dissolved, the pH is adjusted to 7.4 with 1 mM hydrochloric acid (HCl), using a Corning pH meter series 440 series (Corning Inc., NY, USA) on a Cimarec 2 Magnetic Stirrer (Barnstead Thermolyne Co., IA, USA). Once pH has been fixed at 7.4, the buffer volume is completed to 1 L with distilled water. The final Krebs buffer is stored at 4 °C until use.

2.2 Preparation of Other Reagents

The preparation of other reagents such as quinolinic acid (QUIN), ferrous sulfate ($FeSO_4$), hydrogen peroxide (H_2O_2), 3-(4,5-dimet hylthiazol-2-yl)-2,5-diphenyltetrazolium bromide (MTT), acidic alcohol, and propidium iodide is described as follows:

To obtain 5 mL of the QUIN stock (100 μM final concentration in the slice medium), we follow previous specification [2]: 4.2 mg of the reagent are dissolved in 4 mL of deionized water, and pH is then adjusted to 7 with 1 M sodium hydroxide (NaOH) using the pH meter and the magnetic stirrer; then, the solution volume is completed to 5 mL with deionized water. Aliquots of the QUIN solution (200 μl each) are stored at −20 °C until required for experimental purposes.

To prepare 2 mL of the $FeSO_4$ stock (25 μM final concentration in the slice medium), we follow previous specifications [3]: 0.4 mg of the iron salt are dissolved in 2 mL of deionized cold water (4 °C) and stirred until fully dissolved. This solution needs to be used freshly prepared as iron is rapidly oxidized.

To prepare the H_2O_2 solutions at 500 and 800 μM (final concentrations), 5 μL and 8 μL of H_2O_2 (Sigma) are diluted in 1 mL of deionized water for 5 mM or 8 mM stock concentrations, respectively, considering a molecular weight of 34.0147 g/mol, purity of 30%, and density of 1.13 g/cm^3. Then, 50 μL of each stock solution are added to the wells containing the slices (one slice per well) in 500 μL of the buffer solution.

To obtain 2 mL of the MTT stock (5 mg/mL), we follow previous specifications [3, 4]: 10 mg of the reagent are added to 2 mL of deionized water and stirred until completely dissolved. This solution also needs to be used freshly prepared.

To prepare 50 mL of acidic alcohol solution, 643 μL of HCl (0.4 N) are diluted in 25 mL of isopropanol, and then the solution volume is completed to 50 mL and stored at 4 °C.

Propidium iodide (PI; Roche, Switzerland) was used from a 1 mg/mL stock, and the final concentration in the solution was 100 μg/μL.

3 Animal Handling

Obtain your animals from your local suppliers. We obtain adult male Wistar rats weighing 250–300 g from the vivarium of the Instituto Nacional de Neurología y Neurocirugía *Manuel Velasco Suárez*. Animals are housed for 24–48 h in acrylic cages to acclimate. Rats received Rodent Chow Purine and water ad libitum and maintained under standard conditions of temperature (23–26 °C), humidity (50%), and lighting (12:12 light/dark cycle). All experiments are carried out according with the "Guidelines for the Use of Animals in Neuroscience Research" from the Society of Neuroscience, the Local Ethical Committees, and in compliance with of the ARRIVE guidelines [2, 3].

4 Isolation of Cortical Slices

The rat is euthanized by decapitation (small species decapitator; Fig. 1b) after mild anesthesia (inhaled chloroform), and the whole brain is immediately extracted from the cranial cavity [3]. It is accepted that, sometimes, even the lighter anesthesia can produce major neurochemical changes; still, ethical regulations require this step, so we strongly recommend its use. The brain is then placed on a Petri dish with filter paper humidified with Krebs buffer, all on ice. The cortical area (prefrontal and frontal) is separated from other brain structures and placed on the plate of a tissue chopper (McIlwain Tissue Chopper, Campden Instruments, England; Fig. 1a) to obtain by sequential cutting cortical slices of 250–300 μm thickness [3, 4]. This procedure is also practiced to obtain striatal tissue slices. Through the tissue chopper, tissue slices can be obtained thinner or thicker, depending on the experimental purpose. This can be done by adjusting speed and force cutting values. For a correct use of this equipment, the user must collocate a common use razor in the device, procuring that the razor is secured in a vertical position (90° angle) with respect to the chopper plate.

Once collected, the slices are placed on a Petri dish containing approximately 10 ml of the Krebs buffer. Afterward, slices are separated under a stereoscopic microscope light (SCF-11 Series Stereomicroscope, Motic, Hong Kong; Fig. 1c) using a spatula or common slim and soft bristle brushes; subsequently, each slice is placed in a Costar 24-well plate containing 490 μl of the Krebs buffer. The slices are incubated in a CO_2 Incubator (Panasonic MCO-19AIC) for the designed time at 37 °C and 5% of CO_2.

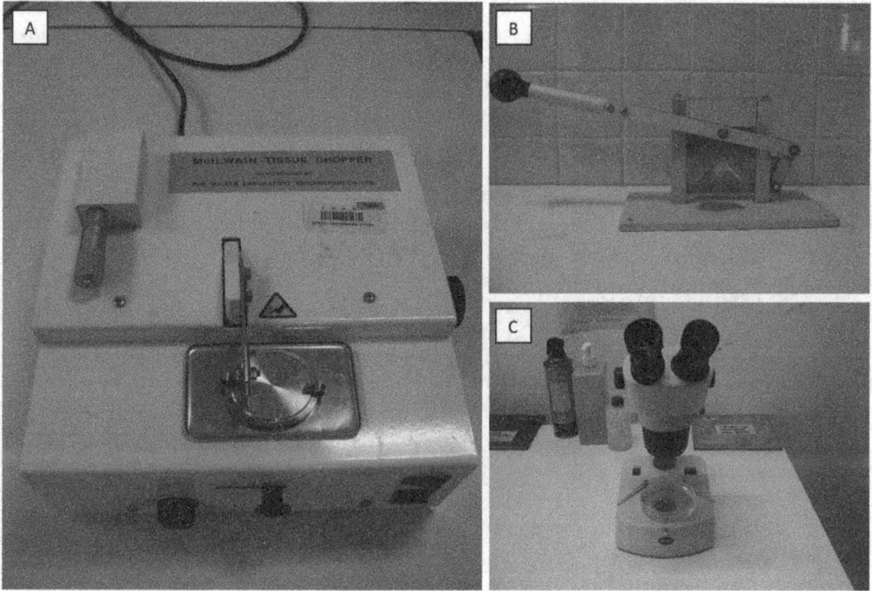

Fig. 1 Exhibit of basic equipment used for the isolation and processing of slices. Panels show the devices as follows: (**a**) tissue chopper; (**b**) rodent decapitator; (**c**) stereoscopic microscope

5 Cell Viability/Mitochondrial Activity Assay

The MTT reduction assay is a colorimetric method capable of assessing the mitochondrial metabolic activity in biological preparation. Mitochondrial oxidoreductase enzymes can indicate the number of viable cells or the number of loss of viable cells. These enzymes reduce the tetrazolium dye MTT to formazan resulting in a purple coloration of cells [5].

To validate this method in tissue slices, 10 μL of QUIN (100 μM, final concentration), FeSO4 (25 μM, final concentration), or H_2O_2 (500 or 800 μM, final concentrations) are added to 490 μl of Krebs buffer containing the incubated slices (Fig. 2a). For us, testing the effects of toxins with different and specific toxic patterns represents an advantage as it allows the investigation of the degree of participation of diverse toxic mechanisms in the same test. For the control conditions, 10 μL of vehicle are added. For each treatment, $n = 6–8$ repeats per experiment are recommended. The slices (one per well) are incubated in the presence of the toxins for 60 min at 37 °C and 5% of CO_2. These conditions are used to validate the experiment as they have been shown to be sufficiently toxic to induce a significant decrease of mitochondrial activity [3]. In addition, in our experience, the 60 min incubation step has been sufficient to induce an acute pattern of toxicity that can be measured by this and other assays.

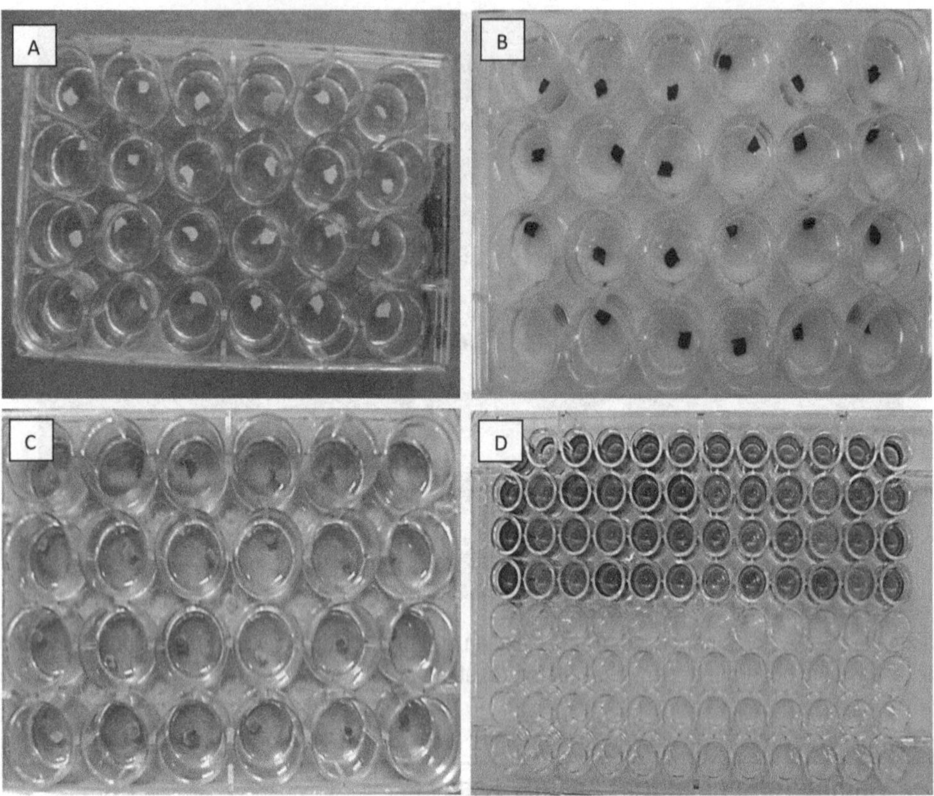

Fig. 2 Exhibit of culture plates at different stages of the experiment. (**a**) Tissue slices in Krebs buffer on 24-well cell culture plates; (**b**) Tissue slices after MTT incubation on 24-well cell culture plates; (**c**) Tissue slices during acidic discoloration; (**d**) Final supernatants on 96-well cell culture plates prior reading (note that the well plate displayed contains two 24 well plates)

Immediately thereafter, the Krebs buffer containing the toxic agents is washed out, and then, 285 μL of new Krebs buffer are added to the tissue slices followed by the addition of 15 μL of the MTT reagent (5 mg/mL), following previous specifications [3]. During this step, slices containing the MTT reagent remain protected from light by being covered with aluminum foil (Fig. 2b). Cortical slices are then incubated again for 60 min at 37 °C and 5% of CO_2. At the end of this step, the medium containing the MTT reagent is withdrawn from the slices. Then, 250 μL of acidic alcohol are added to the samples to obtain the purple chromophore that evidences the mitochondrial metabolism (Fig. 10.2c). The dye is collected after it reacted with the slice, and then, 150 μL of the collected volume are transferred into a 96-well plate (Fig. 10.2d). The optical density from each sample is measured in a Cytation 3 Cell Imaging Multimode Reader (BioTek; VT, USA; Fig. 10.3) at 570 nm (Figs. 3 and 4).

Fig. 3 Multiple reader (Cytation 3 Cell Imaging Multimode Reader; Biotek, VT, USA) and PC used for the analysis of processed samples

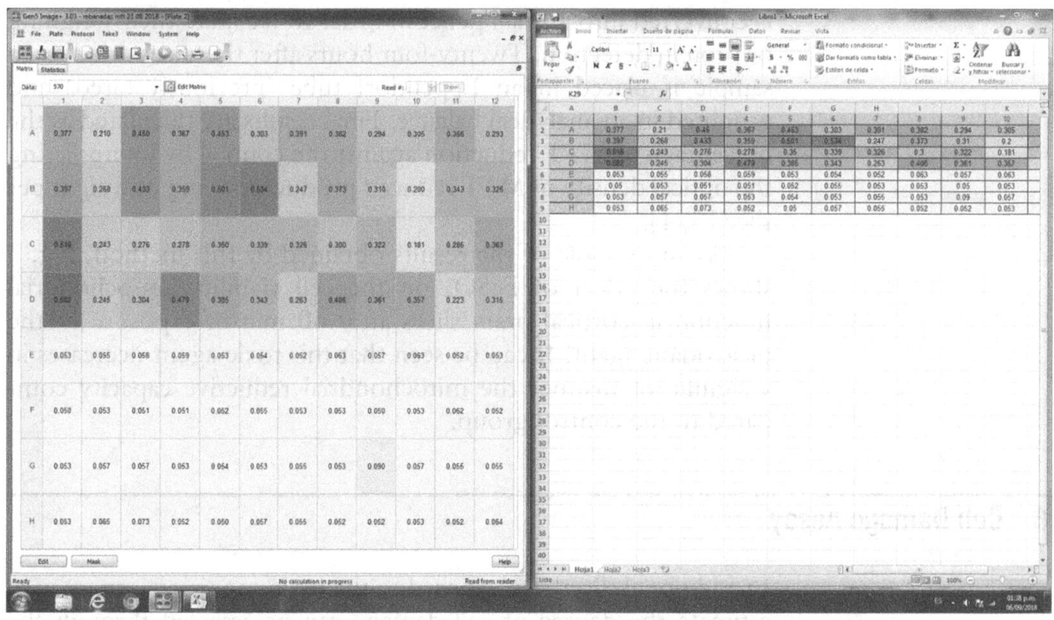

Fig. 4 Screenshot of the multiple reader monitor displaying the Gen5 Image + Software with recordings on the left, and data exported to Microsoft Excel on the right (please notice that not all recordings on the screen represent actual samples)

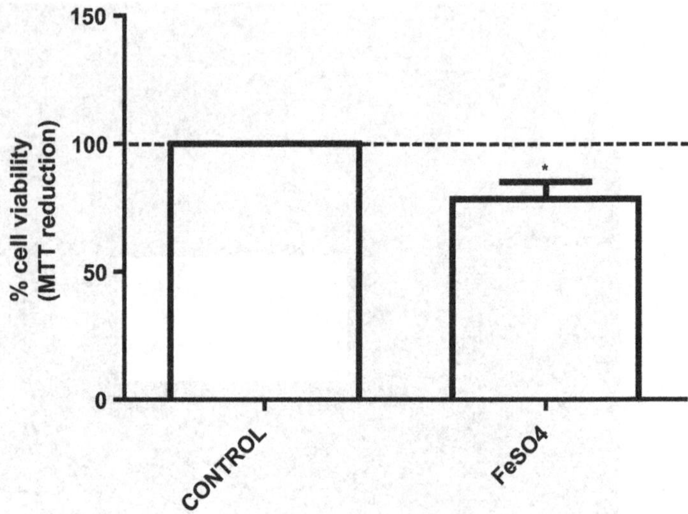

Fig. 5 Graphic representation of the effects of a toxic pro-oxidant agent ($FeSO_4$) on the mitochondrial reductive capacity in rat cortical slices after 60 min of incubation in the presence of the toxin. $FeSO_4$ is different of the control condition at $P \leq 0.05$ (Student's t-test)

The remaining purple supernatant is washed out from each sample. The tissue slices are then allowed to dry, and each of them is converted into a small pellet in order to facilitate their weighing or protein detection. Twenty-four hours after the slices dried, each sample is placed in an Eppendorf tube (previously tared) and weighted in an analytical balance. Final results are calculated as the percentage of MTT reduction against the control by interpolating the optical density of MTT reduced products per mg of weight per slice [3, 4].

As an example of the results obtained by this method, Fig. 5 shows the effect of $FeSO_4$ on the cell viability/mitochondrial function in cortical brain slices after 60 min of exposure to the prooxidant agent. It can be seen that this toxic agent decreases in a significant manner the mitochondrial reductive capacity compared to the control group.

6 Cell Damage Assay

In additional slices also exposed to the toxic conditions, an assay to estimate the degree of cell damage can be assessed through the level of incorporation of the dye PI by cells. For this purpose, all control and treated tissue slices with QUIN (100 μM), FeSO4 (25 μM), or H_2O_2 (500 or 800 μM) are carefully placed into 24-well plates and rinsed with saline buffer at 37 °C using a glass Pasteur pipette to protect the integrity of the brain sections until their analysis, using modifications to previous specifications [6]. Immediately thereafter, the tissue slices are rinsed trice with saline buffer, and the samples from all groups, including the control condition, are incubated in the presence of PI for 10 min in darkness and at room temperature [7]. PI selectively binds to DNA by intercalating with its bases with little or no sequence preference; since this dye is membrane-impermeable, therefore, it is excluded from viable cells whereas stain damaged cells. Then, all samples are washed trice in saline buffer again, fixed in 2% paraformaldehyde in saline buffer for 30 min, and washed trice with 2 mL of saline buffer. After these steps, each slice is carefully placed and extended on glass slides using Cushing Brain Forceps and covered with mounting medium with 4′, 6-diamidino-2-phenylindole (DAPI) for additional nucleic acid staining. DAPI is commonly used to stain fixed cells since the dye is cell-impermeable, though it also labels live cells when used at higher concentrations. The coverslips are sealed around their perimeter with nail polish. All samples are protected from light until analysis. The slides are analyzed in an imager processor (Cytation 3, BioTek) equipped with fluorescent filters for DAPI (358/461 nm) and PI (535/617 nm). Images are analyzed by the software provided by the supplier (Gen5 v3.02.2).

An example of the results obtained by this procedure is shown in Fig. 6, where a prominent staining with DAPI (to label cell nuclei) is evident throughout all the treatments, while a positive staining for PI (to label damaged cells) is only evident in the photomicrographs of toxic treatments (QUIN, $FeSO_4$, and H_2O_2). The merge panel confirms this result. In addition, the densitometric analysis depicted in the same figure reveals the differential intensity of cell damage elicited by each toxic treatment.

7 Detection of Lipid Peroxidation as an Endpoint of Oxidative Damage

Another useful application for the use of brain slices is the employment of simple experimental approaches to detect oxidative damage. In our experience, the detection of lipid peroxidation is a simple method revealing the degree of acute oxidative stress in this biological preparation. Lipid peroxidation is currently determined as the production of thiobarbituric acid-reactive substances (TBARS), following previous specifications [4]. Briefly, each tissue slice is homogenized in 500 μL of HEPES buffer solution (pH 7.0); then, 50 μL-aliquots of each homogenate are added to 100 μL of the TBA reagent. The TBA stock solution contains 0.75 mg of TBA + 15 mg of trichloroacetic acid +2.54 μL of HCl. Samples are then incubated at 94 °C for 20 min, and a pink chromophore is produced in samples visually revealing the amount of peroxidized lipids. Tubes are kept on ice for 5 min and further centrifuged at 3000 g for 15 min. Supernatants are collected and

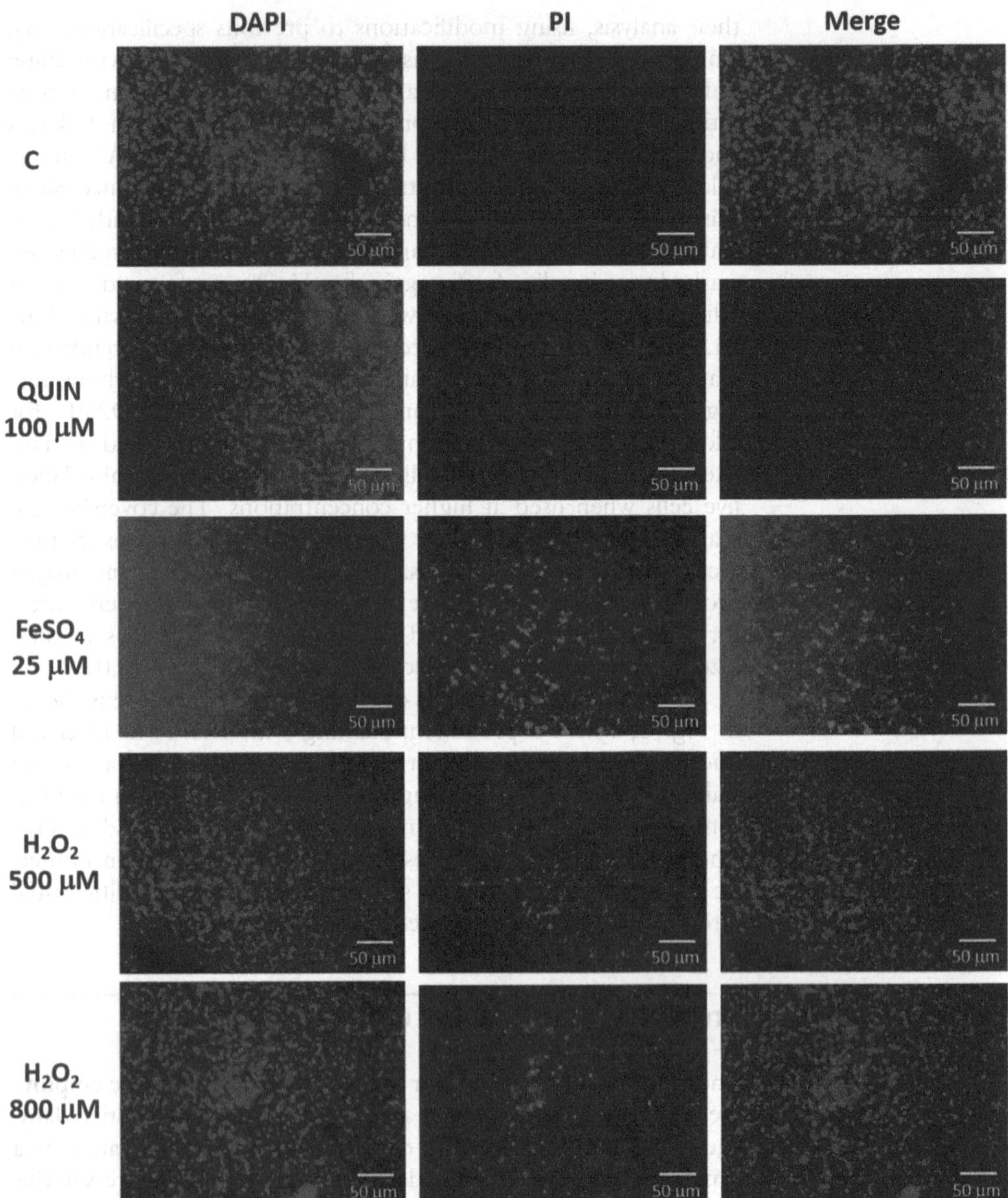

Fig. 6 Cell damage detection assay by propidium iodide (PI) labeling + DAPI staining in rat cortical slices exposed to three toxic agents (QUIN, $FeSO_4$, or H_2O_2) for 60 min. In the upper panels, PI (in red), DAPI (in blue), and co-staining (merge in purple) for slices incubated with the three toxic agents are compared with the control (C) condition. In the bottom panel, the graphic representation of the densitometric analysis depicts the quantitative effects of the three toxic agents. All treatments are different of the control condition at $P \leq 0.05$ (Two-way ANOVA followed by Tukey's test)

their optical density is measured in a multiple reader (Cytation 3, BioTek) at 532 nm. The amounts of TBARS (peroxidized lipids such as malondialdehyde) are calculated by interpolation of values in a MDA standard curve and expressed as nmoles of TBARS per mg of protein.

8 Organotypic Slice Cultures

Despite that the aim of this review is to provide technical information about the processing of brain tissue slices for short-term neurotoxicological assays, the issue of culturing slices for long-term studies constitutes a valuable tool also for this field of research. Also, since this process is in actual progress in our laboratory, here we develop this section as a brief overview to provide the readers basic information on how to proceed with this method.

Among the several advantages of this assay, the slices can be obtained either from adult, young, or neonatal animals; however, it has been suggested that the performance is better when slices are obtained from neonatal animals [8, 9]. Indeed, the organotypic cultures obtained from neonatal rodents maintain up to 50% of cell survival for 12 weeks when appropriate growth factors are added, whereas it is estimated that the cells from organotypic cultures from adult rodents can survive up to 60 days [10, 11]. In addition, it is possible to obtain and preserve slices from mice with specific mutations, though these cultures are not able to survive for long times. It is also suggested for this preparation to have a strict control of survival and/or viability through time curves performed minimally every week. Other functional aspects that can be evaluated regularly are markers of cell communication and axonal and dendritic growth [8].

The most employed method for slice organotypic cultures is the one generated by Stoppini et al. [12] using a membrane. This method is characterized by the thinning of the slices, and during the first days of culture, the slices show a change of color, turning white opaque. Later on, slices turn gray, thus indicating a successful culture. During this last stage, cell survival can be labeled by fluorescence dyes propidium iodide, DAPI, and annexin V; it is also possible to label specific cell types with antibodies (i.e., GFAP for astrocytes or NeuN for neuronal cells). To preserve cell survival in the culture, growing factors for neuronal cells can be added to the horse serum culture medium, maintaining pH at 7.2 [8, 9].

In contrast to the short-term assays, the preparation of the slices for organotypic slice cultures requires that the dissection solution and the culture medium are sterile. The dissection solution is used cold and with a fixed percent of oxygen. Slices are also obtained in a tissue chopper, and the dissection solution consists of 5% CO_2, 1.25 mM KH_2PO_4, 124 mM NaCl, 3 mM KCl, 8.19 mM

$MgSO_4$, 2.65 mM $CaCl_2$, 3.5 mM $NaHCO_3$, 10 mM glucose, 2 mM ascorbic acid, and 39.4 μM ATP, all prepared in ultrapure, sterile, and filtered H_2O adjusted to pH 7.2 [9].

The culture medium consists of 26.6 mM HEPES, 19.3 mM NaCl, 5 mM $NaHCO_3$, 511 μM ascorbic acid, 40 mM glucose, 2.7 mM $CaCl_2$, 2.5 mM $MgSO_4$, 1% glutamine (GlutaMAX), 0.033% (v/v) insulin, 0.5% (v/v) penicillin/streptomycin, and 25% warm horse serum, all in ultrapure, sterile, and filtered H_2O adjusted to pH 7.2 [9]. This medium is a variant to the one used by Stoppini et al. [12], which is also useful and consists of 50% MEM + HEPES (Gibco No. 079-01012), 25% warm horse serum, 25% Hank's solution, 6.5 mg/ml glucose, 2 mM glutamine, and 25% (v/v) penicillin/streptomycin, all in sterile water and adjusted to pH 7.2 [8]. This medium requires pre-warming to 37 °C before used.

The procedure for the isolation (first step) of brain slices is the same than that described above for short-term assays. The rodents are also euthanized by decapitation, and their brains are dissected out. Depending on the specific interest of a given protocol, the dissection of specific brain regions proceeds. The dissected brain regions are maintained in the cold dissection solution to proceed with the slices' separation either in cortical or coronal positions. The slices can be as thick as 200–400 μm, and it is recommended to place three to five slices per well in six-well plates. Each well should contain 1 mL of the sterile culture medium preincubated at 37 °C. The slices are then placed on a 30 mm-diameter translucid membrane with a pore size of 0.4 μm (Millicell-CM). It is necessary for the slices to be placed very carefully on the membrane, taking care of not folding or breaking them [12]. The cultured slices should be maintained in an incubator at 37 °C with 100% humidity and 5% CO_2 for 2 weeks before any exposure to drug treatments. The medium should be replaced twice or trice per week [8, 9, 12, 13].

Once again, this biological preparation is appropriate for several long-term physiological, pharmacological, and toxicological studies, allowing important inferences about how a heterogeneous population of cells behave in terms of their interactions and in response to different stimuli, thus representing a promising interface between in vitro and in vivo translational studies.

9 Conclusion

Here we have described the basic conditions to isolate and prepare rat brain tissue slices from a region (the brain cortex) that is appropriate to study toxic events linked to cell damage, mitochondrial dysfunction, and oxidative stress at the biochemical and molecular levels. By means of quite simple methods, it is possible to establish an integral experimental strategy to perform easy and accurate ex vivo neurotoxicological studies aimed to characterize damaging acute mechanisms in the CNS. Among the several advantages of using this biological preparation, we emphasize three key issues: (1) it is easy to obtain and prepare; (2) it contains all type of cells (neuronal cells, glial cells, endothelial cells, etc.), thus making the inferences derived from its analysis more representative of the complex physiology and physiopathology of the CNS; and (3) the diversity of experimental approaches can be tested, from basic neurochemical and neurophysiological assays to advanced cellular and molecular assessments. Hence, using brain cortical slices for toxicological studies opens a wide variety of possibilities to characterize toxic mechanisms and design therapeutic strategies.

References

1. Lossi L, Merighi A (2018) The use of ex vivo rodent platforms in neuroscience translational research with attention to the 3Rs philosophy. Front Ver Sci 5:1–14
2. Colín-González AL, Sánchez-Hernández S, Lima ME, Ali SF, Chavarría A, Villeda J, Santamaría A (2015) Protective effects of caffeic acid on quinolinic acid-induced behavioral and oxidative alterations in rats. J Drug Alcohol Res 4:1–5
3. Colonnello A, Kotlar I, Lima MA, Ortíz-Plata A, García-Contreras R, Antunes Soares FA, Aschner M, Santamaría A (2018) Comparing the effects of ferulic acid and sugarcane aqueous extract in in vitro and in vivo neurotoxic models. Neurotox Res 34:1–9
4. Colín-González AL, Maya-López M, Pedraza-Chaverrí J, Ali SF, Chavarrí A, Santamaría A (2014) The Janus faces of 3-hydroxykynurenine: dual redox modulatory activity and lack of neurotoxicity in the rat striatum. Brain Res 1589:1–14
5. Berridge MV, Herst PM, Tan AS (2005) Tetrazolium dyes as tools in cell biology: new insights into their cellular reduction. Biotechnol Annu Rev 11:127–152
6. Na L, Xin-Guo D, Shi-Hua Z, Mei-Feng H, Dong-Qing Z (2014) Effects of different concentrations of tetramethylpyrazine, an active constituent of Chinese herb, on human corneal epithelial cell damaged by hydrogen peroxide. Int J Ophthalmol 7:947–951
7. Pietkiewicz S, Schmidt JH, Lavrik IN (2015) Quantification of apoptosis and necroptosis at the single cell level by a combination of imaging flow cytometry with classical Annexin V/ propidium iodide staining. J Immunol Methods 423:99–103
8. Humpel C (2015) Organotypic brain slice cultures: a review. Neuroscience 305:86–98
9. Croft CL, Noble W (2018) Preparation of organotypic brain slice cultures for the study of Alzheimer's disease. F1000Res 7:592
10. Marksteiner J, Humpel C (2008) Beta-amyloid expression, release and extracellular deposition in aged rat brain slices. Mol Psychiatry 13: 939–952
11. Kim H, Kim E, Park M, Lee E, Namkoong K (2013) Organotypic hippocampal slice culture from the adult mouse brain: a versatile tool for translational neuropsychopharmacology. Prog Neuropsychopharmacol Biol Psychiatry 41:36–43
12. Stoppini L, Buchs PA, Muller D (1991) A simple method for organotypic cultures of nervous tissue. J Neurosci Methods 37:173–182
13. Bright R, Raval AP, Dembner JM, Pérez MA, Steinberg GK, Yenari MA, Mochly D (2004) Protein kinase C delta mediates cerebral reperfusion injury in vivo. J Neurosci 24:6880–6888

Chapter 11

Electrophysiological Neuromethodologies

Yukun Yuan and William D. Atchison

Abstract

Neurotoxic chemicals can alter neuronal excitability and disrupt synaptic transmission in the central nervous system to induce a variety of neurological disorders. These neurotoxic effects can be assessed with high sensitivity and in functional tissue and real time by using electrophysiological recording techniques in freshly isolated brain slices, typically in rats or mice. This provides far greater sensitivity than neurochemical measures of neurotransmission and permits assessment of function of both pre- and postsynaptic elements of the synapse, from the same slice. Preparation of brain slices from adult rodents has been a challenge due to poor viability of the tissue post-slicing. This has dramatically hampered use of these techniques in experimental paradigms in which neurophysiologic effects of toxic chemicals can be assessed during chronic exposure of animals over their life span. This chapter presents a simplified method for preparing brain slices from adult animals. It applies voltage- and current-clamp recordings in slices of the brainstem and cerebellum, two brain regions vital to motor function, for which neurotoxic chemicals have been shown to act. However with slight modifications, the principles could be applied to other brain regions such as hippocampus or *corpus striatum* to study effects of neurotoxic chemicals on central nervous system synaptic function.

Key words Acutely prepared brain slice, Adult animals, Whole cell patch-clamp recording, Neuronal excitability, Action potential, Firing pattern, Synaptic transmission, Short-term synaptic plasticity

1 Introduction

Electrophysiological recordings in acutely isolated brain slices have been the method used most commonly for studying synaptic physiology and its disruption by pharmacological and toxic chemicals on neuronal excitability and central synaptic transmission since the application was described by Yamamoto and McIlwain [1]. This is because acute brain slice preparations offer a number of unique advantages over other in vitro or in vivo testing systems. These include:

(a) When cut at an appropriate angle, the brain slices can maintain the normal anatomical relationships and relatively intact synaptic circuits within a certain brain region as compared to acutely isolated neurons or cells in primary cultures.

Michael Aschner and Lucio Costa (eds.), *Cell Culture Techniques*, Neuromethods, vol. 145, https://doi.org/10.1007/978-1-4939-9228-7_11, © Springer Science+Business Media, LLC, part of Springer Nature 2019

(b) The ability to control the recording conditions easily and precisely by modulation of pH, temperature, and osmolarity of the external solutions and application of known concentrations of the chemicals to the bath solution.

(c) The stimulating and recording electrodes can be positioned by direct visual observation in the target areas in slices, avoiding the difficulties of complicated stereotaxic techniques necessary for in vivo experiments.

(d) The stability of recording is significantly improved when compared to in vivo electrophysiological recordings, which are influenced by many variables such as blood pressure, CO_2 concentration, respiration, and heart rate. In our hands, stable recordings (extracellular or intracellular sharp microelectrode or whole cell patch-clamp recordings) in brain slices lasting for 1–4 hours are achieved routinely [2–8], and in the hands of an experienced investigator, the slice can remain viable for periods up to or in excess of 8 hours [2–8].

(e) Unlike in vivo experiments, no anesthetics, paralytics, or sedatives are required during recordings in brain slice preparations, use of which may potentially interfere with the effects of any test chemicals or normal synaptic transmission.

(f) Multiple brain slices can be obtained from a single animal. Therefore, slices from a given animal can be used simultaneously for multiple experiments each with a different purpose, allowing not only for greater cost-effectiveness but also for paired observations and allowing for greater statistical power and rigor.

(g) Unlike primary cell cultures, it only requires a couple of hours to determine the quality of brain slices and obtain electrophysiological recordings in the same day. Moreover, if a given slice is not acceptable from the standpoint of having basic electrical properties, the ability to obtain multiple slices, and maintain them in oxygenated physiological saline, means that an animal, which may have undergone pharmacological treatment, in vivo, is not "lost." Thus this technique is temporally efficient.

In general, brain slices can be prepared from any region of the brain. The most commonly used regions are those with "laminated" architecture or structural landmarks readily visible such as the hippocampus, neocortex, cerebellum, and brainstem. Selection of a brain region for preparing slices should be driven by the purpose of the proposed experiments and knowledge about the potential neurotoxic targets of the tested chemicals, and not simply because it is the "easiest" to record from. For example, if a chemical is known to cause deficits in hearing or central visual function, then slices from the auditory or visual cortex should be considered; if a chemical such as methylmercury is known to cause cerebellar and visual dysfunction, then cerebellar and visual cortical slices should be prepared; if a chemical is suspected to affect learning and short-term memory, then the brain regions associated with certain form of learning and memory such as the hippocampus should be the choice; if a chemical is suspected to interact with the respiratory or cardiovascular center to affect respiratory or cardiac function, then brainstem slices should be used. If the neurotoxicity of a new chemical is unknown, it may make sense first to examine its action using slices that can be prepared easily and reliably and for which there is a wealth of background information on the normal physiology exists, such as the hippocampus, neocortex, or cerebellum. In short, the experimental design should involve selection of the brain region for slicing that is most appropriate for the questions to be asked.

In theory, brain slices for acute electrophysiological recordings can be made from an animal of any age. However, for decades electrophysiologists have relied almost exclusively on brain slices prepared from juvenile or adolescent animals. There are some obvious advantages for using brain slices from young over old animals. These include the following: (a) the brain tissues are more tolerant to anoxic damage during tissue dissection and slice preparation; (b) slices are more light transparent, and cells can be more readily visualized compared to slices from the highly myelinated brain structures of adult animals; and (c) there is less possibility of so-called space-clamp (the inability to control the membrane potential over the region to be recorded from) issues during voltage-clamp recordings because the neuronal axon and dendritic processes are not fully developed at young ages, for example, in cerebellar Purkinje cells. However, many neurological disorders, including the principal neurodegenerative disorders (Alzheimers, Parkinsons, amyotrophic lateral sclerosis), occur only in adults or aged people. Of more relevance, neurotoxic chemicals do not just affect the developing brain; many also cause neurotoxicity in adults. Thus, studies aiming at examining neurotoxicity related to adult exposure require the use of brain slices from adult or aging animals. Unfortunately, one critical limit to preparing acute brain slices from adult or aging animals is the poor viability of neurons in brain slices when prepared using the conventional methods typically used for preparation of acute brain slices from juvenile animals.

Over the past several decades, different strategies for the preparation of brain slices from adult animals have been adopted by different research groups or laboratories. One of the most commonly used strategies is the “protective cutting” method, in which the [NaCl] in the “slicing” solutions is replaced completely by equimolar sucrose [9–13], choline chloride [14, 15], glycerol [15–17], or N-methyl-D-glucamine (NMDG) chloride [15, 18, 19]. The objectives are to reduce overall excitability by omitting

Na^+, as well as to reduce energy requirements placed on the cells by having to pump out Na^+ accumulated in cut cells or those depolarized by released K^+. In most cases, the Ca^{2+} concentration is also reduced, and the Mg^{2+} concentration increased simultaneously to NaCl replacement in slicing solutions [11, 15, 20]. Altered divalent cation concentrations are used to impede release of neurotransmitters, most notoriously glutamate, which could occur during the slicing process. In some cases, ascorbate and/or N-2-hydroxyethylpiperazine-N′-2-ethanesulfonic acid (HEPES) are included to prevent cellular edema [21–23], or transcardial perfusion with cold slicing solutions is performed before the animal is euthanized to minimize ischemic damage [11, 16, 24–27]. In some cases, a combination of the above factors is used in slice preparation [25]. In general, these "protective cutting" methods do improve the health of slices prepared from adult brains, but the quality of slices obtained with protective cutting methods alone is far from ideal in terms of the viability of cells in brain slice for electrophysiological recordings. Interestingly, it has been shown that cutting slices at physiological temperature, as compared to the more commonly used ice-cold temperature, is beneficial for obtaining healthy brain slices from adult animals [28, 29]. Recently, strategies for improving the quality of acute brain slices from adult animals have focused primarily on the post-cutting recovery procedures, in which the NMDG protective recovery method is introduced ([30] and https://www.brainslicemethods.com/). This so-called protective recovery method has been applied successfully to produce healthy acute brain slices from several brain regions of adult rodents by different research groups [24, 26, 27, 30–33]. There is consonance among the observations that the most critical determinant for producing healthy acute brain slices from adult animals is the post-cutting recovery procedures, because adult brain slices undergo edema during the course of recovery in regular artificial cerebrospinal fluid (ACSF) ([30] and https://www.brainslicemethods.com/). In fact, we noted that even brain slices from juvenile animals undergo neuronal swelling or edema after transferring them directly from slicing solution to regular ACSF, suggesting that abruptly changing the solution from low [NaCl] to high [NaCl] concentrations is unhealthy for cells in brain slices. This observation prompted us to make two major changes in slice cutting procedures: first, we replace half, not all, of the NaCl with equimolar sucrose in the slicing solution; second, we added a transient step before incubation of slices in regular ACSF (see Sect. 3). Credited also to the findings that NMDG is a better replacement for NaCl in both protective cutting and recovery solutions ([30] and https://www.brainslicemethods.com/), we confirmed that the short period of NMDG protective recovery step is beneficial for obtaining healthy brain slices from both juvenile and adult animals. However, we did not find that the transcardial perfusion procedure pre-sacrificing really makes much difference in cerebellar or hippocampal slices in terms of the cell viability in slices. In this chapter, we will present a simplified method which includes both the "protective" cutting and recovery procedures for preparation of brain slices from adult mice for electrophysiological examination of neurotoxic chemicals on central synaptic function. We will primarily use brainstem slices as examples but will briefly discuss procedures for preparation of cerebellar, neocortical, and hippocampal slices as well.

Nowadays, all conventional electrophysiological recording techniques including extracellular field potential recordings, intracellular sharp microelectrode recordings, and whole cell patch-clamp including perforated patch recordings and single-channel recording techniques that are used in acutely dissociated or cultured cells or isolated neuromuscular junctions have been routinely applied to brain slice preparations. Again, the choice of a method of electrophysiological recording in brain slices depends principally on the goal of the experiments. For example, if the primary goal is to test whether a chemical induces an effect on central synaptic transmission, neuronal excitability, or potential epileptic seizures (epileptiform activity), then conventional extracellular microelectrode recording of field potentials will be sufficient to answer the question [5, 8, 34–39]. If one wants to test if a chemical disrupts dynamic neural network activity, the substrate-integrated, planar microelectrode array (MEA), a multiple-electrode extracellular recording technique, will be a better choice [40]. However, both conventional extracellular microelectrode and MEA recordings only generate information about the average behavior or synchronized firing activity of a population of neurons in response to effects of a neurotoxic chemical; they cannot provide in-depth information about the precise underlying mechanisms or changes in membrane electrophysiological properties of individual neurons. For this, intracellular sharp microelectrode or whole cell patch-clamp recording techniques will be the method of choice [5–8, 41]. If the cell membrane is too weak for mechanical rupture without causing current leak or if rundown of current is an issue, then the perforated patch-clamp recording will be a better choice [42]. Recent combinations of conventional electrophysiological recordings with newly developed fast voltage-sensitive dye imaging [43–49] and optogenetic techniques [50–56] or conventional single-photon confocal microscopy [7, 8] have provided more powerful tools and options for neurotoxicologists in studying effects of neurotoxic chemicals on central synaptic circuits and function. Nevertheless, electrophysiological recording techniques clearly remain an essential tool for studying chemical-induced neurotoxicity. These methods have been pivotal in our understanding of the neurotoxicity of neurotoxicants such as methylmercury on cerebellar function [5–8] and lead on hippocampal transmission [36, 37, 41, 57–59] and polychlorinated biphenyls (PCBs) on the hippocampus and neocortex [34, 35, 60–62] and others. In this chapter, we will describe the most commonly used whole cell patch-clamp (both voltage- and current-clamp) recording methods for studying effects of neurotoxic chemicals on neuronal excitability, synaptic function, and short-term plasticity in acute brain slices of adult animals.

2 Materials

2.1 Solutions

1. Unlike most protective cutting solutions in which all NaCl is replaced completely by equimolar sucrose [9–13], choline chloride [14, 15], glycerol [15–17], or N-methyl-D-glucamine (NMDG) chloride [15, 18, 19], we replace only half of NaCl with equimolar sucrose. This is because it has been suggested that Na^+- and Cl^--free solutions might cause cellular acidification and swelling, especially at the sliced surfaces [63–65]. We also reduce the extracellular Ca^{2+} concentration from 2 to 0.5 mM. To counter the effects of reduced Ca^{2+} concentration on cell membrane stability, we increase Mg^{2+} concentration to 5 mM. We also reduce the $[K^+]$ to 3.75 from 5 mM, again to reduced membrane excitability during the slicing process. Therefore, our slicing solution contains (in mM) 110 sucrose, 62.5 NaCl, 2.5 KCl, 5 $MgCl_2$, 1.25 KH_2PO_4, 26 $NaHCO_3$, 0.5 $CaCl_2$, and 20 D-glucose (pH 7.35–7.4 when saturated with 95% O_2/5% CO_2 at room temperature of 22–25 °C). We do replace all NaCl with equimolar NMDG-Cl in our protective recovery solution (NMDG-ACSF), which contains (in mM) 125 mM NMDG-Cl, 2.5 KCl, 5 $MgCl_2$, 1.25 KH_2PO_4, 26 $NaHCO_3$, 0.5 $CaCl_2$, 5 mM HEPEs, and 20 D-glucose (pH 7.35–7.4 when saturated with 95% O_2/5% CO_2 at room temperature of 22–25 °C). The NMDG-Cl stock solution is prepared by titrating NMDG to pH 7.4 with HCl. It appears that including 2–5 mM HEPES in the NMDG-ACSF solution may be beneficial for maintaining healthy brain slices. Thus, 20–50 mM HEPES is included in NMDG-Cl stock solution (pH 7.4).
2. The external solution used for whole cell patch-clamp recordings is the regular ACSF containing (in mM) 125 NaCl, 2.5 KCl, 1 $MgCl_2$, 1.25 KH_2PO4, 26 $NaHCO_3$, 2 $CaCl_2$, and 20 D-glucose (pH 7.35–7.4) when aerated continuously with 95% O_2/5% CO_2 at 25 °C as external solution. For internal or pipette solution, we use a potassium gluconate (K-gluconate)-based pipette solution (in mM): 140 K-gluconate, 4 NaCl, 0.5 $CaCl_2$, 10 HEPES, 5 EGTA, 5 phosphocreatine, 2 Mg-ATP, and 0.4 GTP (pH 7.2–7.3 adjusted with KOH).

2.2 Animals

The method described here has been used in both mice and rats, including C57BL/6, BALB/c, and 129 or rats such as Wistar or Sprague-Dawley, either gender, 2–12 months old.

2.3 Equipment and Tools for Brain Slice Preparation

Preparation of healthy brain slices requires a high-quality tissue slicer or vibratome. We use the EMS 5000 Oscillation Tissue Slicer for routine slice cutting. There are many other tissue slicers or vibratomes such as Leica VT1000S or 1200S, CA-5100MZ or CA-5100MZ Plus (World Precision Instruments), or VF-300 series (Precisionary Instruments Inc.) that are commercially available and capable of cutting high-quality brain slices. Other tools include a 95%O_2/5%CO_2 gas mixture cylinder, a pair of large surgical scissors and a pair of small sharp dissecting scissors, a pair of rongeurs, two fine tip tweezers, single-edged and a high-quality double-edged razor blades, at least two homemade slice incubation chambers, a 35–37 °C water bath, and other routine laboratory supplies such as cyanoacrylate glue (Superglue); Whatman #1 or #2 filter papers; 50 or 75 mm plastic petri dishes; a Pasteur pipette and a wide-bore Pasteur pipette (the narrow tip end was broken and inserted into a rubber bulb) for transferring slices; a fine, soft artist's paintbrush; and 5% agar gel block or Sylgard® 184 block.

2.4 Basic Electrophysiological Setup

1. Electrophysiological circuits: The principal component of a functional whole cell patch-clamp system is the patch-clamp amplifier. Although all commercial available patch-clamp amplifiers such as Axopatch 200B should work well, we prefer the computer-controlled, versatile MultiClamp 700B amplifier, which can be used for extracellular field potential recordings, intracellular sharp microelectrode recordings, whole cell voltage- and current-clamp recordings, and single-channel recordings. The other components including data acquisition system such as an Axon Digidata 1440A or 1550A Digitizer or similar, a Windows®-based PC computer system installed with pClamp® 10.x or 11.x software or similar software for data acquisition and analyses, and a hydraulic or step-driven control micromanipulators such as the Burleigh 5000–150 series manipulators are absolutely critical. For recordings of stimulation-evoked synaptic responses, a Narishige YOU-2 coarse/fine mechanical micromanipulator, a Grass S88 stimulator, and SIU5 stimulus isolation unit are needed.
2. Microscope: A NIKON Eclipse FN1 or similar upright microscope equipped with Nomarski optics (at least 40× water immersion objective, 2–3.5 mm long working distance), infrared CCD video camera system, and a black-and-white monitor is essential. Ideally, the microscope is mounted on a fixed and stable platform such as the Gibraltar® platform with an X–Y

stage base. All manipulators and chamber are mounted to the same platform.

3. Recording chamber: A Warner RC26 submersion recording chamber assembled with a SS-slice support and a Warner VC-6 Valve Perfusion Control System or similar are required.
4. For pulling patch electrode, a Sutter P-97 micropipette puller or equivalent and a Narishige MF-900 microforge for firing polishing of pipettes are needed.
5. Others: Ag-AgCl ground electrode (e.g., warner REF-1L reference cell or E206 Ag-AgCl pellet electrode), glass pipette (o.d., 1.5 μm, i.d., 0.75 μm), MicroFil MF28G needle (WPI, Inc., Sarasota, FL), or homemade pipette filler pulled from a 1 ml plastic syringe are required. A concentric bipolar metal electrode or monopolar tungsten electrode (3 mΩ, FHC, Inc., Brunswick, ME) or alternatively a broken tip glass pipette filled with ACSF can be used as the stimulation electrode.

3 Methods

3.1 Prepare Acute Brain Slices from an Adult Mouse

1. Make up 200–400 ml of the slicing solution in a 500 ml glass bottle or beaker (see Sect. 2.1). Put it in a −20 or −80 °C freezer for ~60 min before use. After removing the slicing solution from the freezer, oxygenate it with a 95%O_2/5%CO_2 gas mixture for at least 15 min (pH to 7.35–7.4) while immersing the bottle or beaker in ice. Also, prepare up 50–100 ml of NMDG-ACSF and 1000 ml of ACSF (see Sect. 2.1).
2. Carefully break a double-edged razor blade into two single-edged parts. Clean the blade with alcohol and dry it with a piece of Kimwipe, then insert the broken edge of the half blade into the blade holder of the tissue slicer and secure it tightly. It is recommended that the blade should be placed at a section angle of ~20° to the horizontal. Glue a small (10 × 10 × 10 mm) 5% (w/v) agar or Sylgard® block on the tissue pedestal of the slicer with Superglue®. Clean up the cutting stage of the tissue pedestal and the cutting chamber.
3. The mouse should be sacrificed using the proper procedures approved by an Institutional Animal Care and Use Committee and following the NIH guidelines for animal care and use and then decapitated using a rat/mouse guillotine. Cervical dislocation *is not recommended* as it results in blood pooling in the CNS. Use a pair of large scissors to cut the scalp and a pair of small scissors to cut the skull along the midline. Care should be taken to avoid cutting or damaging the underlying brain tissue by keeping the tip of the scissors against the top inside wall of the skull. Retract the skull with a rongeurs. Pour some

chilled slicing solution over the brain immediately after exposing it. Remove the overlying membrane, detach all nerve and blood vessel connections, and isolate the whole brain. Place the isolated brain immediately on the filter paper moistened with the chilled slicing solution in a petri dish on ice. Wash the brain several times with the chilled and oxygenated slicing solution to remove residual blood. Separate the cerebellum from the rest of the brain by a transverse cut (Fig. 1b, c), or bisect the forebrain along the midline (Fig. 1e) if hippocampal slices are to be prepared (*see* **Note 1**).

4. Prepare transverse (coronal) brainstem and cerebellar slices simultaneously: Cover a small area of the tissue pedestal near the agar block with a thin layer of superglue. Use the fine tweezers to hold the spinal cord end (Fig. 1b–d) loosely and mount the cerebellum and brainstem tissue block on the tissue pedestal of the slicer. Be sure the tissue block is held closely against the agar block with the cerebellum (ventral side) facing the blade and the transection side between forebrain and cerebellum (Fig. 1b–d) on the glue layer. Lower the blade all way down until the desired level of brainstem is attained or the

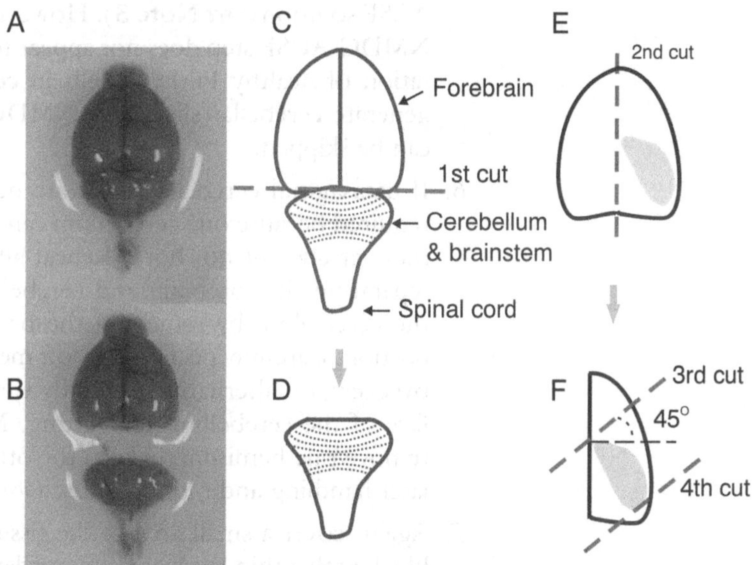

Fig. 1 A diagram of the steps, orientation, and cutting angles for preparing mouse brainstem and hippocampal tissue blocks for slicing. (**a**, **b**) A photograph of dorsal view of the whole mouse brain and after bisection between the forebrain and cerebellum/brainstem. (**c**) A diagram of the bisection of the forebrain and cerebellum and brainstem by the first cut (red dotted line). (**d**) The cerebellum and brainstem tissue block. (**e**) The second cut to separate the brain hemispheres. The pink shadow indicates where the hippocampal formation is located. (**f**) A diagram of the third and fourth cuts to prepare a brain tissue block for slicing transverse hippocampal slices

blade is close to the level of the cerebellum. Adjust the slice thickness to 180–200 μm. Cut the tissue at a very slow rate of advancement and high rate of horizontal oscillation (8–9 Hz). Discard the first few cuts that basically remove the remaining cervical portions of the spinal cord. Collect the brainstem slices at the level as planned. In this case, we collect the portions containing both cerebellar and brainstem slices. Carefully separate the cerebellar and brainstem slices and transfer them with a wide-bore Pasteur pipette to a holding chamber containing the same slicing solution (*see* **Note 2**).

5. Protective recovery incubation: Slices are incubated initially in the same slicing solution for ~30 min at room temperature (directly transferring slices to the next solution by omitting this step also works fine) and then in NMDG-ACSF solution for ~15 min at 35–37 °C. Finally, transfer the slices to a mixture (1:1) of the "slicing" solution and ACSF in another holding chamber aerated continuously with 95% O_2/5% CO_2 at room temperature (25 °C) for at least another 40 min before finally transferring to ACSF for the experiments. We noted that cells in brain slices look healthier and can survive more than 10 hours when incubated in this mixture of slicing and ACSF solutions (*see* **Note 3**). However, we also noted that the NMDG-ACSF step does not appear to be beneficial for preparation of healthy Purkinje cells in cerebellar slices. Thus, to generate cerebellar slices, the NMDG-ACSF incubation step can be skipped.
6. If parasagittal cerebellar slices are desired, another mouse is required because one cerebellum can be cut either sagittally or transversely, but not both. Repeat all 3.1.4 procedures. After separating the forebrain and cerebellum (Fig. 1a, b), isolate the cerebellum by removing the brain stem and spinal cord portion. Carefully peel off the soft membrane with two fine tip tweezers, or, alternatively, slightly trim a small area of the surface of the cerebellum for cutting. Make one sagittal cut to remove one hemisphere; keep the other hemisphere intact for later handling and mounting use (*see* **Note 4**).
7. Again, cover a small area of the tissue pedestal near the agar block with a thin layer of cyanoacrylate glue. Use a pair of fine tweezers to hold the remaining hemisphere loosely, or simply insert the tips of tweezers into the hemisphere tissue, and then mount the cerebellar tissue block on the cutting stage of the tissue pedestal with the ventral side (mostly white matter tissue) of the cerebellar vermis block against the agar block and the sagittal hemisphere cutting end of the cerebellar vermis block against the glue layer.
8. Lower the blade until it is close to the level of the vermis portion. Adjust the slice thickness to 180–250 μm. Note that thin

parasagittal cerebellar slices will cut off most of the parallel fibers. Slice the tissue at a very slow rate of advancement and high rate of horizontal oscillation (8–9 Hz). Discard the first few cuts that basically remove the hemispheric portion. Collect five to eight slices generated from the vermis portion, and transfer them with a wide-bore Pasteur pipette to the holding chamber containing the slicing solution. All other procedures are identical as described for brainstem and transverse cerebellar slices above.

9. If hippocampal slices are desired, all the procedures for cutting and incubation are similar, except that the preparation of the tissue block containing the hippocampus is different. To cut transverse hippocampal slices, the tissue block should be prepared as shown in Fig. 1e–f. Glue the third cut surface of the tissue block on the cutting stage of the tissue pedestal. Depending on the purpose of the experiments, the excitatory and inhibitory synaptic circuits can be better preserved by slightly adjusting the cutting angles. In general, excitatory pathways are better preserved when slices were cut at an angle of 15–30° between the transverse axis and the longitudinal axis of the hippocampus, whereas the recurrent inhibitory pathways are better preserved if the angle is less acute – 30–45°. The slices generated in this way will include both cortical and hippocampal portions.

   Using the procedures described above, we have consistently produced healthy cortical, hippocampal, brainstem, and cerebellar slices (Fig. 2) for electrophysiological recordings in our laboratory. These procedures without the NMDG-ACSF incubation step also produce healthy brain slices from young animals.

3.2 Whole Cell Patch-Clamp Recording in Brain Slices

3.2.1 Whole Cell Current-Clamp Recordings in Brain Slices

Many neurotoxic chemicals alter the passive and active membrane electrical properties of neurons to affect neuronal excitability and action potential firing. These effects can be rigorously assessed using whole cell current-clamp recording techniques in acute brain slices. Here we will use the brainstem slices as an example to show how to use whole cell current-clamp recordings in brain slices to examine the effects of neurotoxic chemicals on neuronal membrane excitability and action potential firing pattern.

1. While waiting for the brain slice recovery incubation, turn on the computer, amplifier, and microscope to pre-warm for 15–30 min. Load the data acquisition programs such as Clampex® of pClamp® 10.x and MultiClamp® 700B Commander. At the same time, prewash the recording chamber with deionized water and then oxygenated ACSF. Adjust the flow rate to 2–4 ml/min. Check if the ground pellet is properly immersed in the bath. Check if the silver wire of

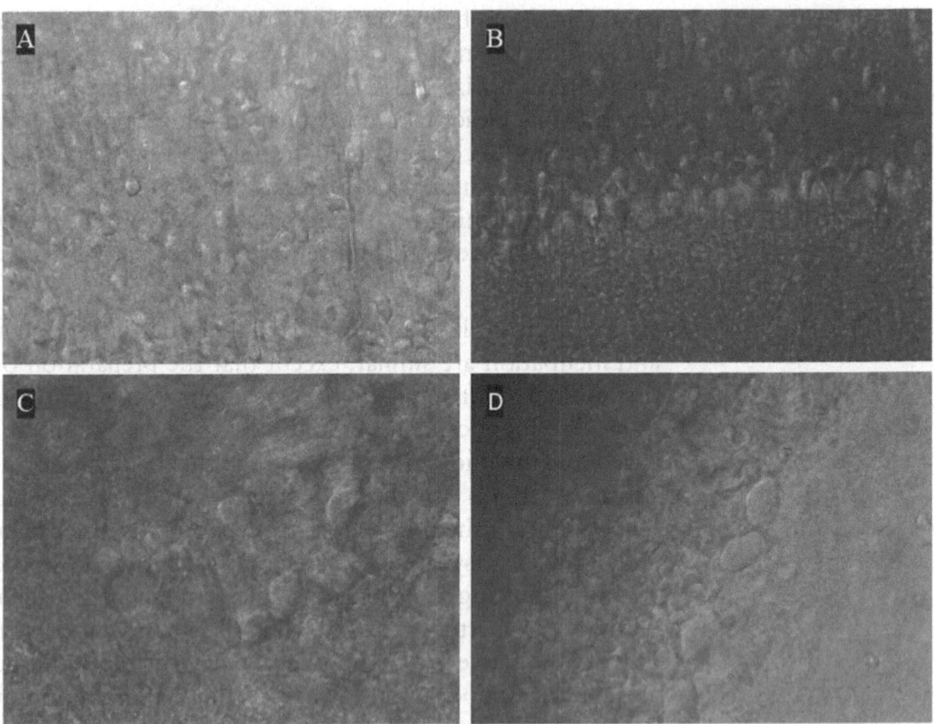

Fig. 2 Photographs of neurons in slices of different brain regions prepared from 3- to 6-month-old mice using our simplified method. (**a**) Cortical layer two-third neurons in a cortical slice. (**b**) CA1 neurons in a hippocampal slice. (**c**) Hypoglossal motoneurons in a brainstem slice. (**d**) Purkinje cells in a cerebellar slice

pipette holder needs to be chlorided (for all electrophysiology-related terms, please see [42, 66] or if the pipette holder needs to be cleaned (*see* **Note 5**).

2. Transfer a brainstem slice from the holding chamber to the recording chamber. Gently put a mesh-containing anchor such as the SHD-26/H2® slice anchor (Warner Instruments, Inc.) over the slice, and then perfuse the chamber continuously with oxygenated ACSF at a rate of 2–4 ml/min (*see* **Note 6**). Lower the 40× water immersion objective lens until it touches the solution to form a "water column." Focus and select a healthy cell (*see* **Note 7**). In brainstem slices, use the central canal or fourth ventricle as a landmark to locate the motoneurons such as those of the nucleus hypoglossal motoneurons. For a better viewing of the neurons, use the video monitor, change the light path to the video camera path of the microscope, and switch to the infrared filter. Adjust the light and zoom on tube to obtain a better image. When done, switch back to the regular light path.
3. Get a freshly pulled and fire-polished pipette, fill it with the K-gluconate-based internal solution, and insert it into the pipette holder which is already attached to the amplifier head-

stage (*see* **Note 8**). Defocus by raising the water immersion objective lens slightly (but don't lose the water column!) to provide more working space for managing the pipette electrodes. Apply gentle positive pressure to the back of the patch electrode using a 1–10 ml syringe (~0.2 ml) and then close the three-way valve to maintain the pressure. Adjust the access angle of the recording electrode to be as large as possible. With the 3.5 mm long working distance and access angle of 45° water immersion objective (NIKON Eclipse FN1), we normally have a 40–45° electrode access angle.

Lower and advance the electrode tip into bath solution; set the LOWPASS BESSEL FILTER at 4–6 kHz and a proper GAIN level on MultiClamp Commander. Run the "Seal Test" program in Clampex and monitor the response of the electrode to a test pulse on the computer screen. One can directly read the electrode resistance from the built-in Seal Test program in Clampex if using pClamp10.x. Recording electrodes should have a resistance of 3–5 MΩ when filled with the potassium gluconate (K-gluconate)-based pipette solution (see Sect. 2.1).

Finding the electrode under 40× water immersion objective lens is somewhat of a challenge for a first-time user. One effective way to do so is first to monitor the advancement of the electrode tip from the side view by visual inspection until the electrode tip approximately reaches the center under the objective. Then look through the eye pieces by changing focus and moving electrode back and forth until the shadow of the electrode is visible. Be sure the tip of the electrode is free of dust or debris. Continue to focus directly on the electrode tip while advancing and lowering it toward the targeted cell (*see* **Note 9**).

4. Just before the electrode tip touches the cell membrane, check and zero the baseline by clicking the Auto Pipette Offset on MultiClamp Commander or manually adjusting the PIPETTE OFFSET of the Axopatch 200B amplifier (null the liquid junction potential of electrode). Use the remote fine control of the manipulator to approach the selected cell soma slowly. One should see the tissue debris on the cell surface being blown away by the wave of ejected solution streaming out of the pipette tip as the tip of the pipette approaches the cell membrane. This helps to clean the surface of the targeted cell (*see* **Note 10**).

   Continue to advance or lower the electrode until it touches the cell and causes a small dimple on the cell membrane. Then immediately release the positive pressure by opening the valve. At this point, the size of square current pulse of "Seal Test" on the screen of the computer should decrease to 1/3–1/2, or the seal resistance should increase. Apply gentle negative pressure by a 1–10 ml syringe (many investigators use mouth) and a negative holding potential of –60 to −80 mV to the pipette by clicking "Patch" until a gigaohm (10^9 Ω) seal forms. Once the gigaohm seal is established, immediately release the negative pressure and holding potential.

Now one can rupture the cell membrane. To rupture the cell membrane, gently apply negative pressure by using the same syringe. Once the cell membrane is ruptured, quickly run a RAMP protocol, which continuously change voltages range from −100 to +60 mV at a rate of 0.3–0.5 mV/ms, at the same time quickly read zero-current potential (i.e., resting membrane potential) value by switching to $I = 0$. One can also examine the membrane capacitance (C_m), membrane resistance (R_m), and electrode access resistance (R_a) using the integral "Membrane Test" program of Clampex®, or immediately perform whole cell capacitance and series resistance compensation. Record these initial parameter values.

5. After capacitance and series resistance compensations have been completed, click or switch the clamp mode to $I = 0$, then open the program protocols designed for current-clamp recordings. Click or switch to "I-C" mode. Does a proper bridge balance if MultiClamp 700B is used? Individual action potentials in neurons can be evoked by injections of a series of 1 ms currents varying from sub- to suprathreshold stimuli in 5–10 pA steps from their resting membrane potentials (Fig. 3a). Repetitive action potential firing can be evoked by injections of a series of 500–1500 ms currents varying from −60 to 250 pA at

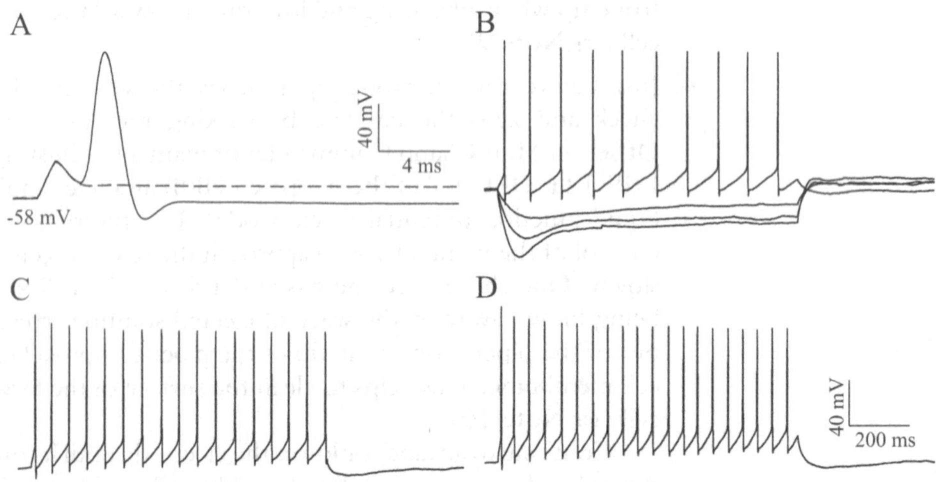

Fig. 3 Representative depiction of action potential and firing pattern of hypoglossal motoneurons in brainstem slice of the mouse. (**a**) An action potential evoked by injection of a short pulse (1 ms) of threshold depolarizing current (solid trace) while the cell was held at resting membrane potential (−58 mV). The dotted line indicates a response evoked by a subthreshold stimulation. (**b–d**) In the same cell, responses were evoked by injections of 1500 ms pulse currents of −60, −40, −20 and +20 pA (**b**), +40 pA (**c**), and +60 pA (**d**)

10 pA-step from their resting membrane potentials (Fig. 3b). Spontaneous action potential firing is recorded at $I = 0$ for 2 min (Fig. 11.3c). If the slices are prepared from animals after in vivo exposure to the test chemicals, the current-clamp experiments are basically completed. One can switch back to $I = 0$ and open protocols for voltage-clamp recordings to continue voltage-clamp experiments if the investigator plans to do so (see next section). If this is an in vitro experiment to examine the effect of a given chemical on neuronal excitability, after stable baseline recordings, one can expose the brain slices to the test chemicals to examine their concentration- and time-dependent effects on membrane electrical properties, action potential, and firing patterns (*see* **Note 11**).

3.2.2 Whole Cell Voltage-Clamp Recording in Brain Slices

Many chemicals can affect either presynaptic transmitter release, or postsynaptic receptor sensitivity, or both, to disrupt synaptic transmission. These effects can be assessed in isolated brain slices by whole cell voltage-clamp recordings of spontaneous or evoked postsynaptic currents and short-term plasticity (STP).

At synapses, postsynaptic responses do not simply follow the pattern or strength of presynaptic signals. Instead, they are modified in a time- and activity-dependent manner by STP, leading to either synaptic enhancement (facilitation) or a decrease (depression) in synaptic strength. Because of its rapid timescale (milliseconds to minutes), STP is thought to serve as a filter in information processing, which selectively strengthens or weakens the presynaptic activity and modifies postsynaptic responses [67–70]. Thus, examining STP will help investigators to understand how the tested chemical disrupts central synaptic function. Here we use cerebellar slices as an example to show how STP is examined. We chose the cerebellar slices because of the unique synaptic circuits of the cerebellar cortex: a single cerebellar Purkinje cell (PC) receives two major excitatory inputs: climbing fibers (CFs) and parallel fibers (PFs). A single CF exclusively innervates proximal dendrites of a PC, whereas hundreds of thousands of PFs selectively innervate the distal dendrites of PCs. Activation of CFs or PFs will generate different excitatory postsynaptic responses.

1. Repeat procedures 3.2.1.1–3.2.1.2, but use a transverse or parasagittal cerebellar slice and find a healthy PC.
2. Connect the concentric bipolar metal stimulation electrode and the Grass S88 stimulator. Raise the 40× water immersion objective lens slightly (but don't lose the water column!) to provide more working space for managing the stimulation electrodes. Again, use the same method to find the stimulation electrode first, and then continue to focus while advancing and lowering the stimulation electrode. If it is a transverse cerebellar slices, place the stimulation electrode in the external

molecular layer. If a parasagittal cerebellar slice is used, place the stimulation electrode in the white matter or granule cell layer (*see* **Note 12**).

3. Once the stimulation electrode is positioned at the desired place, then repeat procedures 3.2.1.3–3.2.1.4. After compensating for membrane capacitance and series resistance, one can first record spontaneous excitatory postsynaptic currents (sEPSCs) in a PC by running a "gap-free" protocol in Clampex with a membrane-holding potential of −60 or −70 mV (Fig. 4). The gap-free protocol should last at least 2 min longer if the frequency of spontaneous synaptic events is too low. Then, open another protocol designed for recording presynaptic stimulation-evoked postsynaptic responses in PCs. To do so, one can stimulate PFs in a transverse (coronal) cerebellar slice or CFs in a parasagittal cerebellar slice to generate PF-Purkinje cell excitatory postsynaptic currents (PF-EPSCs) or CF-Purkinje cell (CF-EPSCs). More specifically, to record PF-EPSCs, turn on the pulse generator to deliver stimuli at a frequency of 0.1 Hz, 0.1 ms and initially a low stimulus intensity with the pre-positioned stimulating electrode in the

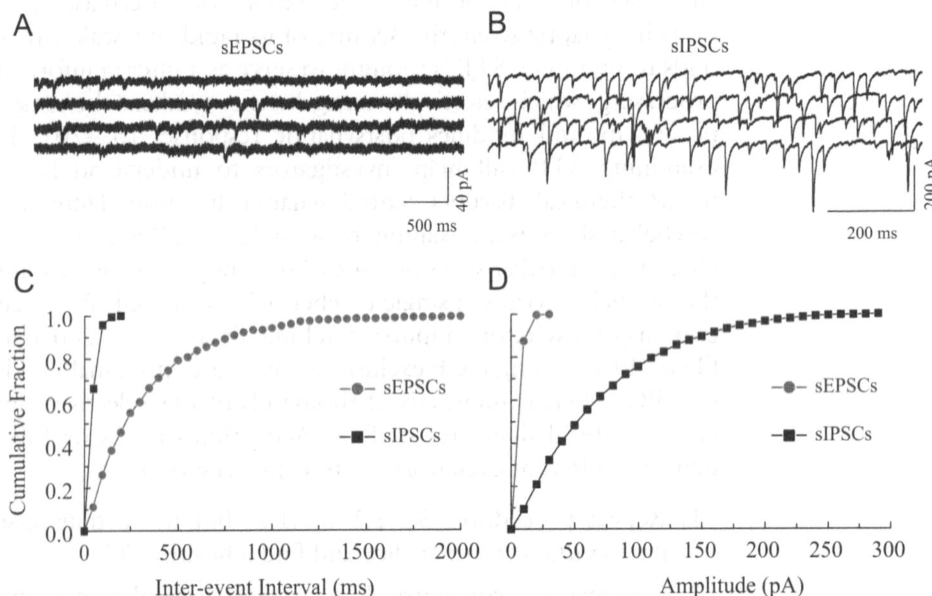

Fig. 4 Representative depiction of spontaneous excitatory or inhibitory postsynaptic currents (sEPSCs or sIPSCs) of Purkinje cells in cerebellar slices from a 12-month-old mouse. (**a**) sEPSCs were recorded in a Purkinje cell of a cerebellar slice from a holding potential of −70 mV in the presence of 10 μM bicuculline. (**b**) sIPSCs were recorded from another Purkinje cell in a different cerebellar slice from a holding potential −70 mV in the presence of 10 μM CNQX and 100 μM APV. sEPSCs and sIPSCs were recorded using K-gluconate- or CsCl-based pipette solution, respectively, in these two cells. (**c**, **d**) Cumulative distribution of inter-EPSC intervals or amplitude of sEPSCs or sIPSCs shown in (**a**) or (**b**)

surface of the external molecular layer. Gradually increase the stimulus intensity until it produces an EPSC amplitude approximately 50–60% of the maximum response for a given cell. The protocol should be triggered externally by stimulus. The cell should be held at a holding potential of −60 or −70 mV. Once the stimulus intensity is set, open a protocol with paired-pulse stimulations. The two stimulus pulses have the same stimulation intensity but with inter-pulse intervals that vary from 20 to 500 ms. Normally, paired-pulse stimulation of PFs generates paired-pulse facilitation (PPF) of PF-EPSCs (Fig. 5, Left). Similarly, to record CF-EPSCs in a Purkinje cell in a sagittal slice, deliver stimuli at a frequency of 0.1 Hz, 0.1 ms and initially a low stimulus intensity to stimulate the CFs with the pre-positioned stimulating electrode in the white matter or granule cell layer. Gradually increase the stimulus intensity until "all-or-none" CF-EPSC responses

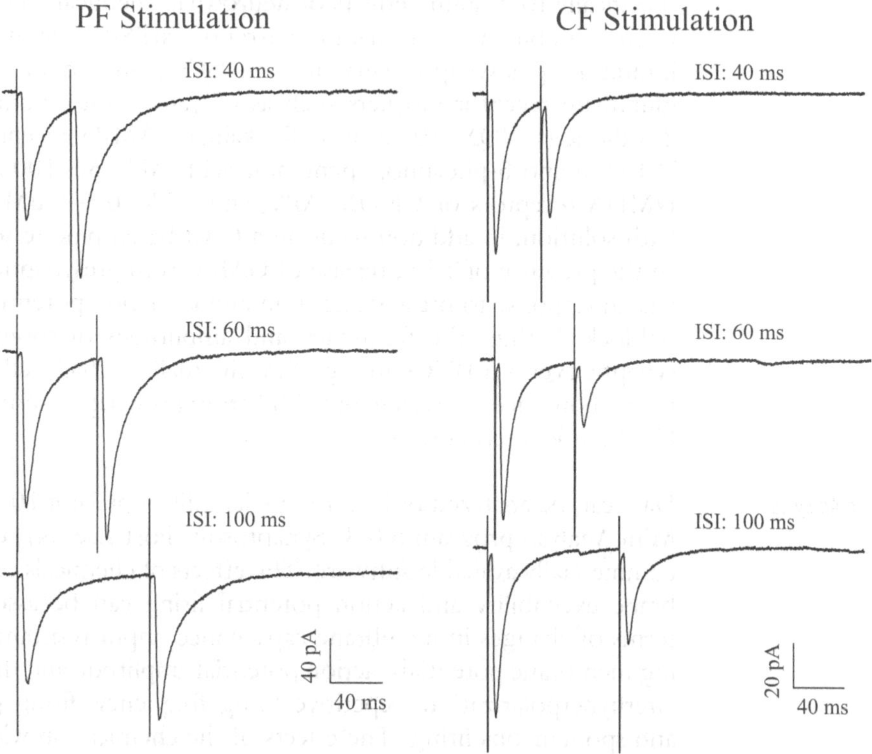

Fig. 5 Representative depiction of paired-pulse evoked excitatory postsynaptic currents (EPSCs) in Purkinje cells of cerebellar slices from 12-month-old mice. *Left*, paired-pulse responses were evoked from a Purkinje cell in a transverse cerebellar slice by stimulation of the parallel fibers (PF stimulation) in the molecular layer at inter-stimulus intervals (ISI) of 40, 60, and 100 ms. Note that the second EPSCs are larger than the first ones, indicating paired-pulse facilitation (PPF). *Right*, paired-pulse responses were evoked from a Purkinje cell in a parasagittal cerebellar slice by stimulation of climbing fiber in the white matter or granule cell layer near the recorded Purkinje cell. Note that the second EPSCs are smaller than the first ones, indicating paired-pulse depression (PPD)

appear. Then use the same paired-pulse protocol to evoke CF-EPSC responses (Fig. 5, right). Unlike responses evoked by paired-pulse stimulation of PFs, paired-pulse stimulation of CFs normally produces paired-pulse depression (PPD) of CF-EPSCs. Again, if the slices are prepared from animals after in vivo exposure to the test chemicals, the experiments are basically completed. If this is an in vitro experiment to examine the effect of a given chemical on neuronal excitability, after stable baseline recordings have been obtained, one can expose the brain slices to the test chemicals to examine their concentration- and time-dependent effects on sEPSCs and STP (*see* **Note 13**).

If one is particularly interested in examining effects of chemicals on spontaneous or miniature EPSCs (sEPSCs or mEPSCs), a $GABA_A$ receptor antagonist such as bicuculline (10–20 μM) or picrotoxin (100 μM) or bicuculline or picrotoxin plus tetrodotoxin (TTX) should be added to the bath solution. Similarly, if one wants specifically to examine effects of neurotoxic chemicals on spontaneous inhibitory postsynaptic currents (sIPSPs) or miniature inhibitory postsynaptic currents (mIPSCs), one should include glutamate receptor blockers such as 6-cyano-7-nitroquinoxaline-2,3-dione (CNQX, 10–20 μM) for kainate/AMPA receptors and D,L-2-amino-5-phosphonopentanoic acid (APV, 50–100 μM) for NMDA receptors or CNQX, APV, and TTX (0.5–1 μM) in the bath solution, in addition to using a CsCl-based pipette solution. In the presence of TTX, release of GABA from presynaptic terminals in response to presynaptic spontaneous action potential firing is blocked; thus, the frequency and amplitudes of spontaneous synaptic events mIPSCs are significantly smaller, which reflexes the true "spontaneous" release of GABA from presynaptic terminals of GABAergic interneurons.

3.2.3 Data Analysis

Data can be analyzed offline using Clampfit of pClamp10.x or the Mini Analysis program 6.0.3 (Synaptosoft, Fort Lee, NJ) or other commercially available software. The effects of chemicals on membrane excitability and action potential firing can be assessed in terms of changes in membrane capacitance, input resistance, resting membrane potentials, action potential amplitude and duration, afterhyperpolarization, repetitive firing frequency, firing pattern, and spontaneous firing. The effects of the chemical on whole cell synaptic currents can be assessed in terms of (a) changes in frequency and amplitude of sEPSCs or mEPSCs; (b) changes in amplitudes of evoked responses; (c) changes in 10–90% rise time, half-width or decay time of synaptic currents; (d) changes in current-voltage (I–V) relationship and reversal potentials; (e) time courses of effects of chemical on PF-EPSCs, CF-EPSC, and MF-EPSCs, and spontaneous synaptic currents; (f) paired-pulse evoked PF-EPSCs and CF-EPSCs, the ratios of the second pulse-evoked eEPSC to the first pulse-evoked eEPSc will be calculated; and (g) concentration- and time-dependent responses and interaction of the test chemical with known relevant pharmacological agonists or antagonists.

4 Notes

1. To avoid damaging the cerebellum and brainstem, we prefer to open the skull from the front part of the head toward the rear.
2. An agar or Sylgard 184 block is very helpful in cutting good slices from adult animals. Be sure that no gap exists between the tissue block and the agar block. Otherwise, the blade will push the tissue and cut slices unevenly.
3. All solutions during cutting and recovery are aerated continuously with 95% O_2/5% CO_2 gas mixture.
4. Leaving one hemisphere uncut will really simplify later handling and gluing of the tissue to the tissue pedestal, and moreover, it will reduce the chances of damaging the cerebellar vermis.
5. The pipette holder should be cleaned regularly after use. If one has problems with making a gigaohm seal, check if the electrode holder has any cracks or leaks.
6. Regardless what perfusion rate is used, it is important to maintain it consistently for every experiment. Otherwise, it will be difficult to compare individual experiments from day to day. This is because the onset of in vitro effects of a given chemical is affected by the perfusion rate.
7. Healthy cells have smooth, soft, clearly visible, and "phase-bright" cell membranes under phase contrast. The healthy cell membrane can be easily dimpled by the ejected solution stream emanating from the tip of the approaching pipette with positive pressure added to the back of pipette. Conversely, unhealthy cells usually have a uniform dark, rough, and "crinkled" appearance with a shrunken cell body or membrane with granular-like patches. Unhealthy cells may also look like "ghost" swelling cells and appear transparent.
8. One trick to fill the pipette is first to drop one small drop of pipette solution on a piece of clean parafilm and dip the tip of pipette in the solution for a few seconds. Then backfill the pipette tip with a MicroFil MF28G needle or homemade pipette filler pulled from a 1 ml plastic syringe. Gently tap the pipette to remove any trapped air bubbles and then fill the pipette up to about 50–70% of its total length. Don't fill it completely. Also, keep the pipette filler on an ice block.

9. Electrode resistance should be monitored during electrode advancement. Any increase in resistance before touching the targeted cell indicates clogging of the electrode tip either by residual particles in the pipette or of bath solution or by slice tissue debris; any of these will certainly interfere with formation of a high resistance seal. Apply more positive pressure to clear the pipette tip or change to a new electrode. This happens when positive pressure is applied to the back of pipette. Alternatively, do not apply positive pressure until the tip is very close to the slice tissue surface.
10. If you don't see the ejected solution stream pushing away the surface tissue debris, perhaps either the pipette tip is clogged or no positive pressure is added to the back of the pipette or the electrode holder leaks. In either case, a good gigaohm seal will not be formed.
11. Always first switch the recording mode to $I = 0$ when changing protocols between voltage-clamp to current-clamp or vice versa. Otherwise, cells will be killed.
12. It is recommended to preset the stimulation electrode in place before establishing whole cell recording configuration, because position of the stimulating electrode may cause loss of whole cell seal, especially when the concentric bipolar electrode is used as it has a big tip.
13. The relative locations of stimulating and recording electrodes are important for evoking synaptic responses, particularly for CF-EPSC recording. Adjustments of stimulating electrode location are often required in order to get evoked CF-EPSC responses.

References

1. Yamamoto C, McIlwain H (1966) Electrical activities in thin sections from mammalian brain maintained in chemically-defined media in vitro. J Neurochem 13:1333–1343
2. Yuan Y, Atchison WD (1993) Disruption by methylmercury of membrane excitability and synaptic transmission of CA1 neurons in hippocampal slices of the rat. Toxicol Appl Pharmacol 120:203–215
3. Yuan Y, Atchison WD (1995) Methylmercury acts at multiple sites to block hippocampal synaptic transmission. J Pharmacol Exp Ther 275:1308–1316
4. Yuan Y, Atchison WD (1997) Action of methylmercury on $GABA_A$ receptor-mediated inhibition is primarily responsible for its early stimulatory effects on hippocampal CA1 synaptic transmission. J Pharmacol Exp Ther 282:64–73
5. Yuan Y, Atchison WD (1999) Comparative effects of methylmercury on parallel-fiber and climbing-fiber responses of rat cerebellar slices. J Pharmacol Exp Ther 288:1015–1025
6. Yuan Y, Atchison WD (2003) Methylmercury differentially affects $GABA_A$ receptor-mediated spontaneous IPSCs in Purkinje and granule cells of rat cerebellar slices. J Physiol Lond 550:191–204
7. Yuan Y, Atchison WD (2007) Methylmercury-induced increase of intracellular Ca^{2+} concentration in presynaptic fibers causes increased frequency of spontaneous synaptic responses in cerebellar slice of rat. Mol Pharm 71:1109–1121
8. Yuan Y, Atchison WD (2016) Multiple sources of Ca^{2+} contribute to methylmercury-induced increased frequency of spontaneous inhibitory synaptic responses in cerebellar slices of rat. Toxicol Sci 150:117–130

9. Aghajanian GK, Rasmussen K (1989) Intracellular studies in the facial nucleus illustrating a simple new method for obtaining viable motoneurons in adult rat brain slices. Synapse 3:331–338
10. Salin PA, Prince DA (1996) Electrophysiological mapping of $GABA_A$ receptor-mediated inhibition in adult rat somatosensory cortex. J Neurophysiol 75:1589–1600
11. Moyer JR Jr, Brown TH (1998) Methods for whole-cell recording from visually preselected neurons of perirhinal cortex in brain slices from young and aging rats. J Neurosci Methods 86:35–54
12. Schmidt-Hieber C, Jonas P, Biochofberger J (2004) Enhanced synaptic plasticity in newly generated granule cells of adult hippocampus. Nature 429:184–187
13. Dougherty KA, Islam T, Johnston D (2012) Intrinsic excitability of CA1 pyramidal neurons from the rat dorsal and ventral hippocampus. J Physiol 590:5707–5722
14. Mainen ZF, Maletic-Savatic M, Shi SH (1999) Two-photon imaging in living brain slices. Methods 18:231–239
15. Tanaka Y, Tanaka Y, Furuta T, Yanagawa Y, Kaneko T (2008) The effects of cutting solutions on the viability of GABAergic interneurons in cerebral cortical slice of adult mice. J Neurosci Methods 171:118–125
16. Ye JH, Zhang J, Xiao C, Kong JQ (2006) Patch-clamp studies in the CNS illustrate a simple new method for obtaining viable neurons in rat brain slices: glycerol replacement of NaCl protects CNS neurons. J Neurosci Methods 158:251–259
17. Lerchner W, Xiao C, Nashmi R, Slimko EM, van Trigt L, Lester HA (2007) Reversible silencing of neuronal excitability in behaving mice by a genetically targeted, ivermectin-gated Cl– channel. Neuron 54:35–49
18. Nashmi R, Velumian AA, Chung I, Zhang L, Agrawal SK, Fehlings MG (2002) Patch-clamp recordings from white matter glia in thin longitudinal slices of adult rat spinal cord. J Neurosci Methods 117:159–166
19. Balthasar N, Mery PF, Magoulas CB, Mathers KE, Martin A, Mollard P (2003) Growth hormone-releasing hormone (GHRH) neurons in GHRH-enhanced green fluorescent protein transgenic mice: a ventralhypothalamic network. Endocrinology 144:2728–2740
20. Richerson GB, Messer C (1995) Effect of composition of experimental solutions on neuronal survival during rat brain slicing. Exp Neurol 131:133–143
21. Rice ME (1999) Use of ascorbate in preparation and maintenance of brain slices. Methods Companion Methods Enzymol 18:144–149
22. Brahma B, Forman RE, Stewart EE, Nicholson C, Rice ME (2000) Ascorbate inhibits edema in brain slices. J Neurochem 74:1263–1270
23. MacGregor DG, Chesler M, Rice ME (2001) HEPES prevents edema in rat brain slices. Neurosci Lett 303:141–144
24. Ting JT, Daigle TL, Chen Q, Feng G (2014) Acute brain slice methods for adult and aging animals: application of targeted patch clamp analysis and optogenetics. Meth Mol Biol 1183:221–242
25. Llano DA, Slater BJ, Lesicko AM, Stebbings KA (2014) An auditory colliculothalamocortical brain slice preparation in mouse. J Neurophysiol 111:197–207
26. Dergacheva O, Dyavanapalli J, Piñol RA, Mendelowitz D (2014) Chronic intermittent hypoxia and hypercapnia inhibit the hypothalamic paraventricular nucleus neurotransmission to parasympathetic cardiac neurons in the brain stem. Hypertension 64(3):597–603
27. Pan G, Li Y, Geng HY, Yang JM, Li KX, Li XM (2015) Preserving GABAergic interneurons in acute brain slices of mice using the N-methyl-D-glucamine-based artificial cerebrospinal fluid method. Neurosci Bull 31:265–270
28. Huang S, Uusisaari MY (2013) Physiological temperature during brain slicing enhances the quality of acute slice preparations. Front Cell Neurosci 7:48. https://doi.org/10.3389/fncel.2013.00048. eCollection 2013
29. Ankri L, Yarom Y, Uusisaari MY (2014) Slice it hot: acute adult brain slicing in physiological temperature. J Vis Exp (92):e52068. https://doi.org/10.3791/52068
30. Ting JT, Lee BR, Chong P, Soler-Llavina G, Cobbs C, Koch C, Zeng H, Lein E (2018) Preparation of acute brain slices using an optimized N-methyl-D-glucamine protective recovery method. J Vis Exp (132):e53825. https://doi.org/10.3791/53825
31. Wang L, Zhang X, Xu H, Zhou L, Jiao R, Liu W, Zhu F, Kang X, Liu B, Teng S, Wu Q, Li M, Dou H, Zuo P, Wang C, Wang S, Zhou Z (2014) Temporal components of cholinergic terminal to dopaminergic terminal transmission in dorsal striatum slices of mice. J Physiol 592(16):3559–3576
32. Dergacheva O (2015) Chronic intermittent hypoxia alters neurotransmission from lateral paragigantocellular nucleus to parasympathetic cardiac neurons in the brain stem. J Neurophysiol 113:380–389
33. Jiang X, Shen S, Cadwell CR, Berens P, Sinz F, Ecker AS, Patel S, Tolias AS (2015) Principles of connectivity among morphologically

defined cell types in adult neocortex. Science 350(6264):aac9462

34. Altmann L, Weinand-Haerer A, Lilienthal H, Wiegand H (1995) Maternal exposure to polychlorinated biphenyls inhibits long-term potentiation in the visual cortex of adult rats. Neurosci Lett 202:53–56
35. Altmann L, Lilienthal H, Hany J, Wiegand H (1998) Inhibition of long-term potentiation in developing rat visual cortex but not hippocampus by in utero exposure to polychlorinated biphenyls. Brain Res Dev Brain Res 110:257–260
36. Hussain RJ, Parsons PJ, Carpenter DO (2000) Effects of lead on long-term potentiation in hippocampal CA3 vary with age. Brain Res Dev Brain Res 121:243–252
37. Carpenter DO, Hussain RJ, Berger DF, Lombardo JP, Park HY (2002) Electrophysiologic and behavioral effects of perinatal and acute exposure of rats to lead and polychlorinated biphenyls. Environ Health Perspect 110(Suppl 3):377–386
38. Dasari S, Yuan Y (2009) Low level postnatal methylmercury exposure in vivo alters developmental forms of short-term synaptic plasticity in the visual cortex of rat. Toxicol Appl Pharmacol 240:412–422
39. Dasari S, Yuan Y (2010) Methylmercury exposure in vivo induces a long-lasting epileptiform activity in layer II/III cortical neurons of cortical slices of rat. Toxicol Lett 193:138–143
40. Kummer KK, El Rawas R, Kress M, Saria A, Zernig G (2015) Social interaction and cocaine conditioning in mice increase spontaneous spike frequency in the nucleus accumbens or septal nuclei as revealed by multielectrode array recordings. Pharmacology 95:42–49
41. Li XM, Gu Y, She JQ, Zhu DM, Niu ZD, Wang M, Chen JT, Sun LG, Ruan DY (2006) Lead inhibited N-methyl-D-aspartate receptor-independent long-term potentiation involved ryanodine-sensitive calcium stores in rat hippocampal area CA1. Neuroscience 139:463–467
42. Penner R (1995) A practical guide to patch clamping. In: Sakmann B, Neher E (eds) Single-channel recording. Plenum, New York, pp 3–30
43. Glover JC, Sato K, Sato Y-M (2008) Using voltage-sensitive dye recording to image the functional development of neuronal circuits in vertebrate embryos. Dev Neurobiol 68:804–816
44. Carlson GC, Coulter DA (2008) In vitro functional imaging in brain slices using fast voltage-sensitive dye imaging combined with whole-cell patch recording. Nat Protoc 3:249–255
45. Chemla S, Chavane F (2010) Voltage-sensitive dye imaging: technique review and models. J Physiol Paris 104:40–50
46. Coulter DA, Yue C, Ang CW, Weissinger F, Goldberg E, Hsu FC, Carlson GC, Takano H (2011) Hippocampal microcircuit dynamics probed using optical imaging approaches. J Physiol (Land) 589:1893–1903
47. Takano H, Coulter DA (2012) Imaging of hippocampal circuits in epilepsy. In: Noebels JL, Avoli M, Rogawski MA, Olsen RW, Delgado-Escueta AV (eds) Jasper's basic mechanisms of the epilepsies [internet], 4th edn. National Center for Biotechnology Information (US), Bethesda
48. Städele C, Andres P, Stein W (2012) Simultaneous measurement of membrane potential changes in multiple pattern generating neurons using voltage sensitive dye imaging. J Neurosci Methods 203:78–88
49. Wright BJ, Jackson MB (2015) Voltage imaging in the study of hippocampal circuit function and plasticity. Adv Exp Med Biol 859:197–211
50. Kramer RH, Fortin DL, Trauner D (2009) New photochemical tools for controlling neuronal activity. Curr Opin Neurobiol 19:544–552
51. Szobota S, Isacoff EY (2010) Optical control of neuronal activity. Annu Rev Biophys 39:329–348
52. Häusser M (2014) Optogenetics: the age of light. Nat Methods. 2014 11:1012–1014
53. Deisseroth K (2015) Optogenetics: 10 years of microbial opsins in neuroscience. Nat Neurosci 18:1213–1225
54. Song C, Knöpfel T (2016) Optogenetics enlightens neuroscience drug discovery. Nat Rev Drug Discov 15:97–109
55. Kim CK, Adhikari A, Deisseroth K (2017) Integration of optogenetics with complementary methodologies in systems neuroscience. Nat Rev Neurosci 18:222–235
56. Rost BR, Schneider-Warme F, Schmitz D, Hegemann P (2017) Optogenetic tools for subcellular applications in neuroscience. Neuron 96:572–603
57. Lasley SM, Gilbert ME (1996) Presynaptic glutamatergic function in dentate gyrus in vivo is diminished by chronic exposure to inorganic lead. Brain Res 736:125–134
58. Lasley SM, Gilbert ME (2000) Glutamatergic component underlying lead-induced impairment in hippocampal synaptic plasticity. Neurotoxicology 21:1057–1068
59. Suszkiw JB (2004) Presynaptic disruption of transmitter release by lead. Neurotoxicology 25:599–604

60. Altmann L, Mundy WR, Ward TR, Fastabend A, Lilienthal H (2001) Developmental exposure of rats to a reconstituted PCB mixture or aroclor 1254: effects on long-term potentiation and [^{3}H]MK-801 binding in occipital cortex and hippocampus. Toxicol Sci 61: 321–330
61. Gilbert ME, Mundy WR, Crofton KM (2000) Spatial learning and long-term potentiation in the dentate gyrus of the hippocampus of animals developmentally exposure to aroclor 1254. Toxicol Sci 57:102–111
62. Gilbert ME (2003) Perinatal exposure to polychlorinated biphenyls alters excitatory synaptic transmission and short-term plasticity in the hippocampus of the adult rat. Neurotoxicology 24:851–860
63. Pitkänen RI, Korpi ER, Oja SS (1985) Cerebral cortex slices in sodium-free medium: ~on of:synaptic vesicles. Brain Res 326:384–387
64. Berdichevsky E, Mufioz C, Riveros N, Cartier L, Orrego F (1987) Neuropathological changes in the rat brain cortex in vitro: effects of kainic acid and of ion substitutions. Brain Res 423:213–220
65. Aitken PG, Breese GR, Dudek FF, Edwards F, Espanol MT, Larkman PM, Lipton P, Newman GC, Nowak TS Jr, Panizzon KL, Raley-Susrnan KM, Reid KH, Rice ME, Sarvey JM, Schoepp DD, Segal M, Taylor CP, Teyler TJ, Voulalas PJ (1995) Preparative methods for brain slices: a discussion. J Neurosci Methods 59:139–149
66. The Axon guide for electrophysiology and biophysics laboratory techniques. Axon Instruments, Union City. https://mdc.custhelp.com/euf/assets/content/Axon%20Guide%203rd%20edition.pdf
67. Abbott LF, Varela JA, Sen K, Nelson SB (1997) Synaptic depression and cortical gain control. Science 275:220–224
68. O'Donovan MJ, Rnzel J (1997) Synaptic depression: a dynamic regulator of synaptic communication with varied functional roles. Trends Neurosci 20:431–433
69. Chance FS, Nelson SB, Abbott LF (1998) Synaptic depression and the temporal response characteristics of V1 cells. J Neurosci 18(12):4785–4799
70. Abbott LF, Regehr WG (2004) Synaptic computation. Nature 431:796–803

Chapter 12

A Method for Sampling Rat Cerebrospinal Fluid with Minimal Blood Contamination: A Critical Tool for Biomarker Studies

Zhen He, John Panos, James Raymick, Tetyana Konak, Li Cui, Diane B. Miller, James P. O'Callaghan, Serguei Liachenko, Merle G. Paule, and Syed Z. Imam

Abstract

Sampling and analysis of cerebrospinal fluid (CSF) is a common clinical practice used in the diagnosis, treatment, and prevention of neurological diseases. A similar interest is the sampling of CSF from rats to bridge the gap between bench-to-bedside work and to foster the development of new CSF biomarkers for clinical use. Here, we describe an improved procedure with an instrument designed in-house, by which rat CSF was successfully collected with indiscernible blood contamination (via the naked eye/surgical microscope amplification). The sampled CSF amounts were over 100 μl regardless of the animal's body weight, hydration status, and symptoms of systemic damage including, but not limited to, seizure, delusion (such as repeated hemorrhagic self-biting), hematuria, and gastrointestinal bleeding. In adult Sprague-Dawley rats above 300 g, the sampled CSF amounts were reliably at 200 μl or above with this method. There were no deaths related to the CSF sampling procedure. In conclusion, the present method provides a reliable and reproducible approach for collecting 200 μl CSF in rats without blood contamination.

Key words Cerebrospinal fluid, In-house designed instrument, Rat

1 Introduction

The two major factors that define the quality of experimental CSF collections are the absence of blood contamination and the amount of sample volume collected. A CSF sample with no blood contamination is necessary because blood contamination can make biomarker analysis in CSF less reliable and the volume of the CSF sample can help in its utilization in exploring new assays, regardless of the limitation of a sample volume for a particular assay.

In the clinic, results of a normal conventional CSF collection and analysis include the criteria related to its appearance: clear, colorless and CSF cell count: 0–5 white blood cells (all mononuclear), and no red blood cells (https://www.nlm.nih.gov/medlineplus/

Michael Aschner and Lucio Costa (eds.), *Cell Culture Techniques*, Neuromethods, vol. 145, https://doi.org/10.1007/978-1-4939-9228-7_12,

ency/article/003428.htm); in addition to the standards related to pressure, total protein amount, gamma globulin, glucose and chloride. We are not aware of the criteria for a normal CSF test in rats covering the same parameters as in humans. Currently, the one-time sampled CSF amount from a rat ranges between 50 and 180 μl [1–5], which could be a limitation for meeting all parameters tested as in human samples. There is always a dilemma in determining quality of a sampled rat CSF without compromising a part of CSF amount for the actual experimental analysis.

The total volume of rat CSF has been estimated to be between 400 and 500 μl [6, 7], and the cisterna magna and the subarachnoid cavity are the major areas containing CSF. As mentioned above, the current literature documents a one-time sampled amount of rat CSF from the cisterna magna to vary from 50 to 180 μl. Recently, an improved method with the ultrasound-guided approach led to a sample volume of 100–200 μl of CSF in male Sprague-Dawley rats (300–400 g), and the 200 μl CSF mark was reached in 2 of 40 rats [8]. In principle, CSF flows from the ventricles to the cisterns and/or subarachnoid space. It is believable that CSF from all surrounding areas will drain to the space of the lowest pressure/vacuum effect, indicating the potential to collect a higher CSF volume in rats.

In this study, we demonstrate a reliable and reproducible method to collect high-quality rat CSF in a sample volume of 200 μl and above via an approach from the cisterna magna. We also discuss the potential hurdles that lead to a smaller volume of CSF collection and blood contamination in relation to the replacement/orientation of the needle bevel into the cisterna magna for CSF collection.

2 Materials and Methods

The CSF collection method was developed for an ongoing project aimed at exploring fluidic biomarkers of central nervous system neurotoxicity in adult rat models of trimethyltin chloride (TMT) exposure [9]. The procedures and instruments were permanently set up at the beginning without any changes. However, during the progress of the project, the orientation of needle bevel (see details later) was modified to test the potential to maximize one-time CSF sampling amount without blood contamination.

Animals

All animal procedures were approved by the National Center for Toxicological Research (NCTR) Institutional Animal Care and Use Committee. For setting up the standard operation procedures and testing the effectiveness of this in-house designed instrument, six male adult Sprague-Dawley rats were obtained from the NCTR breeding colony and used in a trial study. The trial ended in 2 days with encouraging outcomes: all animals survived the surgery with CSF sampled, and five of six animals displayed clear CSF, while one sampled CSF showed blood contamination. Accordingly, the formal experiment was started thereafter. In total, 103 adult male Sprague-Dawley rats (3 months old, Taconic, Inc.) were used in this study: 48 rats were exposed to a single dose of trimethyltin chloride (TMT) (7 mg/kg, ip), and 55 control rats received a similar volume of vehicle (saline) via ip injection [9]. Two, 6, 10, or 14 days posttreatment, CSF was collected.

Needle Bevel Orientations

In a separate project, we casted space of the cisterna magna in few rats (unpublished data) using colored silicone [10, 11]. We verified that the distance between the atlantooccipital membrane and the cisterna magna floor/the dorsal surface of the medulla oblongata was approximately 1 mm (image not shown) with normal physiological status/posture. Accordingly, during sampling, it was hypothesized that the bevel of a needle could be placed in three orientations: Type I, the bevel opening facing the cisterna magna floor; Type II, the bevel opening facing sideways (left or right); and Type III, the bevel opening facing the opposite of the cisterna magna floor (Fig. 1). Using same size of the bevel and the same angle in which a needle is inserted into, Type I was hypothetically favored because it would be less possible for the tip of needle to

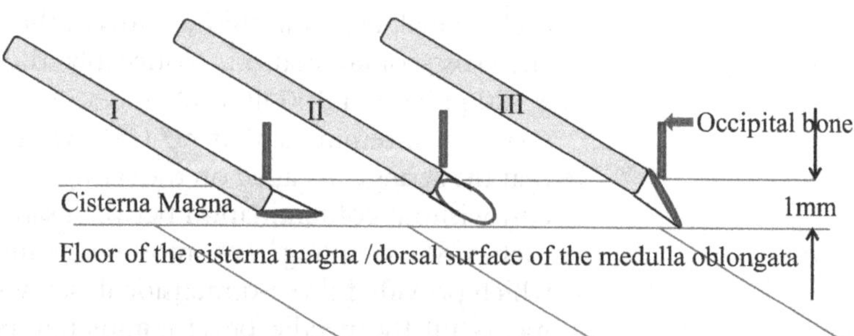

Fig. 1 Orientation of the bevel tip of the sampling needle. Orientation of the bevel tip of the sampling needle was designed based on our unpublished data that the distance between the atlantooccipital membrane and the cisterna magna floor/the dorsal surface of the medulla oblongata was about 1 mm when rats were in the physiological posture. Accordingly, Type I orientation was hypothetically ideal in attempting a complete placement of the needle bevel within the cisterna magna in such a limited space. In contrast, Type II or Type III would risk placing the bevel tail part outside of the cisterna magna without penetrating brain parenchyma in the same sized space. (I) The bevel tip of the sampling needle facing down; (II) the bevel facing sideways; (III) the bevel facing up

penetrate the cisterna magna floor (left panel of Fig. 1). Conversely, Type III would risk penetrating parenchyma of the medulla oblongata with the bevel tip, while the rear side of the bevel might remain the outside the cisterna magna (right panel of Fig. 1). Certainly, the entire bevel head must be placed into the cisterna magna so that the CSF could be effectively sampled by a vacuuming mechanism (see next paragraph). Nevertheless, all three bevel orientations were tested in the current study: Type I was conducted with 6 rats from the NCTR breeding colony in the trial study and 34 rats in the first portion of the TMT study, Type II was used in 34 rats in the middle portion, and Type III was conducted in 35 rats in the last portion of the TMT study.

In-House Designed Instrument for Sampling CSF in Rats

Due to the small physical size of the cisterna magna in rats and a possibility of needle penetration into the medulla oblongata, it could cause blood/tissue contamination to the sampled CSF. Therefore the accuracy, stability, and reliability in all procedures were essential to prevent sudden death/post-surgery paralysis. Accordingly, an in-house designed instrument (Fig. 2a) was developed integrating four parts: (1) a syringe infusion/suction pump set, (2) a 1058 mm length of PE50 tubing, (3) a needle head which was fixed onto (4) the final part, a micromanipulator; all were from commercially available resources. A needle head/the metal pipe part (BD 27G1-1/4) was detached from its tail part, inserted into one end of the PE50 tubing. The other end of the PE50 tubing was then mounted onto the needle of the syringe infusion/suction pump set. The needle head and tubing set was only used once for each animal, ensuring sterile status of the needle head and avoiding any cross-contamination. Noticeably, the tubing was marked with a 100 μl level, a 200 μl level, and a safety guard mark, which here served as a reminder that no CSF would pass across the mark so that the syringe installed on the syringe pump could be used repeatedly without contamination between samples. The needle head for each CSF sampling was mounted onto the micromanipulator which provided three-dimensional, screw-advanced, precise movements till the needle bevel completely penetrated the atlantooccipital membrane at the desired angle (see next paragraph). Finally, the syringe infusion/suction pump provided stable, controllable vacuum, and digitally controlled volume sampling. The in-house designed CSF sampling instrument worked via a volume-change-induced vacuum mechanism provided by the syringe infusion/suction pump.

CSF Sampling

After fur removal by shaving, the rat was placed into a stereotaxic frame (Fig. 1a) under anesthesia of 1–4% of isoflurane. A rhom-

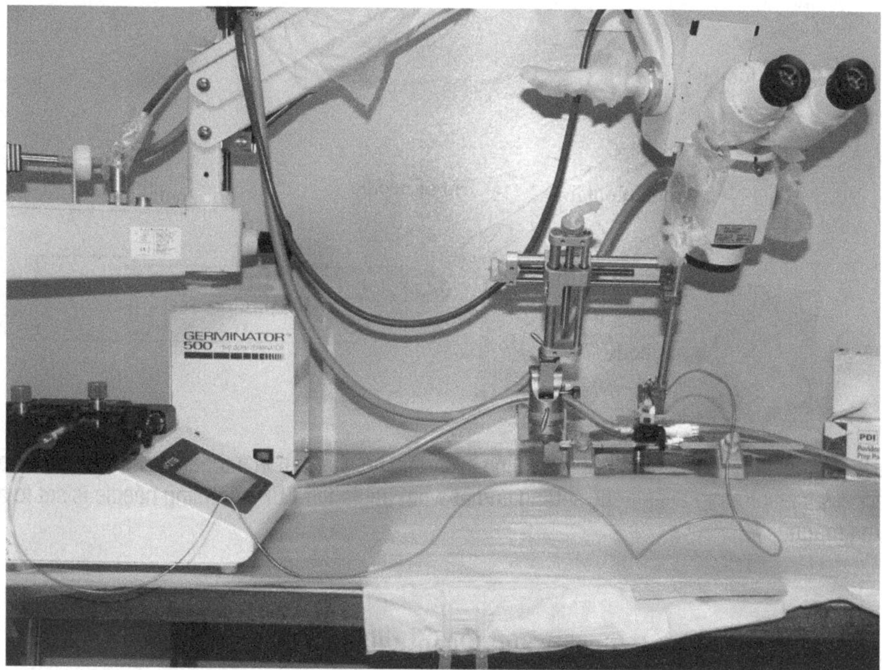

Fig. 2 In-house designed instrument for CSF sampling in rats. The in-house designed instrument (Image **a**) integrates four parts: (I) a syringe installed in an infusion/suction pump and (II) a 1058 mm length of PE50 tubing, one end of which is mounted on the syringe needle and the other end installed on (III) a needle head (pointed by a red arrow head), which is fixed onto the final part, (IV) a micromanipulator, all of which can be purchased from commercial resources. Image (**b**) highlights the two black-ink markers: one close to syringe pump (red arrow) is for the warning sign indicating that the sampled CSF is far over 200 μl level and the sampled CSF may contaminate the syringe which is designed to be repeatedly used without any specific treatment; the other mark indicated by a yellow arrow is for labeling the 200 μl level (see text for detailed calculation per inner diameter of PE50 tubing and length of the tubing). Image (**c**) addresses the needle head-PE50 tubing connection (indicated by a red arrow head on Image (**a**); see text of how needle head was onsite made without compromising its sterile condition). Needle head and PE50 tubing set were used only once per animal ensuring its safety to the animal and preventing from cross-contamination

boid depressed area between the occipital bone and alar vertebrae was exposed under the surgical microscope and a midline incision of the skin, and the first layer of the muscle was followed by blunt separation of the muscles above the atlantooccipital membrane. The in-house designed instrument for CSF sampling was employed, and its needle bevel was advanced to target the atlanto-occipital membrane at a 25° angle with the rat's mouth tilted down 5° (Fig. 3). Thereafter, the syringe pump was switched to withdraw 100–200 μl of CSF at a digitally controlled speed at 400 μl/min. The pump was switched on and off to control the vacuum accumulated within the system. After the predetermined CSF was sampled, the syringe pump was switched to infuse/inject the sampled CSF into a labeled 2 ml Monoject sterile blood collection tube (Tyco Healthcare Group).

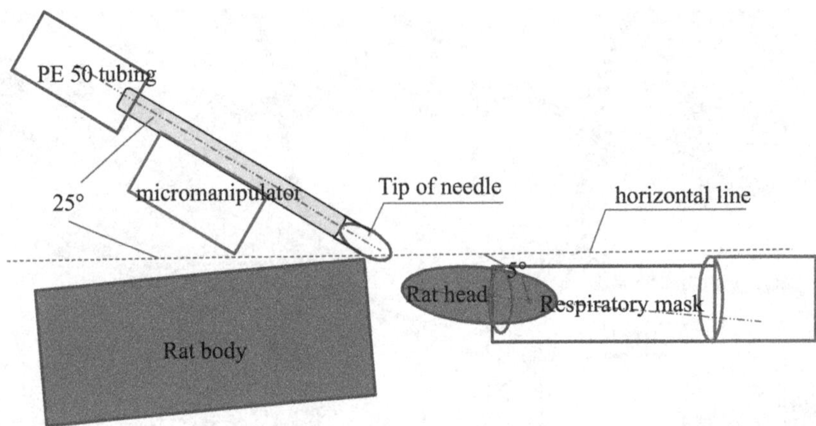

Fig. 3 Illustration of side view of the CSF sampling system. The illustration addresses how the head of a rat is fixed to access a respiratory anesthetic setting and at what angle the CSF-sampling needle is set to approach the cisterna magna

CSF Quality and Quantity Assessment

Quality of the sampled CSF (i.e., visual appearance) was verified by the naked eye and/or with the aid of 10–25× amplification of the surgical microscope to determine whether there was visible blood. The sampled CSF quantity was first estimated by a formula calculating inner volume of the commercially purchased PE50 tubing: 200 μl (desired CSF amount) = $r^2\pi \times L$. The standard inner diameter of the PE50 tubing is 0.58 mm, and accordingly r is 0.29 mm. The supposed PE50 tubing length (L) for 200 μl volume of a CSF sample then is 757 mm per calculation with the formula. To ensure the sampled CSF is 200 μl or above, the eventual 200 μl mark was arbitrarily enforced at the length of 870 mm of each CSF-sampling PE50 tubing, a ~13% increase in addition to a volume in the needle head (~30 mm in length) (see Discussion for reasons of intended increase of length of the sample collection tubing). The 100 μl mark was enacted at the half of the 870 mm tubing length, and the safety guard mark was imposed at the 1058 mm length of the tubing. Theoretically, 123 μl of CSF was sampled at the 100 μl mark of the PE50 tubing, and 238 μl of CSF was sampled at the 200 μl mark of the PE50 tubing per calculation of the formula: volume (100 μl mark)=$(0.29\ \text{mm})^2 \times 3.14159 \times$ (435 mm of tubing length+30 mm of needle head length) and volume (200 μl mark) = $(0.29\ \text{mm})^2 \times 3.14159 \times$ (870 mm of tubing length + 30 mm of needle head length), respectively. To verify the sampled CSF at 100 μl or more and 200 μl or more, a pipetting method was further applied: typically, all the declaimed 200 μl of CSF samples were redistributed into four collection tubes, each with 50 μl or above.

3 Results

Body weights were 409.9 ± 35.7 g in the vehicle control group and 411.1 ± 32.5 g in TMT treatment group at day 0 with no significant difference between groups (Fig. 4). During paired feeding, animals in vehicle control group slightly gained body weight, ranging between 370 and 546 g, while the animals in TMT treatment group displayed progressive body weight loss, varied between 309 and 492 g. Body weight in TMT treatment group was significantly lower than that in vehicle control group at day 12 (443.1 ± 44.5 g vs. 345.9.1 ± 33.8 g). Totally, 103 rats were subjected to the CSF-sampling surgery. The overall success rate (no visible blood contamination) was 89%. If calculating the success rate by examining the sample in the collection tubes, the rate increased to 91% because the visible contaminated blood portion in the sampling tubing in two samples were immediately separated by cutting down the tubing ~20 mm length from the visible blood portion: the bleeding/visible blood occurred after clear CSF was sampled for >200 μl in one rat and >100 μl in another (so called delayed blood contamination). Those samples were marked with questionable quality while awaiting further analysis.

All rats, control or TMT treated, survived the CSF sampling surgery/anesthesia although NCTR's attending veterinarian judged some animals experiencing an unscheduled emergent surgery as "terminal status." Noticeably, three rats with manifestations including but not limited to "signs of recumbency, non-responsiveness, labored respiration, seizures, diarrhea, and significant weight loss" endured the CSF sampling surgery and displayed relatively stable living signs including the rhythmic respiration after withdrawing CSF. One of these three animals experienced temporary respiratory and circulatory arrest during

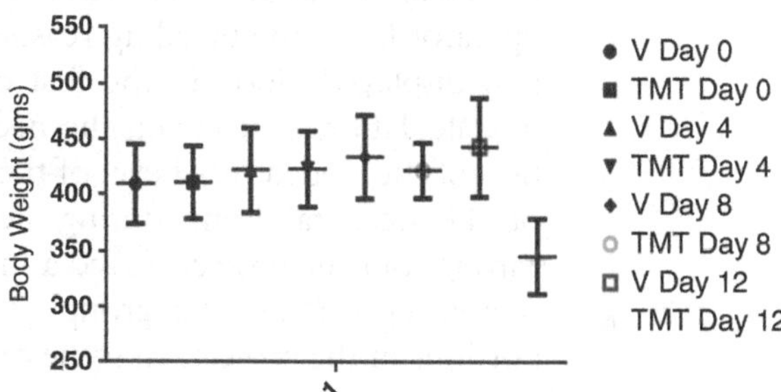

Fig. 4 Time course of changes in body weight following trimethyltin chloride (TMT) exposure. Using software GraphPad Prism6, *t*-test for day 12 shows a significant difference between vehicle (V) group and TMT group, $p < 0.05$. Repeated measures ANOVA cannot be done due to loss of subjects. ANOVA test shows $p < 0.001$

anesthesia induction and recovered after conducting resuscitation (rhythmic chest squeezing) before the CSF sampling surgery. Every drop of CSF was harvested from these three animals as observed by an empty gap displayed in the CSF-sampling tubing after fulfillment of the CSF sampling. The sampled CSF passed the 200 μl mark in two of these three animals, and in the remaining one, ~180 μl CSF was sampled.

4 Discussion

The CSF sampling method described here advances in two aspects compared with the available literature: (1) the instrument established in this study provides a confidence in sampling high-quality CSF with no visible blood contamination and 86–95% success rate, and (2) the present approach ensures a predictable CSF amount sampled: ≥200 μl in normal adult male Sprague-Dawley rats. Noticeably, an average of ~200 μl CSF was sampled even in the "terminal" animals with a significant reduction in body weight (as low as 267 g).

As demonstrated in Fig. 1, there are three bevel orientations of the sampling needle head. It appeared that Type II had a highest successful rate with no visible blood contamination in sampling 100 μl, while no significant difference was detectable between the other two bevel orientations. Type III approach led to sample the highest CSF amount (≥200 μl in 33 of 35 rats) although animals in Type I and Type II groups were targeted at the level of 100 μl. Actually, two CSF samples from Type III group displayed a delayed blood contamination, one after the sampling amount was over the 100 μl mark/123 μl (the calculation amount by tubing size and length plus needle head-hold volume) and another after sampling over 200 μl mark/238 μl per that calculation, indicating that Type II approach might risk similar odds if the sampling amount had been targeted at the level of 200 μl. It is worth mentioning that the operator knew the bleeding reason in one of the two CSF samples that displayed blood in the first drop of CSF in Type III group: miscalculation in selecting the needle penetrating place led a portion of the sloped rear bevel of the needle head to scratch the edge of the occipital bone during needle advance. Such experience throughout the project played a major role in the enhanced success rate in Type II and III groups, which were executed in the later portions of the project. In some cases, minor adjustments of needle penetrations were made to ensure such hemorrhage did not occur. Those adjustments are as follows: (1) the needle advance was halted after entering the atlantooccipital membrane with a length of one and half of its bevel if no CSF was sampled with the syringe pump while advancing the sampling needle head after its bevel passed the membrane; (2) the needle was removed (screwed back), and the penetrating hole was cleaned for removing any fluid/blood; (3) the needle was then advanced again after adjusting/increasing the angle between the animal's trunk and the animal's head. An explanation for these cases is that there is anatomic variation that some animals may have such as an epidural and/or a subdural space before the sampling needle head penetrated the subarachnoid membrane and reached into the cisterna magna. Our unpublished data indicated existence of this cavity when making a model of subarachnoid hemorrhage in rats: a few of the rats failed to display blood deposit on the cranial base (the injected blood went "nowhere"/place not confirmed) after injecting 0.4 ml of fresh blood via penetrating the atlantooccipital membrane.

Techniques to Avoid Blood Contamination

CSF can be contaminated with blood from the injured tissue where blood vessels exist ubiquitously. In contrast to the method via a puncture through skin with or without instrumental guidance such as an ultrasound imaging [8], the present method exposes the atlantooccipital membrane and thus avoids the possibility that the sampling needle will penetrate a vessel between the subcutaneous tissue and the overlaying muscular tissues of the atlantooccipital membrane. Furthermore, vasculature existing within the atlantooccipital membrane can be viewed clearly with the surgical microscope and thus can be circumvented when advancing the needle with the micromanipulator. Nevertheless, the atlantooccipital membrane should be verified free from blood from all the surrounding tissues in the surgical incision/field before advancing the sampling needle through the atlanto-occipital membrane. In general, the blood contamination from the surgical field and the injured atlantooccipital membrane shows a reddish color in the first drops of the sampled CSF within the needle head tubing set. In the present method, the operator was "blind" after the needle tip pierced the atlantooccipital membrane. A beginning blood contamination of the sampled CSF might originate from the brain structures that form the cisterna magna. For example, a hemorrhage might be caused by the injury of the floor of the cisterna magna/the dorsal surface of the medulla oblongata due to an over advanced sample needle (referring to Fig. 1). A reddish color could be displayed in the middle or final drops of the sampled CSF in the needle head tubing set: it occurred three times in the present study when the sampled CSF passed the 100 or the 200 μl mark. We surmised that a high vacuum pressure had accumulated within the tubing set and would account for the bleeding.

Obstacles Affecting CSF Sampling Amounts

(1) Dehydration. The present study demonstrated that there were three animals that appeared to have limited CSF volume (maximum at ~200–238 μl) via the cisterna magna approach in the TMT

treated group. These animals displayed "terminal" signs with significant body weight loss, indicating dehydration as the potential explanation. (2) CSF leakage. The operator did see that CSF leakage occurred around the sampling needle head when advancing the needle across the atlantooccipital membrane. An increased pressure in the cisterna magna or an uneven cutting along the bevel sides (which caused leakage) had been surmised as the potential culprit preventing CSF sampling at the maximum. On the other hand, there were several pragmatic guesses that might lead to a decreased sample volume. First, the vacuuming force was critical. A much higher vacuuming force (a rather "high"-speed vacuum) appeared to lead to a delayed hemorrhage, which interrupted the CSF sampling process. Theoretically, a reduced vacuuming force would also be accountable. The sampled CSF was frequently (not counted but estimated at ~10% of total surgeries) seen with air bubbles in the sampling tubing. As mentioned, an epidural/subdural cavity might exist between the atlantooccipital membrane and the arachnoid membrane as the tent/roof of cisterna magna in some animals. It was hypothesized that a sample needle might be placed with its bevel opening part in the cavity and another part in the cisterna magna, resulting in the bubble-sampled CSF. Also, the bubble-sampled CSF might be the result of an unsealed spot between the sampling needle head and the surrounding tissues (the atlantooccipital membrane). Screening the sampling needle quality before beginning the sample may thus partially avoid the occurrence of bubble-sampled CSF.

In summary, we recommend the following to be able to sample ≥200 µl of CSF without blood contamination:

1. Detailed training in surgical skill.
2. Selecting Type II or III bevel orientation of the sampling needle head. It is possible that a locally increased vacuum/negative pressure may play a role in increased bleeding cases in Type I bevel orientation since the bevel faces the floor of the cisterna magna. On the other hand, Type II or Type III bevel orientation dodges this possibility.
3. Awareness of a possible epidural and/or subdural cavity.
4. Slow suctioning of CSF (arbitrarily ~2 min for sampling 200 µl).

Acknowledgments

This work was supported by the National Center for Toxicological Research/Food and Drug Administration [Protocol # E0758001 to M.G.P. and S.Z.I]. The authors are grateful for the technical expertise provided by the animal care staff of the Priority One Corporation and technical support provided by Susan Lantz and Bonnie Robinson.

Disclaimer

This document has been reviewed in accordance with the US Food and Drug Administration (FDA) policy and approved for publication. Approval does not signify that the contents necessarily reflect the position or opinions of the FDA nor does mention of trade names or commercial products constitute endorsement or recommendation for use. The findings and conclusions in this report are those of the authors and do not necessarily represent the views of the FDA.

References

1. Bouman HJ, Van Wimersma Greidanus TB (1979) A rapid and simple cannulation technique for repeated sampling of cerebrospinal fluid in freely moving rats. Brain Res Bull 4:575–577
2. Consiglio AR, Lucion AB (2000) Technique for collecting cerebrospinal fluid in the cisterna magna of non-anesthetized rats. Brain Res Brain Res Protoc 5:109–114
3. Pegg CC, He C, Stroink AR et al (2010) Technique for collection of cerebrospinal fluid from the cisterna magna in rat. J Neurosci Methods 187:8–12
4. Mahat MY, Fakrudeen Ali Ahamed N, Chandrasekaran S et al (2012) An improved method of transcutaneous cisterna magna puncture for cerebrospinal fluid sampling in rats. J Neurosci Methods 211:272–279
5. Li Y, Zhang B, Liu XW et al (2016) An applicable method of drawing cerebrospinal fluid in rats. J Chem Neuroanat 74:18–20
6. Lai YL, Smith PM, Lamm WJ et al (1983) Sampling and analysis of cerebrospinal fluid for chronic studies in awake rats. J Appl Physiol 54:1754–1757
7. Frankmann SP (1986) A technique for repeated sampling of CSF from the anesthetized rat. Physiol Behav 37:489–493
8. Lu YG, Wei W, Wang L et al (2013) Ultrasound-guided cerebrospinal fluid collection from rats. J Neurosci Methods 215:218–223
9. Imam SZ, He Z, Cuevas E et al (2018) Changes in the metabolome and microRNA levels in biological fluids might represent biomarkers of neurotoxicity: a trimethyltin study. Exp Biol Med (Maywood) 243:228–236
10. He Z, Yamawaki T, Yang SH et al (1999) An experimental model of small deep infarcts involving the hypothalamus in rats: changes in body temperature and postural reflex. Stroke 30:2743–2751
11. He Z, Yang S, Naritomi H et al (2000) Definition of the anterior choroidal artery territory in rats using intraluminal occluding technique. J Neurol Sci 182:16–28

Chapter 13

Transporter Studies: Brain Punching Technique

Cherish A. Taylor and Somshuvra Mukhopadhyay

Abstract

A number of neurological disorders are associated with the dysfunction of transporters in the brain. In order to understand the neural functions of transporters and their roles in disease, it is critical to know where transporters are expressed and active. The brain punching technique provides a method for studying the region-specific functions of transporters in the brain. This chapter describes the materials and procedures for the brain punching technique.

Key words Transporters, Microdissection, Brain punching

1 Introduction

A wide variety of transporters are active in the brain. Their functions range from recycling neurotransmitters released in the synapse to exporting neurotoxicants from the cytosol. Loss of transporter function is implicated in various neurological disorders. For example, loss of function of the manganese efflux transporter SLC30A10 results in manganese-induced parkinsonism [1–3]. An important step in understanding a transporter's function in the brain, and its role in neurological disorders, is characterizing the transporter's pattern of expression and activity. One method for assessing the expression pattern of a transporter would be to visualize expression using immunofluorescence and microscopy. However, specific and high-quality primary antibodies may not be readily available for certain transporters. An alternate method would be to use the brain punching technique to isolate specific brain regions and measure mRNA levels using polymerase chain reaction. Additionally, tissue collected via the brain punching technique could be used for various analyses. Thus, tissue collected from the same sample could be used to assess both expression and activity. For example, to study a metal transporter, both transporter mRNA levels and metal levels could be measured. The brain punching technique allows researchers to study multiple aspects of transporter function.

Michael Aschner and Lucio Costa (eds.), *Cell Culture Techniques*, Neuromethods, vol. 145,
https://doi.org/10.1007/978-1-4939-9228-7_13,

Developed by Miklós Palkovits, the brain punching technique provides a simple method for extracting specific nuclei in the brain [4, 5]. The technique involves slicing frozen brain tissue and using a small, hollow needle to collect tissue from specific regions. Once collected, the tissue can be used in a variety of assays. Here, the authors provide a detailed protocol for the brain punching technique. This protocol lists the steps and materials for extracting tissue for RNA isolation.

2 Materials

1. Brain punch handle and punch of desired size (*see* **Note 1**)
2. Conical tube (15 or 50 mL)
3. Beaker (100 or 150 mL)
4. Beaker (100 mL)
5. Glass bottle (may use any size)
6. Glass microscope slides
7. Microcentrifuge tubes (1.5 mL, *see* **Note 2**)
8. Microtome blade
9. Fine-tipped stainless steel forceps
10. Broad-tipped stainless steel forceps
11. Petri dish (any size)
12. One single-edge industrial razor blade
13. Delicate task wipes
14. Ultrapure water (RNAse and DNAse free)
15. Optimal cutting temperature (O. C. T.) compound
16. Peel-away disposable paraffin embedding molds
17. 2-Methylbutane (isopentane)
18. Dry ice
19. Liquid nitrogen
20. Aluminum foil
21. Two Styrofoam boxes with lids
22. One hundred percent ethanol
23. RNAse decontamination solution (*see* **Note 3**)

3 Procedures

1. Extract whole brain tissue, place in 15- or 50-mL conical tube, and immediately flash freeze in liquid nitrogen; store at −80 °C until ready for use.

2. Set the cryostat temperature to −5 °C, and allow the cryostat to reach this temperature while completing Steps 3–16.
3. Fill the two Styrofoam boxes with finely crushed dry ice until about half full.
4. Place the beaker in one of the boxes filled with dry ice (box 1). Ensure bottom of beaker is surrounded by the dry ice (*see* **Note 4**). In the other box (box 2), create a flat surface with the dry ice; this will be important for later use.
5. Remove frozen brain tissue from liquid nitrogen or −80 °C storage, and immediately place on dry ice in box 1 (*see* **Note 5**).
6. In a biosafety cabinet, fill the beaker inside box 1 with approximately 20 mL of isopentane. Cover beaker with aluminum foil, and place lid on box (*see* **Note 6**). Allow to chill for approximately 15 min (*see* **Note 7**).
7. While the isopentane is chilling, pre-chill the forceps and petri dish by placing them in the same box as the beaker and tissue. Be sure to place the lid back on the box (*see* **Note 8**).
8. While the isopentane, forceps, and petri dish continue to chill, take out a piece of aluminum foil (large enough to fully cover the dry ice in box 2). Spray the foil with RNAse decontamination solution, and wipe with a delicate task wipe. Rinse with ultrapure water and wipe. Spray with 100% ethanol and wipe. Place on top of the flat surface of dry ice in box 2.
9. Place a small piece of dry ice into the beaker filled with isopentane to determine if the isopentane is cold enough. If no bubbling occurs, the isopentane has reached the desired temperature; proceed to *Step 10.* If bubbling occurs, wait another 10 min and repeat *Step 9.*
10. Label the embedding mold (*see* **Note 9**), and fill about half way with O. C. T. Ensure there are no air bubbles in the O. C. T.
11. Remove one of the box lids and place upside down. Take out a few pieces of dry ice from box 1, and place in a flat pile on the lid. Take out the petri dish, and place on the pile of dry ice. Take out both forceps. Empty the frozen brain tissue into the petri dish.
12. Using the broad-tipped forceps, gently pick up the frozen tissue. Place tissue in the embedding mold filled with O. C. T. in the notated orientation (*see* **Note 9**). Using broad-tipped forceps, gently position tissue so that it is as straight as possible. Then use the same forceps to cover the tissue in O. C. T. by dipping the forceps in the O. C. T. and bringing them from one side of the tissue to the other, being careful not to move the tissue (*see* **Note 10**).

13. Use the fine-tipped forceps to pick up the embedding mold, and place it in the cold isopentane (*see* **Note 11**). Cover beaker with aluminum foil, and place lid on box.
14. Allow embedding mold with tissue and O. C. T. to remain in the isopentane until O. C. T. is completely frozen (*see* **Note 12**). This may take approximately 10 min.
15. Remove sample from the isopentane and place on dry ice. The tissue should remain on dry ice until ready for slicing in the cryostat. For long-term storage, store at −80 °C.
16. Repeat Steps 10–15 with any additional frozen brain samples.
17. Remove beaker from dry ice, and pour isopentane into glass bottle. Do not close the bottle. The isopentane must return to room temperature before closing the bottle.
18. While the isopentane warms in the biosafety hood, take the remaining materials to the cryostat. Place the brain tissue frozen in the embedding molds into the cryostat (which should be set to −5 °C). Wait at least 15 min to allow the tissue to acclimate to the new temperature (*see* **Note 13**).
19. As the tissue warms in the cryostat, clean off the fine-tipped forceps with 100% ethanol and a delicate task wipe. Place in cryostat to chill. Clean the microtome blade with 100% ethanol and a delicate task wipe. Place the blade in the cryostat, but do not position until ready to begin slicing.
20. Taking out one microscope slide at a time, label the slide. Spray the slide with RNAse decontamination spray, and wipe with a delicate task wipe. Rinse with ultrapure water and wipe. Spray with 100% ethanol and wipe. Place the slide on top of the aluminum foil in box 2. Repeat with all slides needed (*see* **Note 14**). Close lid and allow slides to chill.
21. Once the tissue has acclimated to the slicing temperature, remove the first sample from the embedding mold (*see* **Note 15**). For coronal sections, orient the sample so that the olfactory bulb is facing up.
22. Squeeze a layer of O. C. T. on the cryostat specimen disc. Place the disc in the cryostat quick freeze shelf. If the cryostat has a Peltier element, activate the Peltier element to cool.
23. As soon as the O. C. T on the disc begins to freeze, place the sample on the disc with the olfactory bulb facing up. The sample should be perpendicular to the disc.
24. Place more O. C. T. around the sample to ensure that it is securely attached to the specimen disc.
25. Allow at least 10 min for the sample to fully freeze to the disc. All of the O. C. T. should be white and opaque.
26. Position the specimen disc in the cryostat platform and secure (*see* **Note 16**).

27. Position the microtome blade in the cryostat blade holder (*see* **Note 16**).
28. Remove labeled microscope slides from the box, and lay out in the cryostat. Only remove the slides for the sample being sliced.
29. Slice one or two 50 μm sections. If the blade slices easily through the O. C. T., increase section to 100 μm. Gradually increase sections in 50–100 μm increments until 300 μm is reached (*see* **Note 17**). The blade should be able to smoothly slice through the O. C. T. and tissue. If the blade is not cutting smoothly, *see* **Note 18**.
30. When a section with a target brain region has been sliced, use the fine-tipped forceps to grab the corner of the slice (only touch the O. C. T.). Place the slice on the appropriately labeled microscope slide (*see* **Note 19**). Use your fingertip to lightly touch the back of the slide, underneath the left and right sides of the section. Do not directly touch the section, and only place fingertips beneath the O. C. T. (*see* **Note 20**).
31. Repeat *Step 30* for remaining sections.
32. When all sections have been collected and secured to the microscope slide, return slides to the dry ice.
33. Repeat *Steps 21–32* for additional tissue. For long-term storage, tissue may be stored at −80 °C in plastic microscope slide folders until ready for punching.
34. Set the cryostat to −15 °C. Place brain punching tool, with inserted punch, inside the cryostat to cool.
35. Fill the 100 mL beaker with dry ice. Ensure a 1.5 mL tube can fit in the center of the beaker and be surrounded by dry ice. Set beaker in the cryostat.
36. Label 1.5 mL tubes. Place labeled tubes in box 1.
37. Place microscope slides for the first sample in the cryostat. Slides for other samples may also be placed in the cryostat if there is room.
38. Once the cryostat has reached −15 °C, take a test brain punch to determine if the tissue is at the correct temperature. Using the punching tool, press the puncher into an unneeded section of tissue. The puncher should get through the tissue without the tissue sticking or cracking. If sticking or cracking occurs, *see* **Note 21**.
39. Take out the appropriately labeled 1.5 mL centrifuge tube, open it, and place it in the center of the beaker filled with dry ice.
40. Use the punching tool to dissect the desired region, and push tissue out into the centrifuge tube (*see* **Note 23**). An example of a punched slice is provided in Fig. 1.

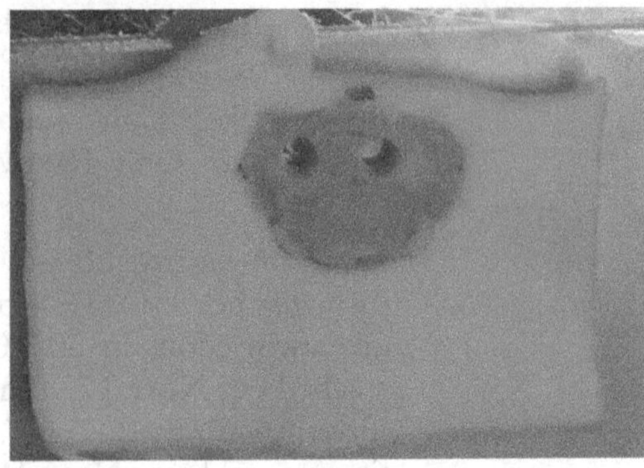

Fig. 1 Image of two punches (1.25 mm diameter) take from a 500 μm slice of mouse cerebellum

41. Return centrifuge tube to box 1.
42. Repeat *Steps 39–41* for remaining punches. Punches may be stored at −80 °C until processing.
43. Return to the biosafety cabinet, and close the bottle of isopentane. Store appropriately.
44. Clean all materials, using only 100% ethanol for cleaning cryostat equipment and items stored in the cryostat.

4 Notes

1. Punch sizes range from 0.25 to 2.00 mm in diameter. Punches can be purchased in a set or individually (see Fig. 2).
2. The total number of tubes will depend on the number of punches collected. Each punch (or set of punches for a particular region) will be placed in a separate tube. All tissue in the same tube will be processed together.
3. If brain punches are not being used for RNA extraction, RNAse decontamination solution is not needed.
4. If desired, during *Step 3*, place a layer of dry ice in one of the boxes; then position the beaker in the center of the box. Tightly pack dry ice completely around the beaker. A small piece of aluminum foil may be placed on the top of the beaker to prevent dry ice from getting inside.
5. To maintain integrity of the tissue, particularly if using for RNA, do not allow the tissue to thaw.
6. *Steps 6–17* should be completed in a biosafety cabinet.
7. The beaker should be filled with only enough isopentane for the embedding mold to be surrounded. If desired, more isopentane may be used. If more isopentane is used, cut a floating foam tube rack to allow the embedding mold to sit firmly inside and float on the isopentane.

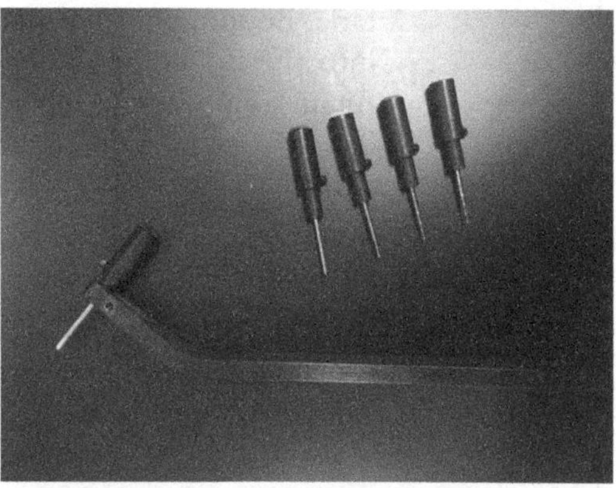

Fig. 2 Image of a brain punching tool and assortment of punch sizes

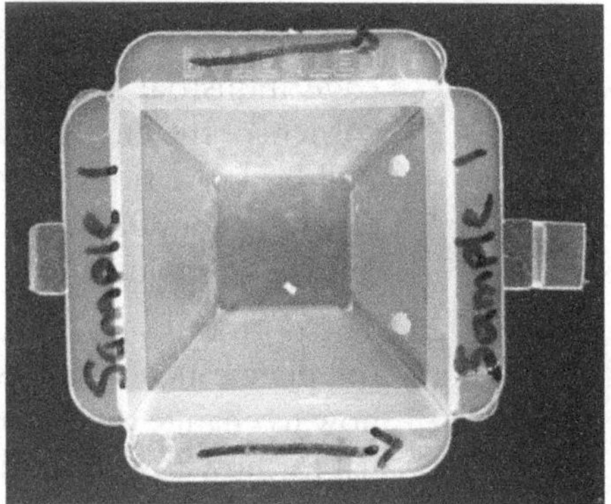

Fig. 3 Example of a labeled embedding mold (*see* **Note 10**)

8. Only place the tip of the forceps in the dry ice.
9. Embedding molds normally have four rims at the top of the mold. Label two of the rims to mark the sample identity. Use the other two rims to note the direction of the tissue. For example, in Fig. 3, the embedding mold is labeled "Sample 1" to note brain number 1. For coronal sections, the arrows indicate the anterior (arrow head) to posterior (arrow end) orientation of the tissue. Be sure to use an ethanol-resistant marker as the isopentane will remove other types of ink.
10. This step must be done quickly to prevent the tissue from thawing. This step must also be done carefully to prevent any possible damage to tissue from handling with the forceps and to ensure proper orientation.
11. As discussed in *Note 7*, there should only be enough isopentane to surround the embedding mold. The isopentane should not get into the embedding mold and touch the O. C. T or tissue. If isopentane does get into the mold, use the forceps to handle the mold, and quickly pour out the isopen-

tane. Be sure not to disturb the tissue. Immediately place back into the isopentane if the O. C. T is not completely frozen. Again, remove any isopentane from the mold if needed. Pour out any excess isopentane from the beaker before working with the next frozen tissue sample.

12. The O. C. T. will be white and opaque when completely frozen. The brain tissue should not be visible.
13. If the tissue was stored in the frozen O. C. T. at −80 °C, additional time may be required to acclimate the tissue to the warmer temperature. Place the tissue in the cryostat at *Step 2*, and allow at least 30 min for the tissue to warm.
14. The total number of slides will depend on the regions being extracted. Each slide fits three to four slices. Estimate the number of slides needed, and prepare at least one extra slide.
15. To remove the sample from the embedding mold, use a single-edge industrial razor blade to make a small cut in the four corners of the embedding mold. Do not cut O. C. T. or tissue. Pull back the sides of the embedding mold. Without directly touching the O. C. T., use fingers to gently push the sample out of the mold.
16. The positions of the platform, specimen disc, blade, and anti-roll plate can all be adjusted. Refer to your cryostat manual for how to change their positions if needed. Also refer to the manual for how to secure the specimen disc to the platform.
17. The provided protocol is based on using the tissue for RNA extraction. For other uses, such as metal analyses, a larger section may be required. Many cryostats are able to slice up to 500 μm sections.
18. If the blade is not cutting smoothly, it is possible the sample is too cold to cut larger sections. Allow the sample more time to warm up to −5 °C and repeat *Step 28*. Depending on the cryostat, the temperature may also need to be increased to −4 °C. It is also possible the sample is not securely attached to the specimen disc. Gently move the sample. If the sample moves, remove the specimen disc from the platform. Gently pull the sample off of the specimen disc. Use the single-edge industrial razor blade to cut away the O. C. T. used to freeze the sample to the specimen disc. Using a new specimen disc, return to *Step 21* and continue. Finally, if the tissue sticks to blade during slicing, the temperature is to warm. Decrease the temperature until the blade is able to smoothly slice the tissue without sticking.
19. It is useful to take a slice before the desired brain region as well. This extra slice can be used to determine the best temperature for punching.
20. Using fingertips on the back of the microscope slide lightly warms the O. C. T. surrounding the tissue slice. This allows the

O. C. T. to melt to the microscope slide, securing the section to the slide while not allowing the tissue to stick to the slide. Only melt one side at a time. Remove fingertip when the O. C. T. is no longer white and slightly translucent. Do not allow the tissue to thaw during this process.

21. If tissue sticks to the puncher or microscope slide, the tissue is too warm for brain punching. Decrease the cryostat temperature, and allow the tissue to chill. If the tissue cracks, the tissue is too cold for brain punching. Increase the cryostat temperature, and allow the tissue to warm. After allowing the tissue to reach temperature, take another test punch. When the tissue is at the desired temperature, proceed to *Step 39.*

22. Multiple punches can be collected in the puncher before needing to push the punches into the centrifuge tube. Again, all punches collected in the same tube will be processed together.

References

1. Tuschl K et al (2012) Syndrome of hepatic cirrhosis, dystonia, polycythemia, and hypermanganesemia caused by mutations in SLC30A10, a manganese transporter in man. Am J Hum Genet 90(3):457–466
2. Quadri M et al (2012) Mutations in SLC30A10 cause parkinsonism and dystonia with hypermanganesemia, polycythemia, and chronic liver disease. Am J Hum Genet 90(3): 467–477
3. Leyva-Illades D et al (2014) SLC30A10 is a cell surface-localized manganese efflux transporter, and parkinsonism-causing mutations block its intracellular trafficking and efflux activity. J Neurosci 34(42):14079–14095
4. Palkovits M (1973) Isolated removal of hypothalamic or other brain nuclei of the rat. Brain Res 59:449–450
5. Palkovits M (1983) Punch sampling biopsy technique. Methods Enzymol 103:368–376

Chapter 14

miRNA as a Marker for In Vitro Neurotoxicity Testing and Related Neurological Disorders

Lena Smirnova and Alexandra Maertens

Abstract

miRNA (miRNAs) are small noncoding RNA molecules, which bind to the 3′UTR of target mRNA and thereby posttranscriptionally regulate gene expression. Thus, miRNA are important fine-tuners of essential processes in the body. In the brain, they regulate neural development and brain homeostasis. Studying miRNA profiling in combination with whole genome transcriptomics after toxicant exposure is a prime way to derive molecular signatures of toxicity. This gives an insight into molecular network perturbations, which underlie systems toxicology. miRNA encapsulated into extracellular vesicles are released into biofluids and in case of in vitro systems into the culture medium as means of intercellular communication but also in response to environmental stress. In addition, miRNA are released into the circulation upon organ injury. Thus, circulating miRNA may serve as potential biomarkers of (brain) injury/toxicity. In this chapter, the importance of miRNA for neural development, neurotoxicity, and neurodegeneration is discussed; the critical steps of miRNA profiling in tissues/cells as well as in biofluids are described; the challenges and options of bioinformatic data analysis are deliberated. The focus of this chapter is on the quality control of miRNA profiling methods.

Key words miRNA, Neurotoxicity, Neurodevelopment, Biomarkers

1 Introduction

1.1 miRNA in Neural Development and Brain Homeostasis

In the last decade, an understanding of posttranscriptional regulation of gene expression has emerged, thanks to the discovery of miRNA (miRNA), small (~22 nt) noncoding regulatory RNA molecules, which bind to specific binding sites in the 3′UTR of target mRNA and repress their translation. About 2000 miRNAs have been identified in humans [1]. More than 50% of all identified miRNAs are expressed in the brain, where they play a particular role in brain development by regulating developmental timing, cell differentiation, proliferation, and cell fate decision, as well as brain morphogenesis [2]. One miRNA may have up to several hundred mRNA targets, while one mRNA may be regulated by several miRNAs. In animal studies, depletion of RNase-III, Dicer, that triggers the loss of miRNA synthesis, results in severe defects in brain devel-

Michael Aschner and Lucio Costa (eds.), *Cell Culture Techniques*, Neuromethods, vol. 145, https://doi.org/10.1007/978-1-4939-9228-7_14,

opment and morphogenesis [3, 4]. The most abundant miRNAs in the CNS, *mir-124* and *mir-9*, promote neuronal identity during differentiation of neural progenitor cells by targeting multiple anti-neuronal factors in neuronal cells [5, 6]. *mir-137* is required for neural differentiation of embryonic stem (ES) cells and modulates differentiation of adult mouse neural stem cells by cross-talking with methylation agents [7, 8]. *mir-138* and *mir-134* contribute to regulation of synaptic development, maturation, and plasticity [9, 10]. Notably, *mir-132* affects synaptic structure, which is modulated by FMRP, the product of the Fmr1 gene responsible for fragile X mental retardation, a syndrome with strong association to autism [11].

Most recently, miRNAs have been detected in biofluids [12]. It has also been shown that these circulating miRNAs can be used as biomarkers for a variety of diseases and organ/tissue injury [13].

1.2 miRNA in Neural Developmental Disorders and Neurodegeneration

Increasing evidence shows that perturbations in miRNA expression patterns have a significant impact on several disorders, including cancer and neurodegenerative and neurodevelopmental disorders (Alzheimer's, Parkinson's, Huntington's disease and autism) [14]. Below are some examples of the role of miRNA in neural developmental disorders (autism) and neurodegeneration (Parkinson's disease, PD).

Autism spectrum disorders (ASD) are a major public health concern in the USA affecting 1 in 59 children ([15] https://www.cdc.gov/ncbddd/autism/data.html). Several miRNAs were shown to be altered in cerebellum and/or blood cells in autism (*mir-21*, *23a*, *128*, *132*, *181b*, *195*, *212*, *219*, *346*, *381*, *495*) (reviewed in [16]). Recent work has observed that among the dysregulated miRNAs in autistic samples, some miRNAs are predicted to target genes, which are known to be involved in genetic causes of autism (e.g., *neurexin*, *SHANK3*, and *PTEN*) [17].

In addition to genetic components (reviewed in [18]) in the etiology of ASD, increasing evidence from epidemiological studies suggests that exposure to drugs and environmental chemicals in the early prenatal period has a substantial impact on ASD risk. For example, valproic acid, thalidomide, misoprostol, lead, and organophosphates may contribute to ASD risk [19–21], but there is limited or no understanding of the mechanisms of environmental impact in the causation of ASD. Thus, there is an urgent need to assess this issue and to identify further DNT compounds and possible miRNA biomarkers of exposures related to ASD.

Several miRNAs regulate functions of dopaminergic neurons (DN) (*mir-133b* [22], *mir-9* [23], *mir-132* [24]). These and some other DN-specific miRNAs were deficient in PD-affected midbrains [25–28]. *miR-7* targets *α-Syn* and inhibits *α-Syn*-mediated DN cell death. In addition, *mir-7* expression is sensitive to MPP^+, an active metabolite of MPTP [27], and rotenone [29], substances found to induce PD-like symptoms in animals and humans. Recently, overexpression of *mir-7* was shown to have protective effects against MPP^+-induced neuronal cell death [30–32]. These findings clearly demonstrate a role of miRNA in development of PD and in MPP^+ neurotoxicity. Moreover, miRNAs play a significant role in mitochondria function [33], including pro-apoptotic *mir-15/16*, anti-apoptotic *mir-21*, and *mir-17–92* cluster. ROS-responsive *mir-210* inhibits cell proliferation and represses the mitochondrial metabolism and respiration by targeting several elements of the TCA cycle [34]. *mir-210* was shown to be regulated by rotenone exposure [35]. Notably, the expression of *mir-210* was shown to be regulated by DNA demethylation induced by hypoxia [36]. Functional studies revealed that overexpression of *mir-338* (brain-specific regulator of cytochrome-c-oxidase-IV) significantly reduced mitochondrial oxygen consumption, metabolic activity, and ATP production [37]. Importantly, there are mitochondria-enriched miRNAs, mitomiRs [38], which may be crucial in managing the mitochondrial response to stress. While these findings demonstrate the role of miRNA in the development of PD and in mitochondria-mediated (e.g., MPTP or rotenone) neurotoxicity, a full, systems-level understanding of the role of miRNA in the cellular response of DN to a mitochondrial toxicant has not been attempted.

1.3 miRNA as Markers for (Developmental) Neurotoxicity

Increasing numbers of studies are addressing the question of whether miRNAs are involved in cellular responses to environmental stress, including chemical exposure (reviewed in [39]); the role of miRNA in DNT, however, is less studied [40, 41]. Taking into account (i) the significant role of miRNA in development of CNS, (ii) high number of targets per one miRNA, and (iii) the fact that mRNA targeted by miRNA are twice as likely to be perturbed following environmental chemical exposure than those which lack miRNA binding sites [42, 43] suggests that miRNA may be mediators of the response to environmental disturbances. miRNA profiling, together with identification of predicted mRNA targets, may have stronger predictive value about the mechanism of action than mRNA alone, as miRNAs were established as crucial developmental regulators that mark developmental timing and cell specification (reviewed in [44]). The main advantage of the miRNome compared to whole transcriptome analysis is the amount of data: miRNA profiling provides more compact data sets to analyze, which can facilitate a statement about substance effects. In addition, miRNAs are phylogenetically conserved, which facilitate the translation of the findings from animal experiments to humans. Some miRNAs are cell-type-/tissue-specific, which help to identify the response to the toxicant from different cell types or tissues. Recently, Fantome 5 program has published a miRNA expression atlas covering human primary cells, which can be used as a basis of miRNA profiling analysis [45]. Identification of miRNAs involved in responses to environmental chemicals in in vitro neural development models, as well as the identification of perturbation in miRNA expression profiles in neural cultures derived from patient induced pluripotent stem cells (iPSC) with neural disorders, can give indications for further research into potential (circulating) biomarkers. Circulating miRNAs are much more stable than mRNA, since they are either bound to a protein complex, called RNA-induced silencing complex (RISC), or are encapsulated into microvesicles or exosomes.

2 Materials

2.1 Materials for mRNA Isolation

A number of commercially available kits can be used for miRNA isolation and purification. Usually, these kits allow to choose from two options: (i) isolation of total RNA (mRNA and small RNA) and (ii) enrichment of small RNA separately from all RNA longer than 200 nt. It is advisable to choose the option (i) and keep both, long and small RNA, in one sample. This will allow to perform both mRNA and miRNA profiling from the same sample, which will increase the accuracy and reproducibility of the experiment. Here it is important to make sure that the columns used from commercially available kits are designed to capture RNA smaller than 200 nt.

In this chapter, we describe how to purify total RNA from cells/tissue material using TRIZOL reagent (Invitrogen) followed by Zymo RNA Clean & Concentrator kit (*see* **Note 1**) as well as from the biofluids using miRCURY RNA isolation kit biofluids (Qiagen). Additional reagents, not provided with the kits, are RNaseZAP solution (Sigma-Aldrich), ethanol (absolute (200 proof) molecular biology grade), isopropanol, RNase-free DNase (optional), RNase-free water, RNase-free tubes, heat block, and microcentrifuge.

2.2 Materials for miRNA Integrity Analysis

NanoDrop 2000 (Thermo Fisher Scientific), Agilent 2100 Bioanalyzer, RNA 6000 Nano or Pico Assay kit, RNase-free water.

2.3 Materials for miRNA Microarray

We use the following materials for microarrays: RNaseZAP solution (Sigma-Aldrich), ethanol, (absolute (200 proof), molecular biology grade), isopropanol, RNase-free DNase (optional), RNase-free water, RNase-free tubes, heat block, microcentrifuge, vortex, and sterile RNase-free small supplies (slide staining dish, Falcon tubes and bottles, forceps, tips, micropipettor).

For Affymetrix microarray we use FlashTag™ Biotin RNA labeling kit (Genisphere); reagents for ELOSA QC assay (Appendix A and C, Affymetrix FlashTag™ Biotin RNA labeling kit guide);

GeneChip® eukaryotic hybridization control kit; GeneChip® hybridization, wash, and stain kit; miRNA array chip (v.4.0, Affymetrix); Laser Tough-Spots® 1/2" and 3/8" diameter (Diversified Biotech); Affymetrix GeneChip® Command Console® software (AGCC); Affymetrix Expression Console™ software (EC) v 1.2 or higher; GeneChip® Fluidics Station 450; GeneChip® Hybridization Oven 645; and GeneChip® Scanner 3000 7G.

For Agilent microarrays we use gene expression wash-buffer kit, miRNA Spike-In kit, Agilent miRNA Complete Labeling and Hyb kit, Agilent SurePrint G3 Human miRNA 8x60K Arrays, hybridization oven with rotator for Agilent Hybridization Chambers, Agilent G2600D SureScan Microarray Scanner, and Agilent Feature Extraction Software.

2.4 Materials for miRNA Bioinformatic Analysis

Several software packages are available for miRNA profiling analysis. GeneChip® Analysis Suite and Affymetrix Microarray Suite are used for array processing. Statistical data analysis and data visualization are performed using Partek Genomics Suite (www.Partek.com for Affymetrix arrays) or GeneSpring software (www.Agilent.com, for Agilent arrays) in combination with an R-package (version 3.4.2) (https://www.R-project.org/) and Bioconductor (version 3.6).

- http://www.mirbase.org
- http://zmf.umm.uni-heidelberg.de/apps/zmf/mirwalk2/
- https://david.ncifcrf.gov/home.jsp
- https://horvath.genetics.ucla.edu/html/CoexpressionNetwork/Rpackages/WGCNA/
- https://pdmap.uni.lu/minerva/

2.5 Materials for miRNA Real-Time PCR

- *TaqMan miRNA assay* (TaqMan miRNA assay contains 5× stem-loop RT-primers, specific for miRNAs of interest, and a 20× real-time probe, containing miRNA-specific forward and reverse primers as well as a miRNA-specific fluorescent MGB probe).
- TaqMan miRNA Reverse Transcription Kit.
- TaqMan Fast Advanced PCR Master Mix for FAST PCR system or TaqMan Universal PCR Master Mix, no AmpErase® UNG for machines operating in standard mode.
- MicroAmp® EnduraPlate™ Optical 96-Well Fast GPLE Clear Reaction Plates with Barcode (Molecular Probes) for fast machines or MicroAmp® Optical 96-Well Reaction Plates for machine operating in standard mode.
- MicroAmp™ Clear Adhesive Films (Molecular Probes).
- Thermocycler.

- Applied Biosystems Real-Time machine: 7300 System, 7500 System, 7900HT System, 7900HT System, 7500 System, StepOnePlus™, StepOnePlus™, ViiA™ 7 System, StepOne™, Mode, StepOne™.
- Standard materials such as Eppendorf tubes, RNase-free water, RNase-away solution, and PCR tubes.

3 Methods for miRNA Detection

3.1 Workflow for miRNA Profiling for (Developmental) Neurotoxicity Testing In Vitro

To study (developmental) neurotoxicity different in vitro neural cell models can be used. Our focus, however, is recently emerging 3D brain organotypic models or microphysiological systems [46]. These systems are human-relevant, since they are derived from human cells, such as induced pluripotent stem cells (iPSC) or human cell lines. 3D organotypic cultures are often better mimicking organ functionality and architecture [47–49]. A panel of functional endpoints can be used to study the cellular response to the potential neurotoxicant, which might give a hint toward substance potency, but not necessarily explain the mechanism of toxicity. Measuring complex molecular responses and perturbations of molecular networking by a combination of omics technologies is a key for systems toxicology and will allow understanding and describing the complex pathways of compound toxicity [50, 51]. miRNA profiling is an essential part of this testing strategy, since miRNAs are essential regulators of gene expression. miRNAs function very often in positive and negative feedback loops in pairs with their targets. Those regulatory loops are crucial for cellular homeostasis. The shift in the feedback loops demonstrates perturbed homeostasis under stress and may answer how cells cope with the toxic insult. Many examples of miRNA/mRNA negative feedback loops were identified for neural development (*mir-124/REST* [5], *mir-9/TLX* [6], *let-7/lin28* [52], and others), demonstrating the usefulness of miRNA profiling accompanied by whole genome transcriptomic analysis to study complex molecular perturbations and derive molecular signatures of exposure. The main miRNA features, important for toxicology, are summarized in Fig. 1. However, isolated analysis of miRNA profiles will provide only limited information about perturbed molecular networks; therefore, a systems toxicology approach – a combination of different -omics techniques with strong bioinformatic tools for data analysis performed in a human-relevant model – should be applied (Fig. 2) [53]. Figure 3 demonstrates a workflow of miRNA profiling. We recently published a step-by-step protocol for miRNA profiling after developmental neurotoxicant exposure, including description of the in vitro neural model, experimental design of exposure to the toxicant, RNA isolation, microarray, RT-PCR, and

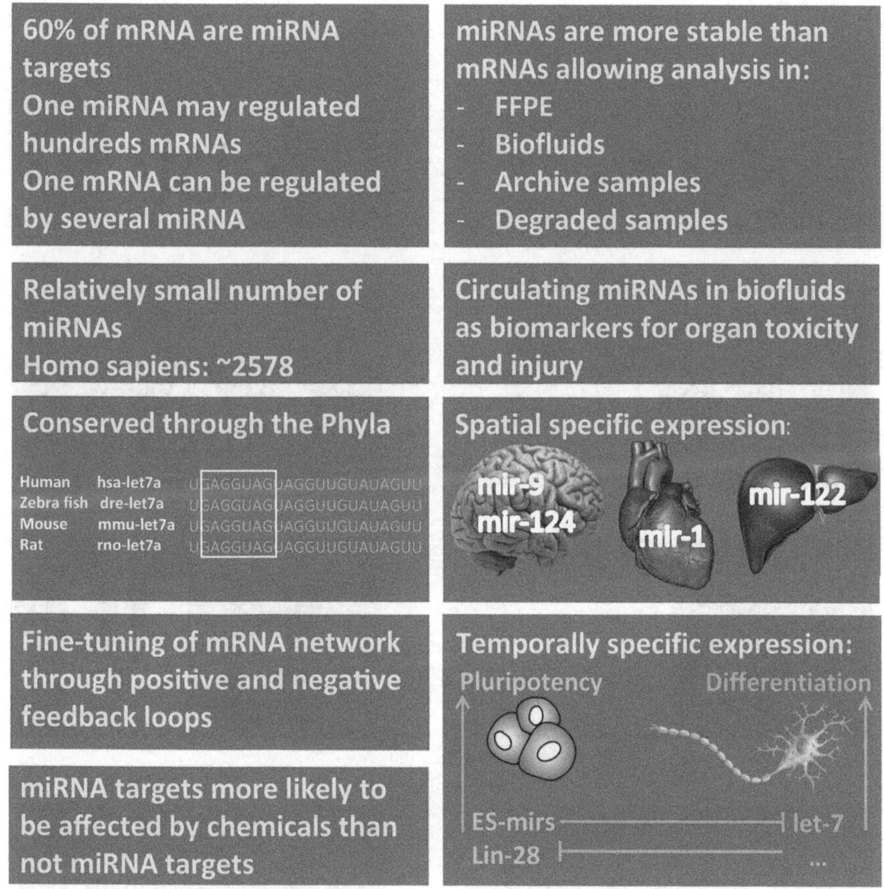

60% of mRNA are miRNA targets
One miRNA may regulated hundreds mRNAs
One mRNA can be regulated by several miRNA
miRNAs are more stable than mRNAs allowing analysis in:
- FFPE
- Biofluids
- Archive samples
- Degraded samples
Relatively small number of miRNAs
Homo sapiens: ~2578
Circulating miRNAs in biofluids as biomarkers for organ toxicity and injury
Conserved through the Phyla
Human hsa-let7a UGAGGUAGUAGGUUGUAUAGUU
Zebra fish dre-let7a UGAGGUAGUAGGUUGUAUAGUU
Mouse mmu-let7a UGAGGUAGUAGGUUGUAUAGUU
Rat rno-let7a UGAGGUAGUAGGUUGUAUAGUU
Spatial specific expression:
mir-9
mir-124
mir-1
mir-122
Fine-tuning of mRNA network through positive and negative feedback loops
Temporally specific expression:
Pluripotency
Differentiation
ES-mirs
let-7
Lin-28
...
miRNA targets more likely to be affected by chemicals than not miRNA targets

Fig. 1 The main points justifying miRNA as a tool for toxicity testing

data interpretation. For details refer to [54]. In addition, a step-by-step protocol for in vitro neuronal 3D model generation, toxicant exposure, and main toxicological endpoints was described by us in [55].

3.2 miRNA Isolation and Integrity Analysis

RNA extraction and sample handling can have a significant impact on the miRNA profiling results. If some cell types or cell lines are easy to handle and give high RNA quality and yield, some cell types or tissues are trickier to get pure RNA samples from. Neuronal cultures are usually on the side of difficult-to-handle cultures. In the tissue samples, the homogenization and lysis are very critical steps influencing the yield of RNA. Sometimes it is difficult to get intact not-degraded RNA from certain tissues, especially from formalin-fixed and paraffin-embedded (FFPE) tissues. Here, miRNA profiling shows again an advantage over whole genome transcriptomics, since miRNAs are not as sensitive to RNase and due to their small size are not degraded. Another consideration to be taken into account when working with tissue samples such as the

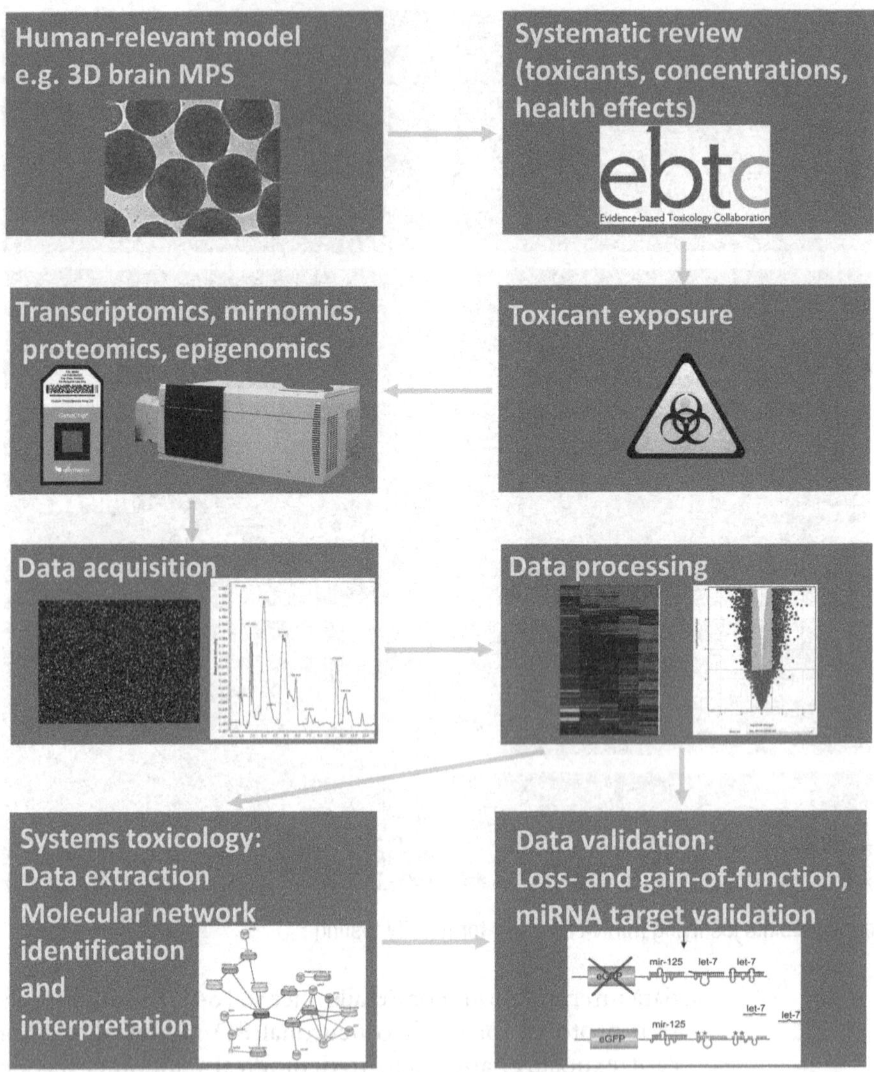

Fig. 2 Generic workflow/systems toxicology approach. Systems toxicology approach for toxicity testing implements use of human-relevant 3D microphysiological systems (MPS) or organotypic cultures in combination with systematic review of existing knowledge about chemical toxicity/human exposure and generation of high-content multi-omics data, which are analyzed in the way to generate molecular networks and signatures of toxicity. Those perturbations identified by omics analysis are undergoing an experimental validation process, to demonstrate relevance to the functionality of the perturbed system

brain is cellular heterogeneity. There are several ways to overcome this problem: Subsequent proper bioinformatic analysis of cell-type-specific miRNA may facilitate the profiling of miRNA in heterogeneous tissue and be able to determine the cell source of the perturbed signal. Cell sorting (e.g., FACS – fluorescent-activated cell sorting) prior to RNA isolation allows studying miRNA profiling in specific cell types (e.g., only astrocytes) and not in the whole tissue. Newly emerged single-cell RNA sequencing [56] may overcome

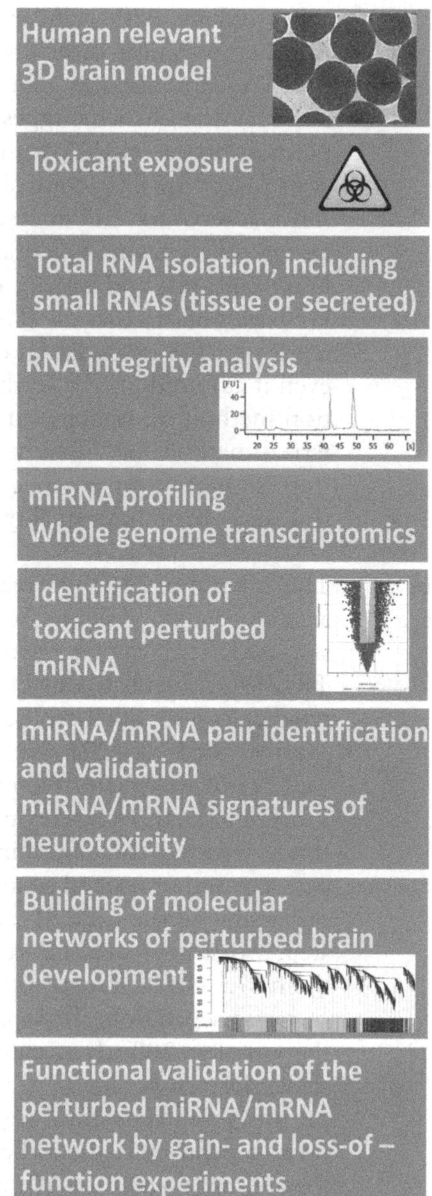

Fig. 3 miRNA profiling for neurotoxicity testing workflow. First, relevant model for the toxicity study should be identified. Next, the exposure scheme should be defined (important to keep the exposure scenario as close as possible to human exposure). After toxicant treatment, total RNA is isolated from the tissue/cells or from the medium in case of circulating biomarker discovery. After RNA integrity check, the miRNome profile is assessed in parallel to the transcriptome profile of protein-coding genes. As the next step, bioinformatic approaches are used to identify significantly perturbed miRNAs and mRNA. Then, miRNA targets are identified computationally, and the relevant ones are confirmed experimentally with miRNA reporter assay. Molecular signatures and networks are derived. Finally, identified networks should be verified experimentally to link the molecular perturbations to functions. In order to do so, the effects of miRNA overexpression or repression on the functionality of the systems are compared with the toxicant effects

the problems with cellular heterogeneity in the tissue samples. Another technical difficulty in RNA purification is isolation of miRNA from biofluids. The amount of circulating miRNA is usually very low, which suggests using specifically designed kits for purification of miRNA from biofluids and always using spike-in controls as quality control and surrogating to determine the miRNA recovery. In our research, we have been working with different cell types and neural models, including 3D organotypic brain cultures, and testing different methods of RNA isolation. We have noticed that the iPSC-derived 3D brain model has higher levels of RNA degradation compared to a single-cell-type model, even if cultured in 3D. In this chapter, we describe an RNA isolation method we established in our laboratory as routine and preferred method for different models. General standard precaution steps should be taken while working with RNA to avoid degradation and obtain high-quality RNA samples (*see* **Note 2**).

3.2.1 miRNA Isolation from the Cells/ Tissue

The amount of TRIZOL reagent used is dependent on the surface area of a culture dish in case of adherent cultures or amount of tissue used. The reagent calculations provided below refer to 1 mL TRIZOL reagent. In our DNT studies, we mainly use 3D organotypic cultures, which are generated in suspension under constant gyratory shaking [29]. We have around 60–100 tissues per one well of a 6-well plate. We resuspend the tissues from one well in 1 mL TRIZOL by vortexing and pipetting up and down several times; we leave the tissues in TRIZOL for 2–3 min at room temperature and resuspend again. We make sure there is no tissue residue visible in the tube. If samples have to be stored after the lysis step, we snap-freeze the samples in liquid nitrogen and transfer them to a –80 °C freezer; otherwise we directly proceed to the step of phase separation. To separate the organic and water phases in the samples, 200 μL of chloroform is added to each sample. We shake the samples for 15 s, incubate them for 3 min at room temperature, and centrifuge at ≥12,000 g at 4 °C for 15 min. If the phase separation is not sufficient, we repeat the centrifugation step for another 5–10 min. We carefully transfer the water phase into the new tube (Note 4.3). At the next step, 1 volume of 100% ethanol is added to the samples, and samples are transferred to the Zymo columns (Zymo RNA Clean & Concentrator kit). After brief (30 s) centrifugation at 10,000 g, 400 μL RNA Prep Buffer is added followed by centrifugation step. Then the Zymo columns are washed with 700 μL Zymo washing buffer (which was diluted with 100% ethanol prior to use). Samples are centrifuged at 10,000 g for 2 min. Finally, RNA is eluted in a new RNase-free tube with 20 μL of RNase-free water. The volume can be reduced to 6 μL, in case the amount of starting material was very low. From this step, samples should be kept on ice, if used for further procedures, or immediately frozen at −80 °C for storage. DNase treatment can be performed during the purification step; however, it is optional, since genomic DNA will not interfere with downstream procedures, if only miRNAs are assessed. In case mRNA will be studied in parallel, then genomic DNA digestion should be performed.

3.2.2 miRNA Isolation from the Biofluids/Cell Culture Supernatant

When studding neurotoxicity in in vitro models, the cell culture supernatant can be used to study the level of secreted miRNA upon toxicant insult as a biomarker of toxicity. Two sets of miRNA can be analyzed: (i) miRNA associated with Argonaute proteins, usually released miRNA from dying cells, and (ii) miRNA, which are incorporated in exosomes for active transport/signaling between cells, which are also increased upon stress conditions. The last one can be analyzed separately from total circulated miRNAs, as there are protocols developed for the extraction of miRNA from the exosomes [57]. Exosomes can be isolated from biofluids or cell culture supernatant by differential ultracentrifugation (first pellet the cells (500 g), then cellular debris precipitation (2000 g), and followed by pelleting of microparticles (10,000 g) and exosomes (100,000 g)). New approaches have recently emerged such as a combination of filtration and ultracentrifugation or size exclusion chromatography and OptiPrep™ density gradient isolation techniques and the high-throughput particle precipitation method (ExoQuick®) [58]. Enrichment of these small fractions of circulating miRNAs is more promising for biomarker discovery than analyzing all circulating miRNAs, because those exosomes carry additional membrane-associated markers of the cell of origin, which can be used to track down the cell the miRNAs was released from. The markers can also be used for enrichment of the miRNA of interest. In case of the CNS, for example, if an abundant neuronal marker is present on the exosome membrane (e.g., NCAM and L1CAM), those exosomes can be separated from the rest of the exosomes originating from other tissues by immunoprecipitation [59].

More and more efforts are put toward the combination of several microphysiological systems (e.g., human-on-a-chip initiative [60]). Identification of circulating miRNAs in cell culture supernatants in such complex systems is a way to measure cell-/organ-type-specific toxicity. For example, release of *mir-122* points to liver toxicity while circulating *mir-124* or *mir-9* will indicate neuronal damage/neural toxicity (Table 1). However, highly standardized

Table 1
Different organ toxicities can be assessed based on the level of circulating miRNA

Brain	Liver	Heart
mir-124, mir-9, mir-384, mir-181	mir-122, mir-192, mir-103a	mir-1, mir-133a, mir-208, mir-34a, mir-499

and reproducible methods delivering highly pure material should be used for these studies, and precaution has to be taken when interpreting the results [61].

The amount of circulating miRNAs is usually extremely low; therefore the methods of miRNA isolation and quantification should be adapted to it. Also, there is no housekeeping gene available, so the normalization methods should be also adapted to this fact. Spike-in controls are essential while working with circulating miRNA [61].

In our laboratory, we successfully used the miRCURY RNA isolation kit for biofluids in combination with glycogen (to increase precipitation) and synthetic ce-mir-39 (*C. elegans* miRNA) as a spike-in control. We followed the steps of the protocol as described below.

The starting volume of medium should be identified based on the size of the culture plates and cellular density. The protocol below describes the procedure with 300 μL starting material. The medium samples are collected, centrifuged at 3000 g for 5 min to pellet any cellular residues. 200 μL are snap-frozen in liquid nitrogen and stored at −80 °C. On the day of isolation, the medium samples are defrosted on ice. Lysis buffer is supplemented with ce-mir-39 oligo (14 pg per sample) and glycogen (10 μg per sample). For example, for 60 μL lysis buffer, we use 2 μL of 1 nM stock of ce-mir-39 oligo and 2 μL of 5 mg/mL glycogen stock. The completed lysis buffer is vortexed for 1 min (*see* **Note 4**). 60 μL of the lysis buffer containing glycogen and ce-mir-39 is added to 200 μL of medium sample, vortexed for 1 min, and incubated for 3 min at room temperature. Then 20 μL of protein precipitation solution is added, and samples are vortexed, incubated for 1 min at room temperature, and centrifuged for 3 min at 11,000 g. Clear supernatant is transferred into new tubes containing 270 μL of isopropanol. The miRCURY RNA isolation kit biofluids protocol is followed for subsequent steps. RNA is eluted from miRCURY columns in 25 μL, snap-frozen in liquid nitrogen, and stored at −80 °C.

3.2.3 Assessment of RNA Quality

The fastest and easiest method to measure RNA purity is the NanoDrop spectrophotometric method. However, this method does not show RNA integrity, since it is based on light absorption by the nucleotides. Thus, degraded RNA samples will still have a signal in NanoDrop. Contamination of RNA sample with genomic DNA cannot be visualized with NanoDrop either, since it cannot distinguish between RNA and DNA nucleotides. Therefore, if NanoDrop is used, subsequent RNA gel electrophoresis is advisable to determine RNA integrity and visualize the presence of genomic DNA. Qubit fluorometric RNA quantification can be used as an alternative to NanoDrop to quantify the presence of genomic DNA. Agilent Bioanalyzer can be used for both RNA quantification and integrity measurement, providing with virtual RNA electrophoresis gel, histogram, and an integrity factor (RIN), which should be between 9 and 10 for a good RNA sample. Examples of good and bad NanoDrop curves are shown in Fig. 4a. An example of good and bad RNA virtual gels/histograms generated by Agilent Bioanalyzer is shown in Fig. 4b.

miRNAs cannot be visualized with either of these methods, since they run on the gel at the same size as degraded mRNA.

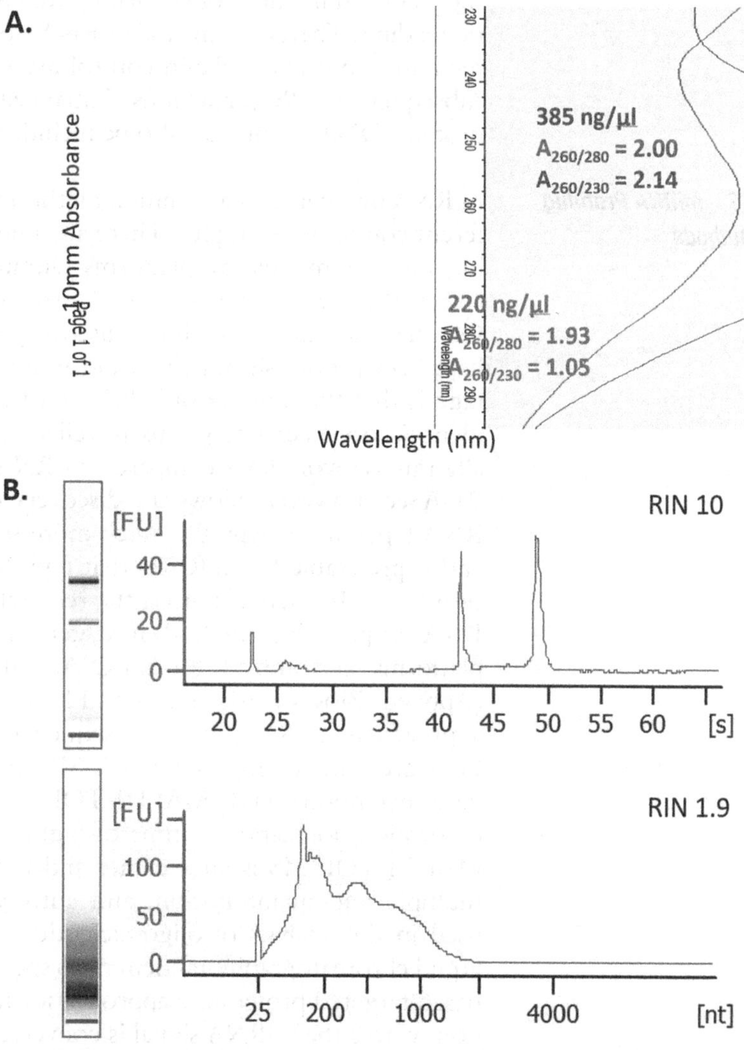

Fig. 4 RNA integrity analysis. Examples of good and bad quality samples. (**a**) NanoDrop graphic report on two RNA samples. Sample 1 has good purity with high $A_{260/280}$ and $A_{260/230}$ ratios, while sample 2 has contamination with organic phase (low $A_{260/230}$ ratio). (**b**) Agilent Bioanalyzer report on intact (RIN 10) and degraded (RIN 1.9) RNA samples. Representative lanes from the virtual electrophoreses gel are depicted on the left, showing the RNA degradation in the sample with RIN 1.9

But if the quality of the total RNA sample is good, miRNA are well preserved in these samples. As mentioned above, if the quality of RNA sample is not good (Bioanalyzer integrity factor is lower than 8, or the NanoDrop ratios are lower than 1.7), it may influence the results of RNA profiling. For example, low $A_{260/230}$ means that the RNA sample is contaminated with organic phase (e.g., phenol). Phenol residues in the RNA samples may inhibit the activity of the reverse transcriptases and polymerases used in subsequent cDNA synthesis and PCR reactions, respectively. It is not possible to assess the yield and integrity of circulating miRNAs by spectrophotometric reading. Therefore, in case of RNA isolation from the medium, medium input and spike-in control are used for normalization in subsequent RT-PCR reactions. Initial cell number (based on total protein/DNA count) can also be included as a normalization step.

3.3 miRNA Profiling Methods

miRNA microarray is a common method to profile miRNAs in different conditions/samples. There are a number of vendors providing miRNA microarray platforms (summarized and reviewed in [62]). We used Affymetrix and Agilent platforms in our studies. On the one hand, this limits discovery to known miRNAs that have been previously described and annotated; however, the advantage is that the number of hybridized RNA probes is limited and identification is unambiguous as well as having a less computationally intense workflows compared to RNA sequencing (RNASeq). RNASeq, however, allows the discovery of new miRNA or other RNA types and is typically much more sensitive in dynamic range and is preferable for miRNAs that typically have very low expression levels. In such instances, the researcher might consider using RNA sequencing method/next-generation sequencing (NGS) platforms customized for miRNA: MiSeq (Illumina), SOLiD (Applied Biosystems), and GS FLX+ (Roche). An alternative approach to miRNA array or sequencing is mass spectrometry. Here are some examples: (i) SPC-SBE (solid phase capture-single base extension) and MALDI-TOF MS (matrix-assisted laser desorption/ionization time-of-flight mass spectrometry). MALDI-TOF MS is an accurate and fast method; is suitable for multiplexing, quantification, and automation; and, therefore, is used in the analysis of oligonucleotides [63]. (ii) A quasi-direct liquid chromatography-tandem mass spectrometry (LC-MS/MS)-based targeted proteomics approach for target-miRNA quantification, where the miRNA signal is converted into the mass response of a reporter peptide via a covalently immobilized DNA-peptide probe [64]. Mass spectrometry methods are time-consuming to establish, require extremely high purity of samples, but are very sensitive and cost-effective to run, once established in the laboratory. They did not get, however, as broad acceptance as NGS or microarrays. In this paragraph, we describe miRNA arrays, as we employed in our own studies. Also, it is very informative for subsequent molecular network and pathway analysis to run whole genome transcriptomics in parallel. By doing so, not only miRNA data can be analyzed in tandem with their mRNA targets (predicated or validated), but also the generated molecular network will provide more nodes of interest for farther building of DNT pathways.

3.3.1 miRNA Microarray

Affymetrix Microarray

For detailed step-by-step protocol for Affymetrix miRNA microarray, refer to the protocol published earlier by the authors [54] as well as to the Affymetrix manuals. An example on how the Affymetrix platform was used for a DNT study is described by Smirnova et al. in 2015 [40]. In this study, we treated mouse embryonic stem cells during the neural differentiation process with two different developmental neurotoxicants, valproic acid and arsenite. Using miRNA profiling, we were able to identify a lineage switch in the neural cultures treated with valproic acid toward myogenesis. This was not observed in arsenite-treated cultures. miRNA results were further confirmed with whole genome transcriptomics and immunohistochemistry.

Briefly, we purify all RNA samples in triplicates as described above and follow Affymetrix GeneChip® Expression analysis user guide for detailed instructions. We use one microgram of RNA for each sample. We use FlashTag™ Biotin RNA labeling kit to add poly(A) to each miRNA and ligate these tailed miRNAs with biotin-labeled DNA in order to get all miRNAs labeled with biotin (for detailed instructions we followed Affymetrix FlashTag™ Biotin RNA labeling kit guide). RNA biotinylation efficiency is analyzed using the enzyme-linked oligosorbent assay (ELOSA) (refer to the FlashTag™ Biotin RNA labeling kit manual, Appendix A for the protocol). After samples are labeled and array information is uploaded into Affymetrix GeneChip® Command Console (AGCC) according to the AGCC user guide, biotin-labeled RNA is hybridized for 18 h at 48 °C to the Affymetrix GeneChip miRNA array using the GeneChip® Scanner 3000 7G System. After hybridization, the arrays are washed, stained with Steptavidin-PE, and scanned according to the GeneChip® Expression wash, stain, and scan user guide for cartridge arrays and Affymetrix GeneChip® Command Console (AGCC) guides. All the data with raw intensity is saved as .CEL files. .CEL files are then imported into microarray-suitable software for data processing such as GeneChip® Analysis Suite and Affymetrix Microarray Suite. Data processing includes background correction, quantile normalization, and summarizing of the log-expression values for each miRNA on each array. Statistical data analysis and data visualization can be performed using Partek Genomics Suite (www.Partek.com) or GeneSpring software (www.Agilent.com) as described below (Sect. 3.4).

Agilent Microarray

As an alternative to Affymetrix microarray, the Agilent platform can be used. The protocols for sample preparation and differential statistical analysis are the same for both methods. For detailed description of Agilent miRNA microarray, please refer to the manufacturer's instructions (https://www.agilent.com/cs/library/usermanuals/public/G4170-90011_miRNA_Protocol_3.1.pdf).

Briefly, 100 nm of total RNA is processed for hybridization to Agilent SurePrint G3 Human miRNA 8x60K Arrays according to Agilent miRNA Complete Labeling and Hyb kit protocol (see link above). The miRNA Spike-In kit is used to prepare and utilize spike-in solutions for controls. The arrays are scanned in the Agilent G2600D SureScan Microarray Scanner using scan protocol AgilentG3_miRNA for miRNA arrays. Agilent Feature Extraction Software is used to assign grids, provide raw image files per array, and generate QC metric reports from the microarray scan data. The QC metric reports are used for quality assessment of all hybridizations and scans. Subsequent statistical analysis is conducted as described in detail below (Sect. 3.4).

3.3.2 miRNA Real-Time PCR

miRNA quantitative real-time PCR (qPCR) is the most sensitive and specific of miRNA profiling methods. This method should also be used for selected miRNAs after total miRNA profiling to confirm RNASeq or microarray results. Several techniques were developed for miRNA qPCR. Most broadly used are TaqMan miRNA assays (Thermo Fisher Scientific) and SYBR Green-based platform – miRCURY LNA qPCR (Qiagen). The design of the assays is described in detail in the manuals and by [62]. In this chapter we describe the TaqMan miRNA assay as we use in our studies.

TaqMan miRNA assay is based on a two-step reaction. In the first step of reverse transcription, a miRNA-specific stem-loop primer is annealed to miRNA, and cDNA is synthetized from it. In the second step of amplification, miRNA-specific forward and reverse primers are used together with a specific probe, which has a fluorescence tag and quencher. The reverse primer is specific to the stem-loop part of the cDNA sequence. The fluorescence tag is quenched on intact probe due to specific proximity between dye and quencher. When DNA polymerase approaches the probe bound to the cDNA, the probe gets cleaved due to DNA polymerase's nuclease activity, and fluorescence dye is released from quencher, which produces a bright signal.

First, we select the TaqMan assay of interest with appropriate housekeeping miRNA (http://www.lifetechnologies.com/us/en/home/life-science/pcr/real-time-pcr/real-time-pcr-assays.html). We select miRNA from the dropdown menu and type the miRNA of interest. Then assay design and species can be selected. It is advisable – if possible – to use inventoried assays, since those are well-established with 100% PCR efficiency.

Housekeeping miRNA should be defined individually for each test system/research project. Usually, three to four housekeeping candidates should be tested and at least two most stable ones kept throughout the experiment. Examples of housekeeping genes for miRNA assay can be found on the Thermo Fisher webpage (e.g., *RNU44*, *U6*, *mir-103*).

For the reverse transcription step, all reagents are provided with miRNA reverse transcription kit except RNase-free water. Stem-loop primers are part of corresponding TaqMan miRNA assays. First, we prepare 30 ng/μL RNA solution by diluting of RNA stock in RNase-free water. We use 2 μL of the 30 ng/μL solution per one RT-PCR reaction. Then, we prepare the reverse transcription master mix. We multiplex up to eight miRNA assays in one reverse transcription reaction as follows: 6 μL of RNase-free water, 3 μL of buffer (10×), 0.6 μL of dNTPs (100 mM each), 0.4 μL of RNase inhibitor (20 U/μL), 2 μL of each stem-loop primer (for eight miRNAs), and 2 μL of reverse transcriptase (50 U/μL). We keep all ingredients and samples on ice for the entire time of sample preparation. We multiplex the volumes provided above by number of samples and add one extra sample to account for the pipetting error. Then we add 28 μL of master mix to preloaded RNA and gently mix without generation of bubbles and start the predesigned cDNA synthesis program on the thermocycler (16 °C for 30 min, 42 °C for 45 min (reverse transcription), 85 °C for 5 min (deactivation of the reverse transcriptase), 4–10 °C until next procedure). Total sample volume is 30 μL. cDNA can be stored at −20 °C or used immediately for the amplification step (real-time PCR).

For real-time miRNA amplification, we follow the TaqMan miRNA assay protocol, but we reduce the volume of one PCR reaction to 10 μL with 1 μL cDNA, 5 μL 5x TaqMan master mix and 0.5 μL of TaqMan miRNA assay, and 3.5 μL of RNase-free water per sample (*see* **Note 5**). It is recommended to run each sample in triplicates and have one water control (no cDNA) for each assay. The following program is used for the amplification reaction in Applied Biosystems Fast 7500 System: first step is 95 °C for 20 s (denaturation and polymerase activation), then the 40 cycles PCR (95 °C for 3 s (denaturation), and 60 °C for 30 s (primer annealing and elongation)). After the reaction is finished, we analyze the PCR curves: for (i) whether thresholds and baselines for each primer are set properly, (ii) artifacts, (iii) whether the water control is negative, and (iv) whether outliers need to be marked or omitted from further analysis (triplicates of a sample should not differ from each other for more than one Ct). Then we exported the Ct values into an Excel table to calculate the fold changes between toxicant-treated samples and controls (see Sect. 3.4.2).

3.4 Data Analysis and Interpretation of the miRNA Profiling Results

3.4.1 Microarray Raw Data Analysis and Differential Statistics

For microarray raw data processing, we import raw data into microarray-suitable software for data processing such as GeneChip® Analysis Suite and Affymetrix Microarray Suite. Data processing includes background correction, quantile normalization, and summarizing of the log-expression values for each miRNA on each array. Statistical data analysis and data visualization are performed using Partek Genomics Suite (www.Partek.com for Affymetrix arrays) or GeneSpring software (www.Agilent.com, for Agilent arrays) in combination with an R-package (version 3.4.2) (https://www.R-project.org/) and Bioconductor (version 3.6). Primary QC by principal component analysis should be conducted to reveal possible batch effects. To eliminate any meaningless probes (expression close to background, duplicates, not-annotated probes), several steps are taken: (1) probes are filtered out if they are not at least 10% brighter than the 95% percentile of the negative control probes on each array on at least three arrays; (2) duplicate probes are summarized; (3) individual probes, which were either labeled by the Feature Extractor Software as not to be used, nonuniform outlier, or population outlier are removed; (4) probes without ID are removed.

Normalization for miRNA remains an open question as there is as yet no agreed-upon approach and there is low correlation between the results from different technologies [65].

For our analysis, differential expression is estimated by empirical Bayes moderation of the standard errors toward a common value (empirical Bayes moderated *t*-test). We calculate fold changes (FC) and *p*-values, corrected for multiple hypothesis testing (e.g., FDR-corrected) for altered miRNAs and genes in treated samples vs. solvent control and set thresholds for FC $\geq |1.5|$ and a false-discovery corrected *p*-value ≤ 0.05 – keeping in mind that the dynamic range of miRNAs is typically less than protein-coding genes and therefore often the signal will be somewhat muted in comparison.

3.4.2 Real-Time PCR Data Analysis

We calculate fold change of miRNA expression in toxicant-treated samples in comparison to solvent-treated controls using the $2^{-\Delta\Delta Ct}$ method as described in [66]. We use the following equation where "I" is miRNA of interest, "HK" is housekeeping miRNA, "Tox" is treated sample, "UC" is control sample, "FC" is fold change, and "Ct" is threshold cycle.

$$2^{-\Delta\Delta Ct} = FC = 2^{-\left[Tox\{Ct(I)-Ct(HK)\}-UC\{Ct(I)-Ct(HK)\}\right]}$$

3.4.3 Interpretation of the miRNA Profiling Results

From a systems biology perspective, there are also multiple challenges, some of which are intrinsic to any experiment that involves transcriptomics and some that are unique to miRNA-focused projects. One of the most fundamental issues specific to miRNAs is that we as yet do not possess a comprehensive catalogue of all miRNAs. Originally, most bioinformatic methods to identify miRNA loci did so by scanning genomic sequences for the characteristic hairpin-forming sequences and filtering this either based on evolutionary conservation (MiRSeeker [67]) or on similarity with known miRNA loci (miRScan [68]). Additional tools that used hierarchical hidden Markov models (HHMMiR) [69] avoided the necessity of evolutionary conservation. While these approaches contributed enormously to our understanding of the ubiquity of miRNA within the genome, all approaches that depend solely on sequence have a high rate of false-positives. More problematic for the purpose of using miRNAs to understand neural developmental, methodologies that depend on similarity to known miRNAs or evolutionary conservation will tend to be biased toward highly conserved miRNAs – and this may miss some relevant aspects that are primate-specific or simply less well-conserved.

NGS has been revolutionary for our understanding of noncoding RNAs, and short RNA sequencing data (RNASeq) is currently the method of choice for the detection and quantification of novel miRNAs. However, there is as yet no workflow that is widely agreed upon. Indeed, both the detection of novel miRNAs and the quantification of differentially expressed miRNAs are each highly dependent on the choice of alignment algorithm, as well as several parameters such as the tolerance allowed in the hairpin alignment, quantification (e.g., how a sequence that can be mapped to multiple miRNAs is counted), and normalization strategies [70]. A recent review applying several of the more common algorithms indicated that both choice of tool and parameter setting can cause substantial variation in both the identified miRNAs and their quantification – although in the presence of a strong signal (in this case, treatment with a toxicant), the treatment effect was preserved with approximately 80% concordance [71]. However, as neurodevelopmental miRNA expression and target gene regulation are often cell-specific (sometimes subcellular compartment-specific), as well as subject to very tight spatiotemporal control, it is likely that very subtle changes can have profound phenotypic consequences [72], but such changes can be very difficult to detect, regardless of choice of tools. Additionally, it is clear that some miRNAs exist as functional variants, known as isomiRs, which can be caused by either imprecise cleavage by Drosha or Dicer and can have phenotypic consequences (e.g., *mir-137* variants were shown to affect synaptogenesis and neuronal transmission) [73]. isomiR detection from RNASeq data is an additional challenge and requires both deep sequencing and an algorithm such as IsomiRage [74] to distinguish canonical miRNA from templated and non-templated isomiRs.

Bioinformatic Approaches of miRNA Profiling in Combination with Whole Genome Transcriptomics

An additional area, where miRNA presents some unique challenges, is in the mechanistic and functional analysis of miRNAs and their target genes. As mentioned above, it is advisable to run whole genome transcriptomics in parallel to miRNA profiling for better understanding of molecular network and genetic feedback loop perturbations. Adding a third -omics, such as metabolomics, will give more comprehensive understanding of the final phenotypical perturbations of the system. However, metabolomics is out of scope of this chapter and will be not described in detail.

Mechanistic interpretation of miRNAs data relies completely on understanding the relationship between miRNAs and target genes, which is the first step of miRNA profiling and whole genome transcriptomic data interpretation. This, however, can be complicated, since very often one mRNA can be regulated by more than one miRNA as well as one miRNA may sometimes have hundreds of targets. Here it is important to remember, however, that those many targets are very often just predicted targets, which were never validated experimentally. There are several approaches available for predicting target genes, which generally look at whether the seed region of the miRNA demonstrates sequence complementarity to the 3′UTR of the target mRNA, as well as conservation of the target site between mRNA of different species, and the thermal stability of miRNA-mRNA duplex [75]. As each algorithm uses a different approach to each aspect of target prediction and assigns different weights to each factor, each result can vary enormously depending on the tool selected, and tools range in their choice of trade-off between sensitivity and specificity [75]. Often, the best choice is to use multiple tools that take different approaches and look for overlap. Additionally, there are several databases such as miRWalk [76] that combine predicted targets with text-mining algorithms to identify co-occurrence between a given miRNA and putative target genes in abstracts, to provide experimental evidence for an interaction. However, such text-mining approaches have to be treated with caution, as similar approaches applied to protein-protein interactions typically have an error rate >30% [77]. Therefore, experimental validation of miRNA/mRNA pairs is the next step before final conclusion on biological significance of perturbed miRNA can be made (see below Sect. 3.4.3.2).

In an experiment with miRNA analysis done in tandem with transcriptomics, there are several ways to extrapolate miRNA gene targets from the data. The most straightforward approach is simply to use pairwise comparisons of miRNA genes using either Pearson correlation or mutual information. Some of the spurious correlations intrinsic to such an approach can be filtered by more sophisticated methods, which take advantage of predicted miRNA gene interactions to prioritize more likely relationships (https://academic.oup.com/bib/article/14/3/263/254884). It is also possible to use additional methods to cluster genes, such as hierarchical clustering or weighted gene correlation network analysis (WGCNA, https://horvath.genetics.ucla.edu/html/CoexpressionNetwork/Rpackages/WGCNA/), to find common regulatory motifs. WGCNA in particular can be very powerful as it combines graph theory and weighted correlations to delineate a network structure that can offer more concrete clues to the interactions among genes.

In terms of understanding the functional consequences of miRNA and their gene targets, generally the most commonly used approach is GO annotations, which have been crucial for the interpretation of transcriptomics of protein-coding genes through "GO term enrichment," which identifies pathways or biological processes in differentially expressed genes from transcriptomic experiments [78]. Owing to relatively lack of GO terms for miRNA, most functional analysis of them looks for the GO terms associated with the genes the miRNA are predicted to regulate [79]. The most straightforward way to do this is typically with DAVID (https://david.ncifcrf.gov/summary.jsp), which can use either microarray probe ID or official gene symbol and will analyze for common pathways as well as GO terms. There are several alternatives to DAVID for miRNA functional annotation, e.g., MiRo, FAME, miRGator, miR2Disease, MAGA, and WebGestalt (described in details in [80]). However, any given miRNA may target up to several thousand mRNA and may do so in a cell-specific manner, and in those instances such an analysis may be uninformative [81]. Annotation for miRNAs and target genes has only lately been approached systematically and will require substantial curation before it can prove as useful to miRNomics as it has to other -omics technologies [79]. It is also important to remember that any approach, which depends exclusively on pathway annotations, can be incomplete or misleading for toxicology, since many cellular responses to a toxicant reflect a combination of pathways that have been repurposed from inflammatory or developmental pathways. Our previous work on a Parkinson's disease model, which used WGCNA to infer transcription factors and target genes, indicated many of the identified transcription factors were not annotated to Parkinson's disease or related pathways [82], thus underscoring the importance of using methods such as WGCNA or hierarchical clustering to look for co-regulated genes and suggest possible pathways – this may prove especially critical if miRNAs are acting to coordinate several different pathways. One way to examine possible pathways of miRNA without depending on annotations is to use miRBase (http://www.mirbase.org), which provides annotations for high-confidence miRNAs as well as text-mining of the literature. In the longer term, it will be critical for our understanding of molecular networks in neural diseases to develop and extend curated maps of all known molecular interactions relevant to a disease, similar to the existing one for Parkinson's disease (https://pdmap.uni.lu/minerva/).

Finally, regardless of the methodology chosen, for any given mRNA expression levels, miRNA regulation is typically only one aspect of regulation, and the other contributors (transcription factors, etc.) can often significantly overshadow the signal of miRNA – optimal understanding of miRNA gene target regulation will likely require both better models of miRNA gene target interactions and larger libraries of tissue-specific miRNA gene target expression data and miRNA perturbation experiments.

Validation of miRNA Targets and miRNA Functions

All the predictions made computationally, as described above, have to be validated experimentally to derive final conclusions about miRNA response to chemical exposure. As for miRNA targets, miRNA and their putative target mRNA have to be expressed in the same cell type to be a valid candidate pair. The most common way to experimentally validate this prediction is to use miRNA reporter assays, where the binding site of the putative mRNA target of a given miRNA is cloned into the 3′UTR of a reporter gene (e.g., GFP or luciferase) and then this reporter vector is transfected into the cells expressing this miRNA of interest. As a control, reporter vector carrying a mutation in the seed region of the binding site and empty reporter vector are used. Alternatively, both miRNA and target gene reporter vector can be transfected into the simple cell line, such as HEK293 cells, where reporter vector expression can easily be quantified. Downregulation of the reporter will indicate interaction with the specific miRNA and confirm the prediction. For example, we followed this approach by transfecting mouse primary neurons and astrocytes with GFP vector carrying the part of 3′UTR of the lin-14 gene with binding sites for mir-125. mir-125 is expressed only in neurons but not in astrocytes; therefore GFP expression was downregulated only in neurons, confirming the functionality of this miRNA [83]. The downside of using HEK293 cells and transfecting both miRNA and reporter vector is the fact that the physiological activity of miRNA of interest in the target cells or tissues is not confirmed by this method. Therefore, it is advisable to use the cell system, where the initial profiling was performed. The schematic representation of a miRNA reporter assay is shown in Fig. 5. The detailed protocols for reporter vector design, cloning strategies, transfection of the cells, and readout are described by [84, 85] and will be not repeated here. The next step to verify miRNA functionality in the system is to run gain- and loss-of-function experiments [86]. By performing gain-of-function experiments, the miRNA of interest is overexpressed via transfection of the cells with synthetic miRNA oligos or miRNA lentiviral constructs (miRNA mimics). By loss-of-function experiments, oligos binding to the miRNA of interest (anti-mirs or miRNA inhibitors) are transfected into the system (e.g., provided by Thermo Fisher Scientific or Qiagen (LNA miRNA inhibitors)). Changes in expression of the endogenous predicted target genes are then assessed by RT-PCR, Western blot, ELISA, or immunocytochemistry. A panel of functional measurements can be conducted, to analyze the role of the miRNA of interest in the system. For example, if, based on the putative targets, miRNA is predicted

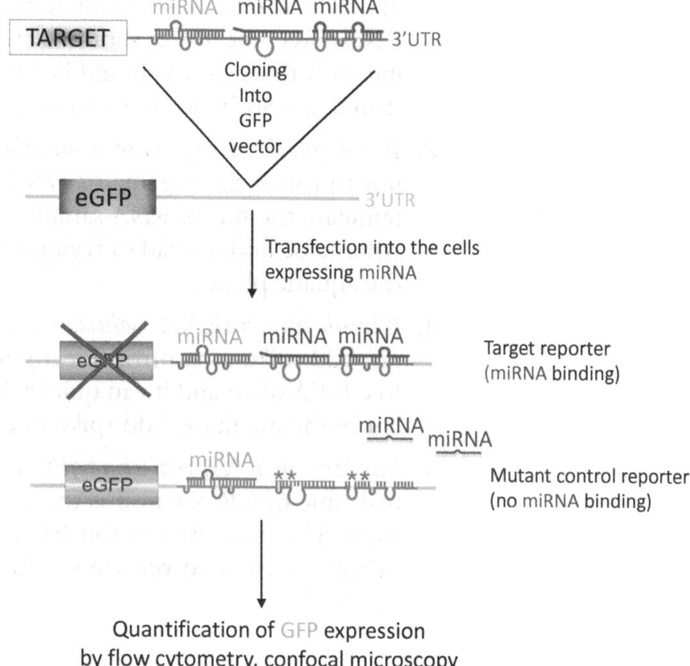

Fig. 5 Schematic representation of experimental validation of miRNA targets miRNA reporter assay. The area with putative miRNA binding sites in the 3′ UTR of the predicted target is cloned into reporter vector, e.g., GFP vector. Then this reporter is transfected into the cells, expressing miRNA of interest. If the miRNA and mRNA are a real biologically relevant pair, miRNA binds to its binding sites and represses the expression of the reporter, which can be then quantified. Control vector, carrying mutations in the binding sites, will be not affected, since miRNA will be not bound

to regulate branching and dendritic spine morphology in neurons, those endpoints can be quantified with high-content imaging after miRNA overexpression or downregulation. Then those results can be compared with the same functional endpoint measured after toxicant treatment, and in case of overlap, this will explain the possible mechanism of chemical toxicity.

4 Notes

1. *RNA purification*. We strongly recommend using column-based cleaning of RNA upon TRIZOL isolation, since it increases RNA purity ($A_{260/280}$ = 1.8–2.0 and $A_{260/230}$ = 2.0–2.2 ratios). From our experience, Zymo RNA Clean & Concentrator allows to obtain highly pure RNA and gives the opportunity to concentrate the RNA, if it was isolated from limited amount of the material. Make sure RNA columns retain small RNA.

2. *Working with RNA*. All materials and reagents have to be RNase-free; the gloves should be worn all the time while working with RNA; RNA should be kept on ice while working and stored at −80 °C for long-term storage.
3. *RNA purification, phase separation step*. It is very important not to touch the organic or DNA phase in order not to contaminate the future RNA sample. It is advisable to leave some volume behind instead of trying to take as much as possible of the aquatic phase.
4. *Circulating miRNA isolation from cell culture medium*. Do not add spike-in control directly to the medium, since this is free RNA oligo and it can quickly be degraded by RNase present in the medium. Add spike-in control to the lysis buffer.
5. *TaqMan miRNA real-time qPCR step*. 10 μL of the sample for real-time qPCR reaction is the minimum volume that can be used. Therefore precaution has to be taken while pipetting; otherwise the end reaction volume can be increased to 20 or 25 μL.

References

1. Kozomara A, Griffiths-Jones S (2011) miRBase: integrating microRNA annotation and deep-sequencing data. Nucleic Acids Res 39:D152–D157
2. Li X, Jin P (2010) Roles of small regulatory RNAs in determining neuronal identity. Nat Publ Group 11:329–338
3. Giraldez AJ, Cinalli RM, Glasner ME et al (2005) MicroRNAs regulate brain morphogenesis in zebrafish. Science 308:833–838
4. Huang TT, Liu YY, Huang MM et al (2010) Wnt1-cre-mediated conditional loss of Dicer results in malformation of the midbrain and cerebellum and failure of neural crest and dopaminergic differentiation in mice. J Mol Cell Biol 2:152–163
5. Conaco C, Otto S, Han JJ et al (2006) Reciprocal actions of REST and a microRNA promote neuronal identity. Proc Natl Acad Sci U S A 103:2422–2427
6. Zhao C, Sun G, Li S et al (2009) A feedback regulatory loop involving microRNA-9 and nuclear receptor TLX in neural stem cell fate determination. Nat Struct Mol Biol 16:365–371
7. Tarantino C, Paolella G, Cozzuto L et al (2010) miRNA 34a, 100, and 137 modulate differentiation of mouse embryonic stem cells. FASEB 24:3255–3263
8. Szulwach KE, Li X, Smrt RD et al (2010) Cross talk between microRNA and epigenetic regulation in adult neurogenesis. J Cell Biol 189:127–141
9. Schratt GM, Tuebing F, Nigh EA et al (2006) A brain-specific microRNA regulates dendritic spine development. Nature 439:283–289
10. Siegel G, Obernosterer G, Fiore R et al (2009) A functional screen implicates microRNA-138-dependent regulation of the depalmitoylation enzyme APT1 in dendritic spine morphogenesis. Nat Cell Biol 11:705–716
11. Edbauer D, Neilson JR, Foster KA et al (2010) Regulation of synaptic structure and function by FMRP-associated microRNAs miR-125b and miR-132. Neuron 65:373–384
12. Weber JA, Baxter DH, Zhang S et al (2010) The microRNA spectrum in 12 body fluids. Clin Chem 56:1733–1741
13. Wang KK, Yuan YY, Li HH et al (2013) The spectrum of circulating RNA: a window into systems toxicology. Toxicol Sci 132:478–492
14. De Smaele E, Ferretti E, Gulino A (2010) MicroRNAs as biomarkers for CNS cancer and other disorders. Brain Res 1338:100–111
15. CDC Report (2018) Prevalence of autism spectrum disorder among children aged 8 years — Autism and Developmental Disabilities Monitoring Network Surveillance Summaries. Morb Mortal Wkly Rep 67(6):1–23
16. Nikolaos Mellios MS (2012) The emerging role of microRNAs in schizophrenia and autism spectrum disorders. Front Psych 3:39

17. Abu-Elneel K, Liu T, Gazzaniga FS et al (2008) Heterogeneous dysregulation of microRNAs across the autism spectrum. Neurogenetics 9:153–161
18. Persico AM, Napolioni V (2013) Autism genetics. Behav Brain Res 251:95–112
19. Landrigan PJ (2010) What causes autism? Exploring the environmental contribution. Curr Opin Pediatr 22:219–225
20. Kuwagata M, Ogawa T, Shioda S et al (2009) Observation of fetal brain in a rat valproate-induced autism model: a developmental neurotoxicity study. Int J Dev Neurosci 27:399–405
21. Geier DA, Kern JK, Garver CR et al (2009) Biomarkers of environmental toxicity and susceptibility in autism. J Neurol Sci 280:101–108
22. Kim J, Inoue K, Ishii J et al (2007) A MicroRNA feedback circuit in midbrain dopamine neurons. Science 317:1220–1224
23. Leucht C, Stigloher C, Wizenmann A et al (2008) MicroRNA-9 directs late organizer activity of the midbrain-hindbrain boundary. Nat Neurosci 11:641–648
24. Yang D, Li T, Wang Y et al (2012) miR-132 regulates the differentiation of dopamine neurons by directly targeting Nurr1 expression. J Cell Sci 125:1673–1682
25. Kim JH, Auerbach JM, Rodriguez-Gomez JA et al (2002) Dopamine neurons derived from embryonic stem cells function in an animal model of Parkinson's disease. Nature 418:50–56
26. Lau P, de Strooper B (2010) Dysregulated microRNAs in neurodegenerative disorders. Semin Cell Dev Biol 21(7):768–773
27. Junn E, Lee K-W, Jeong BS et al (2009) Repression of alpha-synuclein expression and toxicity by microRNA-7. Proc Natl Acad Sci U S A 106:13052–13057
28. Mouradian MM (2012) MicroRNAs in Parkinson's disease. Neurobiol Dis 46: 279–284
29. Smirnova L, Harris G, Delp J, et al (2016) A LUHMES 3D dopaminergic neuronal model for neurotoxicity testing allowing long-term exposure and cellular resilience analysis. Arch Toxicol 90:2725–2743
30. Chaudhuri AD, Choi DC, Kabaria S et al (2016) MicroRNA-7 regulates the function of mitochondrial permeability transition pore by targeting VDAC1. J Biol Chem 291(12):6483–6493
31. Chaudhuri AD, Kabaria S, Choi DC et al (2015) MicroRNA-7 promotes glycolysis to protect against 1-methyl-4-phenylpyridinium-induced cell death. J Biol Chem 290:12425–12434
32. Fragkouli A, Doxakis E (2014) miR-7 and miR-153 protect neurons against MPP(+)-induced cell death via upregulation of mTOR pathway. Front Cell Neurosci 8:182–182
33. Li P, Jiao J, Gao G et al (2012) Control of mitochondrial activity by miRNAs. J Cell Biochem 113:1104–1110
34. Chan YC, Banerjee J, Choi SY et al (2012) miR-210: the master hypoxamir. Microcirculation 19:215–223
35. Kim JH, Park SG, Song S-Y et al (2013) Reactive oxygen species-responsive miR-210 regulates proliferation and migration of adipose-derived stem cells via PTPN2. Cell Death Dis 4:e588
36. Xiong L, Wang F, Huang X et al (2012) DNA demethylation regulates the expression of miR-210 in neural progenitor cells subjected to hypoxia. FEBS J 279:4318–4326
37. Aschrafi A, Schwechter AD, Mameza MG et al (2008) MicroRNA-338 regulates local cytochrome c oxidase IV mRNA levels and oxidative phosphorylation in the axons of sympathetic neurons. J Neurosci 28:12581–12590
38. Bandiera S, Matégot R, Girard M et al (2013) MitomiRs delineating the intracellular localization of microRNAs at mitochondria. Free Radicol Biol Med 64:12–19
39. Smirnova L, Sittka A, Luch A (2012) On the role of low-dose effects and epigenetics in toxicology. EXS 101:499–550
40. Smirnova L, Block K, Sittka A et al (2014) MicroRNA profiling as tool for in vitro developmental neurotoxicity testing: the case of sodium valproate. PLoS One 9:e98892–e98892
41. Pallocca G, Fabbri M, Sacco MG et al (2013) miRNA expression profiling in a human stem cell-based model as a tool for developmental neurotoxicity testing. Cell Biol Toxicol 29:239–257
42. Liu C, Zhao X (2009) MicroRNAs in adult and embryonic neurogenesis. NeuroMolecular Med 11:141–152
43. Wu X, Song Y (2011) Preferential regulation of miRNA targets by environmental chemicals in the human genome. BMC Genomics 12:244
44. Bartel DP (2004) MicroRNAs: genomics, biogenesis, mechanism, and function. Cell 116:281–297
45. de Rie D, Abugessaisa I, Alam T et al (2017) An integrated expression atlas of miRNAs and their promoters in human and mouse. Nat Biotechnol 35:872–878
46. Pamies D, Hartung T (2017) 21st century cell culture for 21st century toxicology. Chem Res Toxicol 30:43–52
47. Paşca AM, Sloan SA, Clarke LE et al (2015) Functional cortical neurons and astrocytes from

human pluripotent stem cells in 3D culture. Nat Chem Biol 12:671–678

48. Lancaster MA, Renner M, Martin C-A et al (2013) Cerebral organoids model human brain development and microcephaly. Nature 501:373–379
49. Alépée N, Bahinski A, Daneshian M et al (2014) State-of-the-art of 3D cultures (organs-on-a-chip) in safety testing and pathophysiology. Altex 31:441–477
50. Kleensang A, Maertens A, Rosenberg M et al (2014) t4 workshop report: pathways of toxicity. ALTEX 31:53–61
51. Hartung T, McBride M (2011) Food for thought ... On mapping the human toxome. ALTEX 28:83–93
52. Rybak A, Fuchs H, Smirnova L et al (2008) A feedback loop comprising lin-28 and let-7 controls pre-let-7 maturation during neural stem-cell commitment. Nat Cell Biol 10:987–993
53. Smirnova L, Kleinstreuer N, Corvi R et al (2018) 3S – systematic, systemic, and systems biology and toxicology. ALTEX 35:139–162
54. Smirnova L, Seiler AEM, Luch A (2015) microRNA profiling as tool for developmental neurotoxicity testing (DNT). Curr Protoc Toxicol 64:20.9.1–20.9.22
55. Harris G, Hogberg H, Hartung T et al (2017) 3D differentiation of LUHMES cell line to study recovery and delayed neurotoxic effects. Curr Protoc Toxicol 73:11.23.1–11.23.28
56. Potter SS (2018) Single-cell RNA sequencing for the study of development, physiology and disease, nature reviews. Nephrology 14:479–492
57. Tang Y-T, Tang Y-T, Huang Y-Y et al (2017) Comparison of isolation methods of exosomes and exosomal RNA from cell culture medium and serum. Int J Mol Med 40:834–844
58. Bhome R, Del Vecchio F, Lee GH et al (2018) Exosomal microRNAs (exomiRs): small molecules with a big role in cancer. Cancer Lett 420:228–235
59. Mustapic M, Eitan E, Werner JK et al (2017) Plasma extracellular vesicles enriched for neuronal origin: a potential window into brain pathologic processes. Front Neurosci 11:278
60. Marx U, Andersson TB, Bahinski A et al (2016) Biology-inspired microphysiological system approaches to solve the prediction dilemma of substance testing. ALTEX 33:272–321
61. Witwer KW (2015) Circulating microRNA biomarker studies: pitfalls and potential solutions. Clin Chem 61:56–63
62. Pritchard CC, Cheng HH, Tewari M (2012) MicroRNA profiling: approaches and considerations. Nat Rev Genet 13:358–369
63. Kim S, Park J, Na J et al (2016) Simultaneous determination of multiple microRNA levels utilizing biotinylated dideoxynucleotides and mass spectrometry. PLoS One 11:e0153201
64. Liu L, Xu Q, Hao S et al (2017) A quasi-direct LC-MS/MS-based targeted proteomics approach for miRNA quantification via a covalently immobilized DNA-peptide probe. Sci Rep 7:5669
65. Schwarzenbach H, da Silva AM, Calin G et al (2015) Data normalization strategies for MicroRNA quantification. Clin Chem 61:1333–1342
66. Schmittgen TDT, Livak KJK (2008) Analyzing real-time PCR data by the comparative C(T) method. Nat Protoc 3:1101–1108
67. Lai EC, Tomancak P, Williams RW et al (2003) Computational identification of Drosophila microRNA genes. Genome Biol 4:R42
68. Lim LP, Lau NC, Weinstein EG et al (2003) The microRNAs of *Caenorhabditis elegans*. Genes Dev 17:991–1008
69. Kadri S, Hinman V, Benos PV (2009) HHMMiR: efficient de novo prediction of microRNAs using hierarchical hidden Markov models. BMC Bioinform 10(Suppl 1):S35
70. Bortolomeazzi M, Gaffo E, Bortoluzzi S (2017) A survey of software tools for microRNA discovery and characterization using RNA-seq. Brief Bioinform
71. Bisgin H, Gong B, Wang Y et al (2018) Evaluation of bioinformatics approaches for next-generation sequencing analysis of microRNAs with a toxicogenomics study design. Front Genet 9:22
72. Rajman M, Schratt G (2017) MicroRNAs in neural development: from master regulators to fine-tuners. Development (Cambridge, England) 144:2310–2322
73. Strazisar M, Cammaerts S, van der Ven K et al (2015) MIR137 variants identified in psychiatric patients affect synaptogenesis and neuronal transmission gene sets. Mol Psychiatry 20:472–481
74. Muller H, Marzi MJ, Nicassio F (2014) IsomiRage: from functional classification to differential expression of miRNA isoforms. Front Bioeng Biotechnol 2:38
75. Riffo-Campos ÁL, Riquelme I, Brebi-Mieville P (2016) Tools for sequence-based miRNA target prediction: what to choose? Int J Mol Sci 17:1987
76. Dweep H, Gretz N, Sticht C (2014) miRWalk database for miRNA – target interactions. In: Alvarez ML, Nourbakhsh M (eds) RNA mapping: Methods Mol Biol. 1182:289–305

77. Krallinger M, Valencia A, Hirschman L (2008) Linking genes to literature: text mining, information extraction, and retrieval applications for biology. Genome Biol 9(Suppl 2): S8–S8
78. Huang DW, Sherman BT, Lempicki RA (2009) Systematic and integrative analysis of large gene lists using DAVID bioinformatics resources. Nat Protoc 4:44–57
79. Huntley RP, Sitnikov D, Orlic-Milacic M et al (2016) Guidelines for the functional annotation of microRNAs using the gene ontology. RNA 22:667–676
80. Liu B, Li J, Cairns MJ (2014) Identifying miRNAs, targets and functions. Brief Bioinform 15:1–19
81. Bleazard T, Lamb JA, Griffiths-Jones S (2015) Bias in microRNA functional enrichment analysis. Bioinformatics 31:1592–1598
82. Maertens A, Luechtefeld T, Kleensang A et al (2015) MPTP's pathway of toxicity indicates central role of transcription factor SP1. Arch Toxicol 89:743–755
83. Smirnova L, Gräfe A, Seiler A et al (2005) Regulation of miRNA expression during neural cell specification. Eur J Neurosci 21:1469–1477
84. Jin Y, Chen Z, Liu X et al (2013) Evaluating the microRNA targeting sites by luciferase reporter gene assay. Methods Mol Biol 936:117–127
85. Aldred SF, Collins P, Trinklein N (2011) Identifying targets of human microRNAs with the LightSwitch Luciferase Assay System using 3′UTRreporter constructs and a microRNA mimic in adherent cells. J Vis Exp JoVE:28(55) e3343–e3343
86. Kuhn DE, Martin MM, Feldman DS et al (2008) Experimental validation of miRNA targets. Methods Companion Methods Enzymol 44:47–54

Chapter 15

Application of Non-Animal Methods to More Effective Neurotoxicity Testing for Regulatory Purposes

Anna Bal-Price and Francesca Pistollato

Abstract

The identification of chemicals that have the potential to induce developmental or adult neurotoxicity is currently entirely based on animal testing. At the regulatory level, systematic testing of developmental (DNT) and adult (NT) neurotoxicity is not a standard requirement within the EU legislation of chemical safety assessment, except for pesticides where NT testing is required. Both DNT and NT evaluation are only performed when triggered based on structure-activity relationships or evidence of neurotoxicity in systemic adult, developmental, or reproduction studies. However, these triggers are rarely used as to date only a limited amount of chemicals have been tested for either DNT or NT effects. Furthermore, the animal-based regulatory DNT or NT studies are unsuitable for screening large number of chemicals, since they are low throughput and costly and use large number of animals. Therefore, new, reliable, and efficient screening and assessment tools are needed for better identification, prioritization, and evaluation of chemicals with potential to induce neurotoxicity. A new framework is proposed for the development of a mechanistically informed IATA (Integrated Approaches to Testing and Assessment) which would integrate various sources of information (e.g., in vitro approaches, in silico modeling, mechanistic knowledge built in the relevant adverse outcome pathways, etc.) as well as in vivo animal and human data, speeding up the evaluation of thousands of compounds present in industrial, agricultural, and consumer products that lack safety data on NT and DNT potential.

Key words Neurotoxicity, In vitro methods, Adverse outcome pathways, Regulatory context

1 Introduction

The specialized metabolic and physiological features of the nervous system convey unique vulnerabilities to toxic compounds that may act on multiple sites through different toxicity pathways. The same substance can potentially affect the central (CNS) or peripheral (PNS) nervous system causing a variety of toxic effects with different impact on function depending on the life stage at which exposure takes place during development, adulthood, or aging.

Neurotoxicity (NT) occurs when exposure to neurotoxins alters the normal activity of mature CNS or PNS due to changes of neuronal and/or glial cell structure, chemistry, and function

Michael Aschner and Lucio Costa (eds.), *Cell Culture Techniques*, Neuromethods, vol. 145,
https://doi.org/10.1007/978-1-4939-9228-7_15,

causing neuroanatomical, neurophysiological, or behavioral changes. Functional neurotoxic effects can induce adverse changes in somatic/autonomic, sensory, motor, and cognitive function. Symptoms may appear immediately after exposure or be delayed [1, 2]. Developmental neurotoxicity (DNT) is defined as pathological alterations of the nervous systems (CNS or PNS) induced by exposure to a xenobiotic during development. These adverse effects may be expressed at any time during the life span of the exposed individual [3]. The developing nervous system is known to be more vulnerable to chemical exposure as compared to the adult nervous system. The higher vulnerability to toxicity of the developing brain stems from the complex developmental processes, such as the commitment and differentiation of neuronal progenitor cells followed by glial and neuronal cell proliferation, migration, differentiation into various neuronal and glial subtypes, synaptogenesis, pruning, myelination, networking, and maturation of terminal functional neuronal and glial cells [4–9]. Therefore, a particular challenge in DNT testing is that the neurodevelopmental outcome depends not only on the kind of exposure (dose, duration) but also on the developmental stage of the brain at the time of exposure [7]. Moreover, the blood-brain barrier (BBB) is not completely formed, permitting the entrance of a chemical into the fetus/neonatal brain [10].

Currently, at the regulatory level, there is a recognized need for neurotoxicity evaluation [11, 12]; however, systematic testing for DNT or NT is not a mandatory requirement in Europe for chemical safety assessments. Both NT and DNT testing are performed only as higher-tiered tests triggered based on structure-activity relationships or evidence of neurotoxicity in standard adult, developmental, or reproduction studies [11, 13, 14] after acute exposure (OECD TGs 402, 403, 420, 423, 425, 436) [15–20], repeated dose toxicity (OECD TGs 407 and 408) [21, 22], or chronic exposure (OECD TG 452) [23]. NT testing is required in the USA and Europe for pesticide evaluation only [24].

Currently, for regulatory purposes, the identification of chemicals with neurotoxic potential is entirely based on the use of animal tests since there are no regulatory accepted alternative methods for this purpose. In vivo test methods used for neurotoxicity evaluation of the developing (OECD TG 426) [25] as well as the adult nervous system (OECD TG 424) [26] are based on neurobehavioral evaluation of cognitive, sensory, and motor functions accompanied by morphometric and histopathological studies.

Additionally, OECD TG 418 [27] and OECD TG 419 [28] are applied to assess delayed neurotoxicity after respectively acute and repeated exposure to organophosphorus compounds and carbamates in laying hens.

Additional testing for offspring that have been exposed to chemicals in utero and during early lactation includes sensory function testing, sexual maturation (OECD TG 426 and OECD TG 443 extended one-generation reproductive toxicity study [29]), and assessments of behavioral ontogeny and learning and memory (OECD TG 426) [25]. When guideline neurotoxicity studies are conducted, more advanced tests of nervous system function can be applied (as described in the OECD Guidance Document for Neurotoxicity Testing (2004) [30]) but are seldom conducted in practice. For example, guidance on the US EPA Neurotoxicity Screening Battery for Pesticides and Chemicals (OPPTS 870.6200 [31] and 40 CFR 799.9620 [32], respectively), which is closely aligned to the OECD TG 424 [26], notes that "there is no clear consensus concerning the use of specific behavioural tests to assess chemical-induced sensory, motor, or cognitive dysfunction in animal models," leaving large flexibility for study design and interpretation of results that is often based on expert judgment. Additionally, the rodent guideline studies result in high uncertainties due to extrapolation of results to humans, including differences in timing of brain development, as well as possible differences in toxicokinetics.

At the same time, these guidelines are complex and very resource intensive in terms of animals, time, and overall cost [33, 34] and have been used only for a limited number of pesticides and industrial chemicals [35, 36]. Currently, approximately less than 200 chemicals [37], mostly pesticides, have been evaluated [24], and only a few of these studies contributed to risk assessment [14]. This highlights the pressing need for developing reliable and efficient alternative methodologies that can more rapidly and cost-effectively screen and assess large numbers of chemicals for their potential to cause DNT or NT [11, 38]. Therefore the efforts should be focused on the development of a testing strategy based on NT and DNT non-animal approaches, including in vitro batteries of test methods.

Taking into consideration the state of science, a testing battery of alternative DNT methods is already available and could be used in a fit-for-purpose manner for either [1] screening and prioritization or [2] as a first step to conduct targeted testing in a tiered testing approach in the process of hazard identification and characterization for specific chemical risk assessment. To meet both these needs, a battery of DNT in vitro assays combined with in silico approaches (QSAR, read-across, computational modeling) and possibly incorporating nonmammalian animal models (e.g., zebrafish, medaka, or *C. elegans*) is required.

These multiple sources of information including also available in vivo and human data should contribute to development of Integrated Approaches to Testing and Assessment (IATA) designed in a fit-for-purpose manner (in relation to the chemical/class of chemicals) for screening and prioritization, hazard identification and characterization, and/or safety assessment.

Several recently developed DNT and NT AOPs (AOP-Wiki: https://aopwiki.org/; [39, 40]) could be used as a framework to facilitate the development of AOP-informed IATA, which should be based on in vitro assays anchored to the key events (KEs) identified in the AOPs at the molecular, cellular, or tissue level. The identification within an AOP of the causative link between the molecular initiating event (MIE), the KEs, and the adverse outcome (AO) provides mechanistic understanding and increases the scientific confidence on the relevance of the in vitro testing battery for regulatory purposes.

However, the current number of available AOPs is limited, and while the development of further specific DNT- and NT-related AOPs will take time, this should not delay the development and implementation of testing strategies. Therefore, it is suggested that besides the KEs already identified in the existing DNT and NT AOPs, additional critical adult (e.g., axonal transport, enzyme activities, synaptic neurotransmission, receptor and ion channels function, etc.) and neurodevelopmental processes (e.g., cell proliferation, migration, differentiation, synaptogenesis, neuronal network formation and function, etc.) should be considered as KEs, since it has been shown that impairment of any of these processes will result in adverse outcomes specific for NT or DNT, respectively. At the same time, chemical testing across a potential testing battery could in the future inform further AOP-building.

Furthermore, mechanistic information on pathways of toxicity specific for DNT and NT identified as KEs in the existing AOPs, as well as critical developmental (DNT) and adult (NT) processes, should guide the design of a testing strategy/IATA, composed of a fit-for-purpose testing battery of in vitro assays. These test methods should be consistent with in vivo human biological and physiological data. Currently most of the DNT and NT in vitro assays can be performed using human in vitro models derived from human pluripotent stem cells, decreasing the need of cross-species extrapolation for animal-based findings. Moreover, some of these assays are amenable to implementation on high-throughput screening (HTS) platforms, permitting the testing of large numbers of chemicals. Incorporation of an in vitro battery of NT and DNT test methods could be supportive across different regulatory domains according to the specific need, e.g. as the first tier approach referring to the relevant OECD TGs. The follow-up animal testing (if necessary) could be refined based on the data obtained from the proposed integrated testing strategy, resulting in targeted, well designed in vivo experiments, delivering required information and leading to more efficient hazard and risk assessment of chemicals.

2 Current Non-animal and Non-testing Methods for Developmental and Adult Neurotoxicity Assessment

2.1 In Vitro Models

So far in vitro models are generally used to study the mechanisms of NT rather than to detect NT for prediction of hazards to human health. One of the main reasons is that available single in vitro test systems cannot fully mimic the immense complexity of human brain development, maturation, and senescence.

Recently EFSA has published a detailed report on the evaluation of in vitro models and endpoints as well as other alternative testing approaches (in silico modeling, read-across, nonmammalian models, etc.) for DNT [41]. This systematic review identified a variety of methods covering early and late stages of neurodevelopment that have the ability to predict DNT effects of chemicals.

Human cell-based systems are recommended as the most relevant in the context of the twenty-first century approach in toxicity testing to reduce the uncertainty in extrapolation of results. The use of human cell-based models should result in better prediction of human toxicity [42–44]. Numerous human neuronal models are currently available ranging from various neuroblastoma lines to stem cell-derived systems [11, 45]. Cancer cell lines, such as human neuroblastoma, present some disadvantages, as the expression profiles of these cells indicate upregulation of several tumor growth-related genes that may affect cell response upon chemical exposure.

Alternative human in vitro neuronal cultures derived from neural progenitor cells (NPCs) have been studied intensively over the past decade as they are primary, of human origin, and expandable although not immortalized [46–48]. NPCs can be obtained from two major types of pluripotent stem cells (PSCs), human embryonic (hESCs) or human-induced pluripotent stem cells (hiPSCs). They can be differentiated into different types of postmitotic neurons as well as astrocytes. However, in hESC- or hiPSC-derived mixed neuronal cultures, the amount of glial cells (astrocytes, oligodendrocytes, and microglia) is still too low in comparison with the in vivo situation. Therefore, differentiation protocols need to be further optimized since glial cells, such as oligodendrocytes (responsible for myelin formation), microglia (involved in inflammatory response), and astrocytes (antioxidant capacity, release of pro-survival factors, uptake of glutamate, ion balance, etc.), play a critical role in chemical-induced mechanisms of neurotoxicity [49]. Therefore, mixed neuronal/glial (preferably organotypic, 3D) cultures of human origin are currently regarded as the most relevant test systems for the evaluation of human developmental or adult neurotoxicity, as they provide a higher level of functional integration between different neuronal and glial cell types that is hampered in monolayer cultures.

Taking into consideration the ethical issues and the differences in national legislation regulating the generation and use of hESCs, hiPSC-derived neuronal and glial models are strongly recommended.

The in vitro neuronal and glial models derived from different types of human stem cells are mainly suitable for DNT [50], rather than adult NT testing, since neurons derived from pluripotent (induced or embryonic) human stem cells do not generally reach full, terminal morphological and functional differentiation, as shown by morphological characterization and measurements of neuronal electrical activity (e.g., [51, 52]) even after a long-term (3 months) culture [53]. Therefore, further efforts are needed to optimize differentiation protocols that will improve neuronal and glial differentiation, resulting in terminal maturation of neurons, astrocytes, oligodendrocytes, and microglia. Establishment of in vitro models of mixed, terminally differentiated cell populations, in proportions similar to the in vivo situation, will be relevant to assess adult NT testing.

The use of hiPSC-derived neuronal models for DNT testing has been described in several studies. Human iPSCs (Fig. 1a) can be used to form neuroectodermal cells (rosettes) (Fig. 1b), resembling neural tube formation in vitro. NPCs derived from rosettes (Fig. 1c) can be expanded in vitro, and their migration can be assessed by phase contrast microscopy (Fig. 15.1d). NPCs can be further differentiated into various neuronal subtypes and glial cells (Fig. 1e, f), and their neuronal network formation and function can be evaluated by multielectrode array (MEA) analysis (Fig. 1f).

At the Joint Research Centre (JRC), a relatively fast and robust protocol for the expansion, cryopreservation, and neuronal and glial differentiation of IMR90-hiPSC-derived NPCs has been optimized [54]. HiPSC-derived NPCs expressing nestin, a neural stem cell marker (Fig. 1c), can be differentiated into heterogeneous cultures of glutamatergic neurons expressing the vesicular glutamate transporter 1 (VGlut1), GABAergic neurons expressing glutamic acid decarboxylase 67 (GAD67) and gamma-aminobutyric acid (GABA), and tyrosine hydroxylase $(TH)^+$ dopaminergic neurons (Fig. 1f, inserts), along with a discrete proportion of glial fibrillary acidic protein $(GFAP)^+$ astrocytes (~20% of total cells) (Fig. 1e) [51, 55, 56]. Differentiated neuronal cells show spontaneous electrical activity (~200 spikes/min, after 21 days of differentiation) (Fig. 1f, g). HiPSC-derived neuronal and glial cell cultures can be used to identify chemically induced signaling pathway perturbations. For instance, differentiated neuronal and glial cells express the nuclear factor (erythroid-derived 2)-like 2 (Nrf2) signaling pathway [51, 56], which regulates endogenous cellular

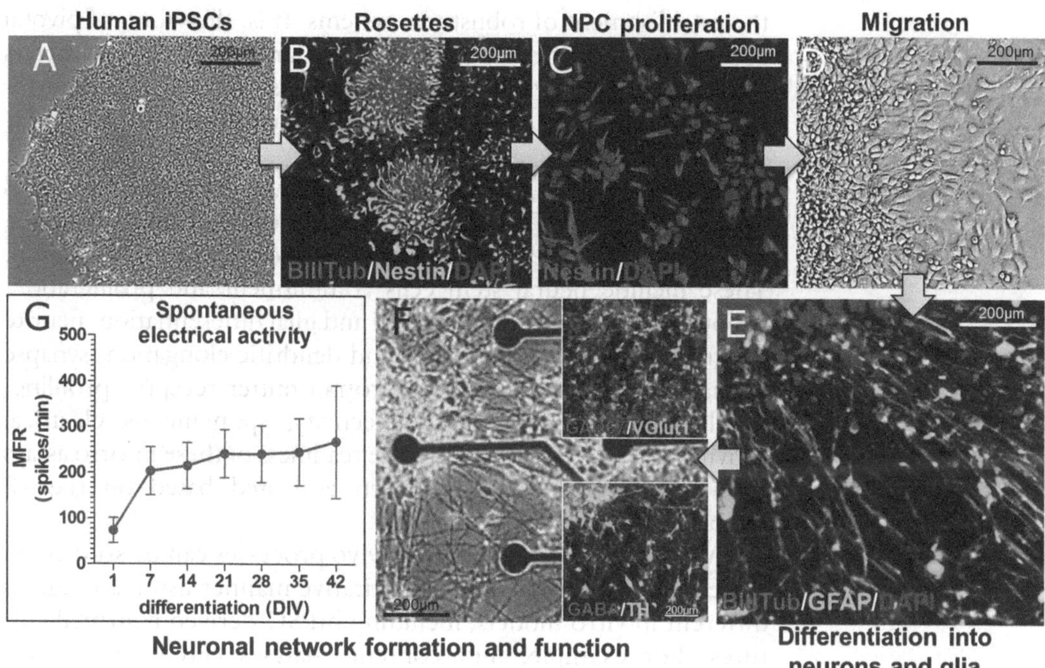

Fig. 1 Human-induced pluripotent stem cell (hiPSC)-derived neural progenitors cells (NPCs) as a model for DNT testing. IMR90-HiPSCs (**a**) can be used to form rosette (**b**), and neural progenitor cells (NPCs) expressing nestin (red; blue DAPI shows nuclear staining) can be derived from rosettes and expanded in vitro (**c**). NPC migration can be measured by phase contrast microscopy (**d**), and NPCs can be further differentiated (**e**) into neuronal cells expressing β-III-tubulin (red) and astrocytes expressing glial fibrillary acidic protein (GFAP, green) indicative of astrocytes. The main neuronal subtypes are present in this mixed culture as evaluated by immunostaining (**f**) for vesicular glutamate transporter 1 (VGlut1), indicative of glutamatergic neurons; glutamic acid decarboxylase 67 (GAD67); gamma-aminobutyric acid (GABA), expressed in GABAergic neurons; and tyrosine hydroxylase (TH), expressed by dopaminergic neurons (see inserts in **f**). NPCs undergoing differentiation on multielectrode array (MEA) (**f**, phase contrast image) show modest increase of mean firing rate (MFR, number of spikes/minute) indicative of spontaneous electrical activity generation measured by MEA analysis (**g**, showing mean ± SEM of nine biological replicates). (**e**) and (**f**) panels show cells differentiated for 42 days

defense against oxidative stress [57], thus allowing to capture oxidative stress response induced by chemicals (e.g., rotenone). Moreover, the cAMP response element binding (CREB) protein pathway, involved in the formation of long-term memories and the regulation of neuronal plasticity and protection [58], results activated in this neuronal/glial model, enabling assessment of CREB pathway perturbation in relation to cellular effects upon treatment with CREB pathway inhibitors, such as 2-naphthol-AS-E-phosphate (KG-501) [55].

It is important to mention that growing stem cell lines in a stable state and delivering reliable and well-defined cultures for toxicity assessments require a high level of standardization of both undifferentiated and differentiated cell cultures, in order to ensure the establishment of robust test systems. It is, therefore, of pivotal importance to define and internationally agree on crucial parameters to judge the quality of the cellular models before using them for toxicity testing [59], especially those derived from PSCs [60, 61].

2.2 *In Vitro Assays*

Regarding DNT in vitro testing, there are well-established methods available for the evaluation of key neurodevelopmental processes, known to be specific for normal brain development and maturation. These include neural stem cells commitment and proliferation, apoptosis, cell migration, neuronal and glial differentiation, neurite outgrowth, myelination, axonal and dendritic elongation, synapse formation, synapse pruning, neurotransmitter receptor profiling, development of neuronal connectivity, spontaneous electrical activity, etc. [41, 62]. Recently, the readiness of these in vitro assays for regulatory purposes has been evaluated based on the 13 established criteria [40].

Most of the DNT-specific in vivo processes can be studied by these in vitro methods in a quantitative manner using a range of different in vitro models, including hiPSC-derived neuronal cultures. For example, high-content analyses (HCA) were performed by the US EPA to assess the effects on neurite outgrowth of approximately 300 chemicals [63], using hiPSC-derived neuronal culture (for 80 chemicals) [64, 65], assessing neural proliferation [63, 66], and synaptogenesis [67]. Biochemical biomarkers of some of the differentiation processes were also studied using primary cultures and neural stem cell-derived cultures [68] based on gene [4, 69] and protein expression [70], as well as OMICs analysis performed in mixed neuronal/glial cultures [71].

Similarly, biological processes characteristic for mature (adult) brain including axonal transport, enzyme activities, synaptic neurotransmission, vesicular release, receptor and ion channel function, neuronal-glial interaction, astrocytic glutamate uptake, receptor activation or blocking, etc. can be studied using well-established in vitro methods [1, 2, 72]. Neuronal networking have been investigated in different cell models [70] by measuring electrical activity using MEA analysis [69, 73–76], including hiPSC-derived neuronal models (e.g., [51, 53, 56]). As shown in a multi-laboratory ring trial studies [74, 76], this technology holds great promise in providing an overall assessment of the chemically induced disruption of critical neuronal function, especially suitable for acute neurotoxicity testing [71], for both developmental and adult neurotoxicity testing.

Summing up, in vitro human as well as rodent neuronal models and relevant endpoints are currently available to study in a quantitative manner (concentration-dependent) the majority of the key biological processes and pathways of toxicity that are specific for both the developing and adult brain. However, most of these methods are currently low throughput, and further efforts could be made in order to scale them up to medium- or high-throughput level.

2.3 Nonmammalian Species

Evaluation of brain development using alternative (nonmammalian) species has revealed that the mechanisms underlying the development and function of the nervous system are well conserved across the phylogenic tree. Many of the basic molecular developmental processes are identical in mammals and in nonmammalian species. In the last 10 years, several alternative species (e.g., small fish models, including *Danio rerio*, medaka, etc.) have been used as vertebrate models for screening neurodevelopmental toxicants [77–79], mainly for behavioral studies. Nonmammalian models are relevant for DNT studies mainly for to three reasons: (1) molecular biology has revealed the basic concordance of cellular events in a wide range of small fish species to that in mammalian species, including humans; (2) the concordance has been verified with advances in genetics and pathway analyses; and (3) the size and speed of development of small fish makes their use particularly ideal for high-throughput assays, including evaluation of behavioral changes (impossible to study using in vitro methods only).

The most investigated model is the zebrafish embryo, which is considered as an alternative high-throughput model for traditional in vivo DNT screening [80].

2.4 In Silico Approaches

There are a few (Q)SAR studies/models that have described the effects of chemicals on the CNS and PNS through the modeling based on in vivo toxicity data [81]. SAR study discussed 14 different triazole fungicides that cause hyperactivity in rats [82]. A QSAR for polychlorinated biphenyl (PCB) neurotoxicity [83], based on data for 28 ortho-substituted PCBs, and building on earlier work [84], revealed a relationship between electronic descriptors (ELUMO, EHOMO, the ELUMO•EHOMO gap, and molecular polarizability) and the binding affinity of PCBs to the aryl hydrocarbon (Ah) receptor. In particular, impairment of the developing nervous system by PCBs has been linked to their ability to alter the spatial and temporal fidelity of Ca^{2+} signaling in muscle and nerve cells through one or more receptor-mediated processes [85]. Prediction of organophosphorus acetylcholinesterase inhibition has been evaluated using three-dimensional quantitative structure-activity relationship (3D-QSAR) methods [86]. Multivariate toxicity profiles and QSAR modeling of 21 non-dioxin-like PCBs have been also determined based on 17 different in vitro screening assays on specific endpoints related to neurotoxicity [87].

Besides QSAR models, another approach applied for filling data gaps is read-across. Read-across is based on the identification of similar compounds taking into consideration chemical structure to make a prediction of a new chemical activity [88]. The read-across tools can be used to search for analogs of known DNT or NT substances; however, none of these tools is specific for the DNT or NT endpoints.

A new chemical, with no information on BBB penetration or transport, first of all has to be assessed with respect to whether it crosses the BBB and whether it reaches brain concentrations that may trigger neurotoxicity. Estimation of a chemical penetration and its quantitative uptake across BBB is currently determined using in vitro systems, including human models [71, 89], or various in silico approaches ranging from physicochemical rules to QSAR models (around 130 publications). The QSAR modeling has been proposed for many different classes of chemicals including diverse organic compounds [90], drug-like molecules [91], a phenyl-amidine class of NMDA receptor antagonists [92], or agonists of the ryanodine receptor type 1 channel [93].

A supervised artificial neural network model has been developed for the accurate prediction of BBB partitioning of a structurally diverse set of 108 compounds using simplified molecular input line system (SMILES) code. The model is able to correctly predict the behavior of a very heterogeneous series of compounds in terms of BBB passage. The results indicate that this approach may represent a useful tool for the prediction of absorption, distribution, metabolism, excretion, and toxicity (ADMET) properties [94] of neurotoxicants.

Recently, a novel, hybrid in silico/in vitro approach and an in silico screening model for the effective evaluation of a chemical permeability coefficient and P-glycoprotein efflux ratio was proposed based on the studies of 176 compounds belonging to different classes of chemicals [95]. These are only a few examples of a large set of QSAR studies that should be carefully assessed to determine their readiness for routine use in testing strategies.

In the case of exposure during pregnancy, a DNT chemical has also to pass the placental barrier in order to reach the developing fetus. The current QSAR models are based on the physicochemical and structural properties of the substance facilitating placental transport, and not on the DNT effect itself. If a substance is not able to cross the placental barrier, such substance is not recognized as a potential DNT chemical. Hewitt and colleagues developed five QSAR models for placental transfer based on different data sets of drugs [96]. Placental transfer is expressed as clearance index or transfer index, i.e., the ratio between the clearance of a substance and the reference substance antipyrine that is easily transported across the placenta via passive diffusion. In general, placental transfer is positively correlated with lipophilicity, expressed by the octanol-water partition coefficient of compounds [97].

3 Application of Adverse Outcome Pathway Concept to Support Development of DNT Testing Strategies

The OECD has developed and endorsed an AOP framework as a tool to apply mechanistic understanding of toxicity pathways into regulatory decisions [98]. AOP describes a sequence of measurable KEs, starting from an MIE in which a chemical interacts with a biological target(s), followed by a sequential series of KEs at the cellular, tissue, and organ level, resulting in an AO relevant to risk assessment [39, 99, 100]. By design, AOPs are simplified representations of disease pathways with KEs at various levels of biological organization which inform purpose-specific application in regulatory context. AOPs are not intended to provide full mechanistic molecular descriptions of causality. Illustrated as a linear series of KEs (Fig. 2), in reality, AOPs represent interdependent networks of events with feedback loops in which disease outcomes are initiated or modified [100].

Due to the complexity of the CNS, development of AOPs relevant to DNT and NT is challenging [101]. A major concern is a general lack of understanding of the MIEs that are causally responsible for triggering KEs leading to a linear cascade of events to the AO in humans. Taking into consideration the possible multiple MIEs for the same AO and variety of potential pathways involved in DNT and NT, numerous AOPs integrated into networks have to be developed, and this will take time. Moreover, an approach based on individual AOPs presents the limitation of being able to identify only a small number of positive "hits" (neurotoxic compounds) eliciting toxicity through the AOP-specific pathways. Therefore, it has been proposed to identify "Converging Key Events" that are common to many individual AOPs. Following this recommendation, networking of the existing AOPs relevant to DNT or NT and identification of the common key events may serve as a base for first-choice in vitro assay selection, leading to the possibility to identify an increased number of neurotoxicants, even if toxicity is triggered by different MIEs, mediated through various pathways and resulting in diverse AOs.

It is well documented in the existing literature that learning and memory processes rely on physiological functioning of the N-methyl-D-aspartate (NMDA) receptor as has been demonstrated in both animal and human studies [102]. It is interesting to notice that disturbance in NMDA receptor function through binding of antagonist (Fig. 2a) or agonist (Fig. 2b) triggers different cascade of KEs but results in the same AO, defined as impairment of learning and memory.

The AOP concept requires description of the mechanistic, causative relationships between the MIE, the KEs, and the AO that should be of regulatory concern. If this is supported by a

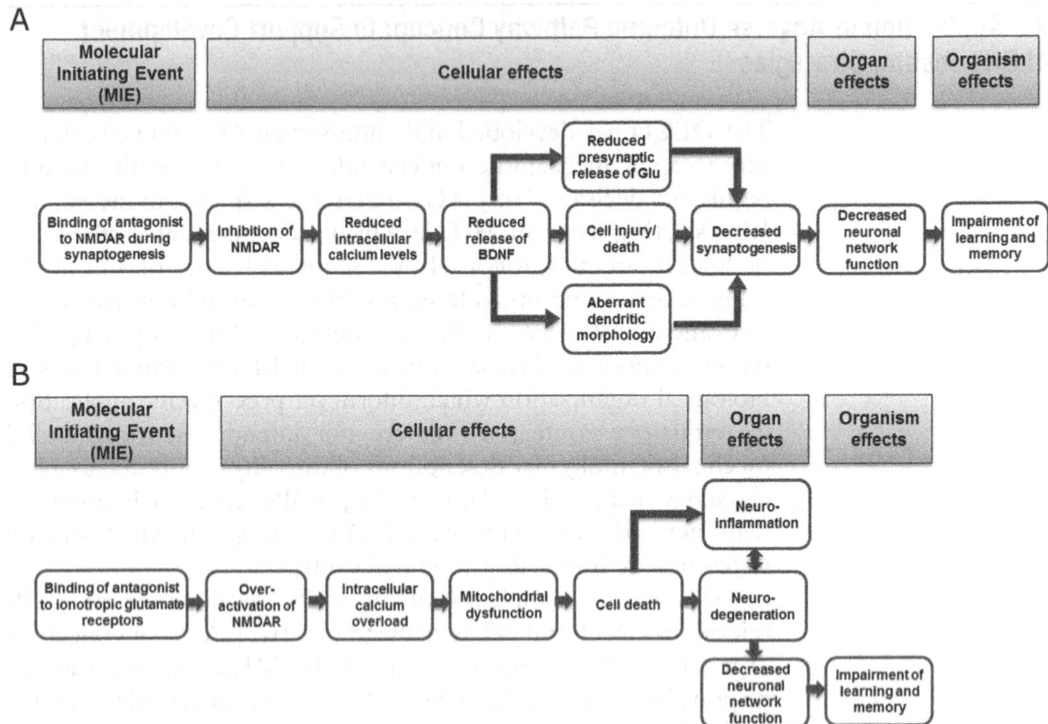

Fig. 2 DNT (**a**) and NT (**b**)-relevant AOPs. (**a**) Chronic binding of antagonist to N-methyl-D-aspartate receptors (NMDARs) during brain development induces impairment of learning and memory abilities (https://aopwiki.org/wiki/index.php/Aop:13). (**b**) Binding of agonist to ionotropic glutamate receptors in the adult brain causes excitotoxicity that mediates neuronal cell death, contributing to learning and memory impairment (https://aopwiki.org/wiki/index.php/Aop:48)

strong weight of evidence, the identified KEs could serve as a base for assay development and be used as an important component of IATA/TS construction for detecting neurotoxicants with potential to cause the human AO, especially when based on human test systems that avoid extrapolation of results. Understanding the likelihood of effects (e.g., initiation of a toxicity pathway) at lower levels of biological organization through, e.g., in vitro testing or structure-activity relationships, can help to inform whether testing at higher levels of biological organization (i.e., in vivo) is warranted [103].

In conclusion, AOP framework provides the biological context that permits development of AOP-informed IATA or TS for regulatory decision-making and facilitates a shift toward a mechanistic knowledge-driven paradigm for chemical hazard identification and characterization (e.g., [104]) and, where feasible, risk assessment. These considerations will influence the construction of IATA in terms of different data requirements, types of testing (e.g., in vitro, in chemico, in vivo), non-testing methods (QSAR, read-across), data integration approaches, and acceptable levels of uncertainty.

4 Strategy to Refine and Reduced Animal Use in Developmental and Adult Neurotoxicity Testing

Based on an analysis of the regulatory requirements for DNT and NT testing, the efforts should be focused on the development of a mechanistically, AOP-informed IATA that uses various sources of information (non-testing methods, in vitro approaches, in vivo and human data) delivering data for the hazard identification and characterization and possibly safety assessment if information on a chemical exposure and ADME are available. The in vitro assays in such IATA should refer to KEs identified in the relevant existing AOPs as well as key biological processes critical for the adult brain (NT) or the brain during development (DNT). The in vitro testing battery anchored to critical cellular processes specific for DNT and NT is already available and could be used in a fit-for-purpose manner. Moreover, the upscaling of these in vitro test methods to HTS platforms and the use of human PSC-derived neuronal models will lay the foundation for a more efficient and human-relevant testing for safety assessment, allowing the screening of a large number of chemicals.

Taking into account both the current state-of-the-art in the field of DNT and NT and considerations for a new paradigm shift in toxicology, IATA should be developed based on combination of carefully selected human models (where possible) and assays amenable for HTS addressing mechanisms of toxicity referring to (i) the identified MIEs and cellular/tissue KEs in the developed AOPs (https://aopwiki.org; [101]), as well as those identified in the literature, (ii) key biological processes specific for brain development (DNT) or maturation (NT), and (iii) if necessary nonmammalian, complementary models (e.g., zebrafish). Combination of these assays should be carefully built into an IATA according to a defined regulatory purpose.

The AOP-informed IATA could be used for an initial chemical screening to identify those with neurotoxicity potential leading to refined, targeted, and more effective further in vivo testing when needed. Moreover, if produced data are combined with available exposure and ADME information relevant to the chemical, such IATA could deliver information useful not only for hazard characterization but possibly risk assessment. Finally, such IATA could be incorporated into existing OECD TGs for DNT (426) or NT (424) not only as complementary testing but also as a possible replacement strategy of the selected in vivo endpoint(s), resulting in reduction of animal testing, leading to more efficient and faster hazard and possibly risk assessment [44, 101].

References

1. Harry GJ, Billingsley M, Bruinink A et al (1998) In vitro techniques for the assessment of neurotoxicity. Environ Health Perspect 106(Suppl 1):131–158
2. Harry J, Kulig B, Ray D et al (2001) Neurotoxicity risk assessment for human health: principles and approaches. Available from: http://www.inchem.org/documents/ehc/ehc/ehc223.htm#_223318000
3. Coecke S, Goldberg AM, Allen S et al (2007) Workgroup report: incorporating in vitro alternative methods for developmental neurotoxicity into international hazard and risk assessment strategies. Environ Health Perspect 115(6):924–931
4. Hogberg HT, Kinsner-Ovaskainen A, Coecke S et al (2010) mRNA expression is a relevant tool to identify developmental neurotoxicants using an in vitro approach. Toxicol Sci 113(1):95–115
5. Hogberg HT, Kinsner-Ovaskainen A, Hartung T et al (2009) Gene expression as a sensitive endpoint to evaluate cell differentiation and maturation of the developing central nervous system in primary cultures of rat cerebellar granule cells (CGCs) exposed to pesticides. Toxicol Appl Pharmacol 235(3):268–286
6. Krug AK, Balmer NV, Matt F et al (2013) Evaluation of a human neurite growth assay as specific screen for developmental neurotoxicants. Arch Toxicol 87(12):2215–2231
7. Rice D, Barone S Jr (2000) Critical periods of vulnerability for the developing nervous system: evidence from humans and animal models. Environ Health Perspect 108(Suppl 3):511–533
8. Stiles J, Jernigan TL (2010) The basics of brain development. Neuropsychol Rev 20(4):327–348
9. Yang D, Kania-Korwel I, Ghogha A et al (2014) PCB 136 atropselectively alters morphometric and functional parameters of neuronal connectivity in cultured rat hippocampal neurons via ryanodine receptor-dependent mechanisms. Toxicol Sci 138(2):379–392
10. Adinolfi M (1985) The development of the human blood-CSF-brain barrier. Dev Med Child Neurol 27(4):532–537
11. Bal-Price AK, Coecke S, Costa L et al (2012) Advancing the science of developmental neurotoxicity (DNT): testing for better safety evaluation. ALTEX 29(2):202–215
12. Smirnova L, Hogberg HT, Leist M et al (2014) Developmental neurotoxicity – challenges in the 21st century and in vitro opportunities. ALTEX 31(2):129–156
13. Bal-Price AK, Hogberg HT, Buzanska L et al (2010) In vitro developmental neurotoxicity (DNT) testing: relevant models and endpoints. Neurotoxicology 31(5):545–554
14. Makris SL, Raffaele K, Allen S et al (2009) A retrospective performance assessment of the developmental neurotoxicity study in support of OECD test guideline 426. Environ Health Perspect 117(1):17–25
15. OECD (1981) Test guideline 403. OECD guideline for testing of chemicals. Acute Inhalation Toxicity
16. OECD (1987) Test guideline 402. OECD guideline for testing of chemicals. Acute Dermal Toxicity
17. OECD (2002) Test guideline 420. OECD guideline for testing of chemicals. Acute Oral Toxicity – Fixed Dose Procedure
18. OECD (2002) Test guideline 423. OECD guideline for testing of chemicals. Acute Oral Toxicity – Acute Toxic Class Method
19. OECD (2008) Test guideline 436. OECD guideline for testing of chemicals. Acute Inhalation Toxicity – Acute Toxic Class Method
20. OECD (2008) Test guideline 425. OECD guideline for testing of chemicals. Acute Oral Toxicity – Up-and-Down Procedure
21. OECD (1998) Test guideline 408. OECD guideline for testing of chemicals. Repeated Dose 90-day Oral Toxicity Study in Rodents
22. OECD (2008) Test guideline 407. OECD guideline for testing of chemicals. Repeated Dose 28-day Oral Toxicity Study in Rodents
23. OECD (2009) Test guideline 452. OECD guideline for testing of chemicals. Chronic Toxicity Studies
24. Bjorling-Poulsen M, Andersen HR, Grandjean P (2008) Potential developmental neurotoxicity of pesticides used in Europe. Environ Health 7:50
25. OECD (2007) Test guideline 426. OECD guideline for testing of chemicals. Developmental Neurotoxicity Study
26. OECD (1997) Test guideline 424. OECD guideline for testing of chemicals. Neurotoxicity Study in Rodents
27. OECD (1995) Test guideline 418. OECD guideline for testing acute neurotoxicity of organophosphorus substances in laying hens
28. OECD (1995) Test guideline 419. OECD guideline for testing delayed neurotoxicity of organophosphorus substances in laying hens
29. OECD (2011) Test guideline 443. Extended one-generation reproductive toxicity study
30. OECD (2004) Series on testing and assessment number 20, Guidance document for neurotoxicity testing
31. EPA (2009) OPPTS 870.6200 Neurotoxicity Screening Battery [EPA 712–C–98–238]. Available from: https://www.

regulations.gov/document?D=EPA-HQ-OPPT-2009-0156-0041

32. U.S. Government Publishing Office (2018) 40 CFR § 799.9620 – TSCA neurotoxicity screening battery. Available from: https://www.govinfo.gov/app/details/CFR-2018-title40-vol35/CFR-2018-title40-vol35-sec799-9620
33. Rovida C, Hartung T (2009) Re-evaluation of animal numbers and costs for in vivo tests to accomplish REACH legislation requirements for chemicals – a report by the transatlantic think tank for toxicology (t(4)). ALTEX 26(3):187–208
34. Tsuji R, Crofton KM (2012) Developmental neurotoxicity guideline study: issues with methodology, evaluation and regulation. Congenit Anom (Kyoto) 52(3):122–128
35. Aschner M, Ceccatelli S, Daneshian M et al (2017) Reference compounds for alternative test methods to indicate developmental neurotoxicity (DNT) potential of chemicals: example lists and criteria for their selection and use. ALTEX 34(1):49–74
36. Fritsche E, Grandjean P, Crofton KM et al (2018) Consensus statement on the need for innovation, transition and implementation of developmental neurotoxicity (DNT) testing for regulatory purposes. Toxicol Appl Pharmacol 354:3–6
37. Grandjean P, Landrigan PJ (2006) Developmental neurotoxicity of industrial chemicals. Lancet 368(9553):2167–2178
38. Crofton KM, Mundy WR, Lein PJ et al (2011) Developmental neurotoxicity testing: recommendations for developing alternative methods for the screening and prioritization of chemicals. ALTEX 28(1):9–15
39. Bal-Price A, Crofton KM, Sachana M et al (2015) Putative adverse outcome pathways relevant to neurotoxicity. Crit Rev Toxicol 45(1):83–91
40. Bal-Price A, Hogberg HT, Crofton KM et al (2018) Recommendation on test readiness criteria for new approach methods in toxicology: exemplified for developmental neurotoxicity. ALTEX 35(3):306–352
41. Fritsche E, Alm H, Baumann J et al (2015) Literature review on in vitro and alternative Developmental Neurotoxicity (DNT) testing methods. External Scientific Report. EFSA supporting publication. EN-778
42. Gassmann K, Abel J, Bothe H et al (2010) Species-specific differential AhR expression protects human neural progenitor cells against developmental neurotoxicity of PAHs. Environ Health Perspect 118(11):1571–1577
43. NRC (2007) In: NRCN (ed) Toxicity testing in the 21st century: a vision and a strategy. The National Academies Press, Washington, D.C.
44. Bal-Price A, Pistollato F, Sachana M et al (2018) Strategies to improve the regulatory assessment of developmental neurotoxicity (DNT) using in vitro methods. Toxicol Appl Pharmacol 354:7–18
45. Fritsche E, Gassmann K, Schreiber T (2011) Neurospheres as a model for developmental neurotoxicity testing. Methods Mol Biol 758:99–114
46. Moors M, Cline JE, Abel J et al (2007) ERK-dependent and -independent pathways trigger human neural progenitor cell migration. Toxicol Appl Pharmacol 221(1): 57–67
47. Moors M, Rockel TD, Abel J et al (2009) Human neurospheres as three-dimensional cellular systems for developmental neurotoxicity testing. Environ Health Perspect 117(7):1131–1138
48. Breier JM, Gassmann K, Kayser R et al (2010) Neural progenitor cells as models for high-throughput screens of developmental neurotoxicity: state of the science. Neurotoxicol Teratol 32(1):4–15
49. Aschner M, Costa L (2005) Role of glia in neurotoxicity. Vol. RC347.5.R65. CRC Press, New York
50. Fritsche E, Cline JE, Nguyen NH et al (2005) Polychlorinated biphenyls disturb differentiation of normal human neural progenitor cells: clue for involvement of thyroid hormone receptors. Environ Health Perspect 113(7):871–876
51. Pistollato F, Canovas-Jorda D, Zagoura D et al (2017) Nrf2 pathway activation upon rotenone treatment in human iPSC-derived neural stem cells undergoing differentiation towards neurons and astrocytes. Neurochem Int 108:457–471
52. Yla-Outinen L, Heikkila J, Skottman H et al (2010) Human cell-based micro electrode array platform for studying neurotoxicity. Front Neuroeng 3:111
53. Amin H, Maccione A, Marinaro F et al (2016) Electrical responses and spontaneous activity of human iPS-derived neuronal networks characterized for 3-month culture with 4096-electrode arrays. Front Neurosci 10:121
54. Pistollato F, Canovas-Jorda D, Zagoura D et al (2017) Protocol for the differentiation of human induced pluripotent stem cells into mixed cultures of neurons and glia for neurotoxicity testing. J Vis Exp. 124. https://doi.org/10.3791/55702
55. Pistollato F, Louisse J, Scelfo B et al (2014) Development of a pluripotent stem cell derived neuronal model to identify chemically induced pathway perturbations in relation to neurotoxicity: effects of CREB

pathway inhibition. Toxicol Appl Pharmacol 280(2):378–388
56. Zagoura D, Canovas-Jorda D, Pistollato F et al (2017) Evaluation of the rotenone-induced activation of the Nrf2 pathway in a neuronal model derived from human induced pluripotent stem cells. Neurochem Int 106:62–73
57. Yang Y, Jiang S, Yan J et al (2015) An overview of the molecular mechanisms and novel roles of Nrf2 in neurodegenerative disorders. Cytokine Growth Factor Rev 26(1):47–57
58. Sakamoto K, Karelina K, Obrietan K (2011) CREB: a multifaceted regulator of neuronal plasticity and protection. J Neurochem 116(1):1–9
59. Coecke S, Balls M, Bowe G et al (2005) Guidance on good cell culture practice. A report of the second ECVAM task force on good cell culture practice. Altern Lab Anim 33(3):261–287
60. Pamies D, Bal-Price A, Simeonov A et al (2017) Good cell culture practice for stem cells and stem-cell-derived models. ALTEX 34(1):95–132
61. Pistollato F, Bremer-Hoffmann S, Healy L et al (2012) Standardization of pluripotent stem cell cultures for toxicity testing. Expert Opin Drug Metab Toxicol 8(2):239–257
62. Harrill JA, Freudenrich T, Wallace K et al (2018) Testing for developmental neurotoxicity using a battery of in vitro assays for key cellular events in neurodevelopment. Toxicol Appl Pharmacol 354:24–39
63. Mundy WR, Radio NM, Freudenrich TM (2010) Neuronal models for evaluation of proliferation in vitro using high content screening. Toxicology 270(2–3):121–130
64. Druwe I, F TM, Wallace K et al (2016) Comparison of human induced pluripotent stem cell-derived neurons and rat primary cortical neurons as in vitro models of neurite outgrowth. Applied In Vitro Toxicology 2(1):26–36
65. Ryan KR, Sirenko O, Parham F et al (2016) Neurite outgrowth in human induced pluripotent stem cell-derived neurons as a high-throughput screen for developmental neurotoxicity or neurotoxicity. Neurotoxicology 53:271–281
66. Breier JM, Radio NM, Mundy WR et al (2008) Development of a high-throughput screening assay for chemical effects on proliferation and viability of immortalized human neural progenitor cells. Toxicol Sci 105(1):119–133
67. Harrill JA, Freudenrich TM, Robinette BL et al (2011) Comparative sensitivity of human and rat neural cultures to chemical-induced inhibition of neurite outgrowth. Toxicol Appl Pharmacol 256(3):268–280
68. Kuegler PB, Zimmer B, Waldmann T et al (2010) Markers of murine embryonic and neural stem cells, neurons and astrocytes: reference points for developmental neurotoxicity testing. ALTEX 27(1):17–42
69. Hogberg HT, Sobanski T, Novellino A et al (2011) Application of micro-electrode arrays (MEAs) as an emerging technology for developmental neurotoxicity: evaluation of domoic acid-induced effects in primary cultures of rat cortical neurons. Neurotoxicology 32(1):158–168
70. Mundy WR, Robinette B, Radio NM et al (2008) Protein biomarkers associated with growth and synaptogenesis in a cell culture model of neuronal development. Toxicology 249(2–3):220–229
71. Schultz L, Zurich MG, Culot M et al (2015) Evaluation of drug-induced neurotoxicity based on metabolomics, proteomics and electrical activity measurements in complementary CNS in vitro models. Toxicol In Vitro 30(1 Pt A):138–165
72. Costa LG (1998) Neurotoxicity testing: a discussion of in vitro alternatives. Environ Health Perspect 106(Suppl 2):505–510
73. Frank CL, Brown JP, Wallace K et al (2018) Defining toxicological tipping points in neuronal network development. Toxicol Appl Pharmacol 354:81–93
74. Novellino A, Scelfo B, Palosaari T et al (2011) Development of micro-electrode array based tests for neurotoxicity: assessment of interlaboratory reproducibility with neuroactive chemicals. Front Neuroeng 4:4
75. Valdivia P, Martin M, LeFew WR et al (2014) Multi-well microelectrode array recordings detect neuroactivity of ToxCast compounds. Neurotoxicology 44:204–217
76. Vassallo A, Chiappalone M, De Camargos Lopes R et al (2017) A multi-laboratory evaluation of microelectrode array-based measurements of neural network activity for acute neurotoxicity testing. Neurotoxicology 60:280–292
77. Geier MC, James Minick D, Truong L et al (2018) Systematic developmental neurotoxicity assessment of a representative PAH superfund mixture using zebrafish. Toxicol Appl Pharmacol 354:115–125
78. Ruszkiewicz JA, Pinkas A, Miah MR et al (2018) *C. elegans* as a model in developmental neurotoxicology. Toxicol Appl Pharmacol 354:126–135
79. Sipes NS, Padilla SKnudsen TB (2011) Zebrafish: as an integrative model for twenty-first century toxicity testing. Birth Defects Res C Embryo Today 93(3):256–267
80. Noyes PD, Haggard DE, Gonnerman GD et al (2015) Advanced morphological – behavioral test platform reveals neurodevelopmental defects in embryonic zebrafish exposed to comprehensive suite of halogenated and organophosphate flame retardants. Toxicol Sci 145(1):177–195

81. Lapenna S, Fuart-Gatnik M, Worth A (2010) Review of QSAR models and software tools for predicting acute and chronic systemic toxicity. JRC Scientific and Technical report
82. Crofton KM (1996) A structure-activity relationship for the neurotoxicity of triazole fungicides. Toxicol Lett 84(3):155–159
83. Pessah IN, Hansen LG, Albertson TE et al (2006) Structure-activity relationship for noncoplanar polychlorinated biphenyl congeners toward the ryanodine receptor-Ca2+ channel complex type 1 (RyR1). Chem Res Toxicol 19(1):92–101
84. Nevalainen T, Kolehmainen E (1994) New QSAR models for polyhalogenated aromatics. Environ Toxicol Chem 13(10):1699–1706
85. Pessah IN, Cherednichenko G, Lein PJ (2010) Minding the calcium store: ryanodine receptor activation as a convergent mechanism of PCB toxicity. Pharmacol Ther 125(2):260–285
86. El Yazal J, Rao SN, Mehl A et al (2001) Prediction of organophosphorus acetylcholinesterase inhibition using three-dimensional quantitative structure-activity relationship (3D-QSAR) methods. Toxicol Sci 63(2):223–232
87. Stenberg M, Hamers T, Machala M et al (2011) Multivariate toxicity profiles and QSAR modeling of non-dioxin-like PCBs – an investigation of in vitro screening data from ultra-pure congeners. Chemosphere 85(9):1423–1429
88. ECHA (2008) Guidance on information requirements and chemical safety assessment. Chapter R, 8. Available from: https://echa.europa.eu/guidance-documents/guidance-on-information-requirements-and-chemical-safety-assessment
89. Paradis A, Leblanc DDumais N (2016) Optimization of an in vitro human blood-brain barrier model: application to blood monocyte transmigration assays. MethodsX 3:25–34
90. Zhang L, Zhu H, Oprea TI et al (2008) QSAR modeling of the blood-brain barrier permeability for diverse organic compounds. Pharm Res 25(8):1902–1914
91. Kortagere S, Chekmarev D, Welsh WJ et al (2008) New predictive models for blood-brain barrier permeability of drug-like molecules. Pharm Res 25(8):1836–1845
92. Abreu PA, Castro HC, Paes-de-Carvalho R et al (2013) Molecular modeling of a phenylamidine class of NMDA receptor antagonists and the rational design of new triazolylamidine derivatives. Chem Biol Drug Des 81(2):185–197
93. Rayne S, Forest K (2010) Quantitative structure-activity relationship (QSAR) studies for predicting activation of the ryanodine receptor type 1 channel complex (RyR1) by polychlorinated biphenyl (PCB) congeners. J Environ Sci Health A Tox Hazard Subst Environ Eng 45(3):355–362
94. Guerra A, Páez JACampillo NE (2008) Artificial neural networks in ADMET modeling: prediction of blood–brain barrier permeation. Mol Inform 27(5):586–594
95. Zhang YY, Liu H, Summerfield SG et al (2016) Integrating in silico and in vitro approaches to predict drug accessibility to the central nervous system. Mol Pharm 13(5):1540–1550
96. Hewitt M, Madden JC, Rowe PH et al (2007) Structure-based modelling in reproductive toxicology: (Q)SARs for the placental barrier. SAR QSAR Environ Res 18(1–2):57–76
97. Giaginis C, Zira A, Theocharis S et al (2009) Application of quantitative structure-activity relationships for modeling drug and chemical transport across the human placenta barrier: a multivariate data analysis approach. J Appl Toxicol 29(8):724–733
98. OECD (2013) Series on testing and assessment no. 184. Guidance document on developing and assessing adverse outcome pathways
99. Ankley GT, Bennett RS, Erickson RJ et al (2010) Adverse outcome pathways: a conceptual framework to support ecotoxicology research and risk assessment. Environ Toxicol Chem 29(3):730–741
100. Bal-Price A, Meek MEB (2017) Adverse outcome pathways: application to enhance mechanistic understanding of neurotoxicity. Pharmacol Ther 179:84–95
101. Bal-Price A, Crofton KM, Leist M et al (2015) International STakeholder NETwork (ISTNET): creating a developmental neurotoxicity (DNT) testing road map for regulatory purposes. Arch Toxicol 89(2):269–287
102. Sachana M, Rolaki ABal-Price A (2018) Development of the Adverse Outcome Pathway (AOP): chronic binding of antagonist to N-methyl-d-aspartate receptors (NMDARs) during brain development induces impairment of learning and memory abilities of children. Toxicol Appl Pharmacol 354:153–175
103. OECD (2017) Guidance document on the reporting of defined approaches to be used within integrated approaches to testing and assessment OECD series on testing and assessment. Vol. 255
104. OECD (2017) Guidance document on the reporting of defined approaches and individual information sources to be used within Integrated Approaches to Testing and Assessment (IATA) for skin sensitisation OECD series on testing and assessment. Vol. 256

Index

Michael Aschner and Lucio Costa (eds.), *Cell Culture Techniques*, Neuromethods, vol. 145,
https://doi.org/10.1007/978-1-4939-9228-7,